PROGRESS IN

Nucleic Acid Research and Molecular Biology

Volume 34

PROGRESS IN

Nucleic Acid Research and Molecular Biology

edited by

WALDO E. COHN

Biology Division
Oak Ridge National Laboratory
Oak Ridge, Tennessee

KIVIE MOLDAVE

University of California
Santa Cruz, California

Volume 34

ACADEMIC PRESS, INC.

Harcourt Brace Jovanovich, Publishers

Orlando San Diego New York Austin
Boston London Sydney Tokyo Toronto

Academic Press, Inc.
San Diego, California 92101

United Kingdom Edition published by
ACADEMIC PRESS INC. (LONDON) LTD.
24-28 Oval Road, London NW1 7DX

Library of Congress Catalog Card Number: 63-15847

ISBN 0-12-540034-9 (alk. paper)

PRINTED IN THE UNITED STATES OF AMERICA
87 88 89 90 9 8 7 6 5 4 3 2 1

Contents

Transport of mRNA from Nucleus to Cytoplasm

Heinz C. Schröder, Michael Bachmann, Bärbel Diehl-Seifert, and Werner E. G. Müller

Foreign Gene Expression in Plant Cells

Paul F. Lurquin

Epstein–Barr Virus Transformation

Samuel H. Speck and Jack L. Strominger

Proteins Covalently Linked to Viral Genomes

Andrei B. Vartapetian and Alexei A. Bogdanov

Abbreviations and Symbols

All contributors to this Series are asked to use the terminology (abbreviations and symbols) recommended by the IUPAC-IUB Commission on Biochemical Nomenclature (CBN) and approved by IUPAC and IUB, and the Editors endeavor to assure conformity. These Recommendations have been published in many journals (*1*, *2*) and compendia (*3*) and are available in reprint form from the Office of Biochemical Nomenclature (OBN); they are therefore considered to be generally known. Those used in nucleic acid work, originally set out in section 5 of the first Recommendations (*1*) and subsequently revised and expanded (*2*, *3*), are given in condensed form in the frontmatter of Volumes 9–33 of this series. A recent expansion of the one-letter system (5) follows.

SINGLE-LETTER CODE RECOMMENDATIONS[a] (5)

Symbol	Meaning	Origin of designation
G	G	Guanosine
A	A	Adenosine
T(U)	T(U)	(ribo)Thymidine (Uridine)
C	C	Cytidine
R	G or A	puRine
Y	T(U) or C	pYrimidine
M	A or C	aMino
K	G or T(U)	Keto
S	G or C	Strong interaction (3 H-bonds)
W[b]	A or T(U)	Weak interaction (2 H-bonds)
H	A or C or T(U)	not G; H follows G in the alphabet
B	G or T(U) or C	not A; B follows A
V	G or C or A	not T (not U); V follows U
D	G or A or T(U)	not C; D follows C
N	G or A or T(U) or C	aNy
Q	Q	Queuosine (nucleoside of queuine)

[a] Modified from *Proc. Natl. Acad. Sci. U.S.A.* **83**, 4 (1986).
[b] W has been used for wyosine, the nucleoside of "base Y" (wye).

Enzymes

In naming enzymes, the 1984 recommendations of the IUB Commission on Biochemical Nomenclature (*4*) are followed as far as possible. At first mention, each enzyme is described *either* by its systematic name *or* by the equation for the reaction catalyzed *or* by the recommended trivial name, followed by its EC number in parentheses. Thereafter, a trivial name may be used. Enzyme names are not to be abbreviated except when the substrate has an approved abbreviation (e.g., ATPase, but not LDH, is acceptable).

REFERENCES

1. *JBC* **241,** 527 (1966); *Bchem* **5,** 1445 (1966); *BJ* **101,** 1 (1966); *ABB* **115,** 1 (1966), **129,** 1 (1969); and elsewhere.† General.
2. *EJB* **15,** 203 (1970); *JBC* **245,** 5171 (1970); *JMB* **55,** 299 (1971); and elsewhere.†
3. "Handbook of Biochemistry" (G. Fasman, ed.), 3rd ed. Chemical Rubber Co., Cleveland, Ohio, 1970, 1975, Nucleic Acids, Vols. I and II, pp. 3–59. Nucleic acids.
4. "Enzyme Nomenclature" [Recommendations (1984) of the Nomenclature Committee of the IUB]. Academic Press, New York, 1984.
5. *EJB* **150,** 1 (1985). Nucleic Acids (One-letter system).

Abbreviations of Journal Titles

Journals	*Abbreviations used*
Annu. Rev. Biochem.	ARB
Annu. Rev. Genet.	ARGen
Arch. Biochem. Biophys.	ABB
Biochem. Biophys. Res. Commun.	BBRC
Biochemistry	Bchem
Biochem. J.	BJ
Biochim. Biophys. Acta	BBA
Cold Spring Harbor	CSH
Cold Spring Harbor Lab	CSHLab
Cold Spring Harbor Symp. Quant. Biol.	CSHSQB
Eur. J. Biochem.	EJB
Fed. Proc.	FP
Hoppe-Seyler's Z. physiol. Chem.	ZpChem
J. Amer. Chem. Soc.	JACS
J. Bacteriol.	J. Bact.
J. Biol. Chem.	JBC
J. Chem. Soc.	JCS
J. Mol. Biol.	JMB
J. Nat. Cancer Inst.	JNCI
Mol. Cell. Biol.	MCBiol
Mol. Cell. Biochem.	MCBchem
Mol. Gen. Genet.	MGG
Nature, New Biology	Nature NB
Nucleic Acid Research	NARes
Proc. Nat. Acad. Sci. U.S.	PNAS
Proc. Soc. Exp. Biol. Med.	PSEBM
Progr. Nucl. Acid. Res. Mol. Biol.	This Series

† Reprints available from the Office of Biochemical Nomenclature (W. E. Cohn, Director).

Some Articles Planned for Future Volumes

The Structural and Functional Basis of Collagen Gene Diversity
P. BORNSTEIN

UV-Induced Crosslinks in Nucleoproteins
E. I. BUDOWSKY

Molecular Structure and Transcriptional Regulation of the Proline-Rich Multigene Protein Complexes of the Salivary Gland
D. M. CARLSON

The Chromatin Structure and Organization of Genes during Transcription
R. CHALKLEY

Cloning and Sequencing of IGF-II Receptor
M. CZECH

Ferritin Gene Expression
J. W. DRYSDALE

Physical Monitoring of Meiotic and Mitotic Recombination
J. E. HABER

Modulation of Gene Expression by Oncogenes
M. W. LIEBERMAN

Molecular Biology of β-Glucuronidase Regulation
K. PAIGEN

Early Signals and Molecular Steps in the Mitogenic Response
E. ROZENGURT

Mechanisms of the Antiviral Action of Interferons
C. E. SAMUEL

Use of Mutagenesis to Interpret the Role of Reactive Groups Identified by Chemical Modification
P. SCHIMMEL AND A. PROFY

Eukaryotic mRNA 5′ Cap and Secondary Structure: Effectors of Translational Regulation
N. SONENBERG

Structural Organization and Regulation of Human Histone Genes
G. STEIN AND J. STEIN

Damage to Mammalian DNA by Ionizing Radiation
J. F. WARD

Messenger RNA Capping Enzymes from Eukaryotic Cells

KIYOHISA MIZUMOTO AND
YOSHITO KAZIRO

Institute of Medical Science
University of Tokyo
Tokyo 108, Japan

Eukaryotic mRNAs are synthesized in the nucleus in the form of precursor molecules (pre-mRNA or hnRNA) 10 to 20 times larger than the mature mRNA. The primary transcripts undergo various posttranscriptional processings, such as 5′-terminal "capping," internal methylation, cleavage, 3′-terminal polyadenylation, and splicing, before the translatable mRNAs are transported to the cytoplasm. Since mRNAs are synthesized, translated, and possibly degraded in a 5′-to-3′ direction, studies on the synthesis and the function of their 5′-modified structures are of considerable interest.

Most of the cellular as well as viral mRNAs in eukaryotes contain a 5′-terminal "cap" structure, m^7GpppN, in which the terminal 7-methylguanosine residue is linked to the 5′ position of the penultimate nucleotide through a 5′–5′ triphosphate bridge (*1, 2*). Ubiquitous in nature, the cap structure is required for efficient initiation of translation (*3–5*) and for protection of mRNAs against nuclease attack (*6, 7*). It has also been suggested that the synthesis of the methylated cap structure and/or the association of the capping enzyme system with RNA polymerase may play an important role in the initiation of transcription of viral (*8, 9*) and cellular (*10*) mRNAs. Furthermore, the cap structure is required for RNA splicing (*11, 12*). The cap structure

Progress in Nucleic Acid Research
and Molecular Biology, Vol. 34

is synthesized at the initial stage of transcription by RNA polymerase II[1] (*13–16*) and is conserved at 5′ termini of RNAs while they are processed in the nucleus and transported to the cytoplasm (*17, 18*). Elucidation of the mechanism of synthesis of the cap structure is important for understanding the molecular mechanism of eukaryotic gene expression and its regulation.

The mechanism of cap formation was first studied with viruses replicating in the cytoplasm. These viruses possess their own capping enzyme system (in virions) together with their own RNA polymerase. Under appropriate conditions, some of these viruses can efficiently synthesize *in vitro* the mRNAs having the methylated cap structures. The fundamental mechanism of cap formation has been studied using the *in vitro* transcription systems derived from cytoplasmic polyhedrosis virus (CPV) (*8, 19, 20*), reovirus (*21, 22*), vaccinia virus (*23, 24*), and vesicular stomatitis virus (VSV) (*25, 26*). Cap formation is catalyzed by a series of enzymes including mRNA guanylyltransferase (EC 2.7.7.50), mRNA (guanine-7-)methyltransferase (EC 2.1.1.56), and mRNA (nucleoside-$O^{2'}$-)methyltransferase (EC 2.1.1.57) (Fig. 1). In reovirus, CPV, and vaccinia virus, the cap structure is formed by the transfer of the GMP moiety from GTP to the 5′-diphosphate end of the acceptor RNA. On the other hand, in VSV, the cap formation is thought to proceed through the transfer of a GDP moiety from GTP to the 5′-monophosphate end of RNA (*25*) and the (ribose-O^2-)methylation seems to precede the (guanine-7-)methylation (*26*).

A capping enzyme complex containing mRNA guanylyltransferase and mRNA (guanine-7-)methyltransferase activities has been solubilized and highly purified from vaccinia virus (*27, 28*). The complex had an M_r of 127,000 and consisted of two nonidentical subunits of M_r 95,000 and 31,400 (*27*) in a molar ratio close to 1 : 1. The former polypeptide contains the catalytic center of the guanylyltransferase (*29*). An mRNA (nucleoside-$O^{2'}$-)methyltransferase activity that catalyzes the methylation of the 2′ OH of the penultimate nucleoside of capped RNA has also been isolated from vaccinia virus (*30*). The M_r of the enzyme was estimated to be 38,000.

Biochemical studies on the capping mechanism of cellular mRNAs was initially hampered by the fact that no *in vitro* transcription system as efficient as viral ones, either with isolated nuclei or soluble systems, had been developed, and the primary transcripts are processed

[1] The differentiation of RNA polymerases I, II, and III is set out definitively by R. A. Roeder in "RNA Polymerase" (R. Losick and M. Chamberlin, eds.), p. 285. Cold Spring Harbor Laboratory, Cold Spring Harbor, N.Y., 1976. [Eds.]

1. Capping

a. $\overset{\alpha\beta\alpha}{pppG} + \overset{\beta'\alpha'}{ppN}\text{-} \xrightarrow[\text{transferase}]{\text{mRNA guanylyl-}} \overset{\alpha\beta\alpha'}{GpppN}\text{-} + \overset{\beta\gamma}{PPi}$

(reovirus, cytoplasmic polyhedrosis virus, vaccinia virus)

b. $\overset{\alpha\beta\alpha}{pppG} + \overset{\alpha'}{pN}\text{-} \longrightarrow \overset{\alpha\beta\alpha'}{GpppN}\text{-} + \overset{\gamma}{Pi}$

(vesicular stomatitis virus, spring viremia of carp virus[1])

2. Methylation[2]

$GpppN\text{-} + AdoMet \xrightarrow[\text{methyltransferase}]{\text{mRNA (guanine-7-)-}} m^7GpppN\text{-} + AdoHcy$

$m^7GpppN\text{-} + AdoMet \xrightarrow[\text{methyltransferase}]{\text{mRNA (nucleoside-}O^{2'}\text{-)-}} m^7GpppNm\text{-} + AdoHcy$

FIG. 1. Mechanism of viral mRNA cap formation. (1) The 1.a mechanism is operative in some conditions (*65*). (2) (Guanine-7-)methylation does not necessarily precede ribose-O^2-methylation in VSV (*26*).

in a more complicated manner than in the viral systems. The isolated nuclei from HeLa cells (*31*) and mouse L cells (*17, 32*) can synthesize methylated cap structures *de novo* under conditions that allow α-amanitin-sensititive transcription. However, the low efficiency of cap formation and the complexity of the reactions that occur in these crude systems were inadequate to analyze the precise mechanism of capping.

To understand the molecular mechanism of mRNA capping in eukaryotic cells in detail, it is necessary to purify and characterize the enzymes involved. Efforts have been made to isolate these enzymes from various eukaryotic cells including HeLa cells (*33–39*), rat liver (*40–43*), *Artemia salina* (*44, 45*), calf thymus (*46*), wheat germ (*44, 47, 48*), *Neurospora crassa* (*49*), and yeast (*48, 50–54*). So far, six enzymatic activities that participate in the modification of the 5′ terminus of RNA to form fully modified "cap-II" structures ($m^7GpppN^1m\text{-}N^2m\text{-}$) (Table I) have been identified. Each enzyme activity listed in Table I can be assayed independently with appropriate substrates *in vitro* as a posttranscriptional modification activity. Among these, the activities of four enzymes, mRNA (guanine-7-)methyltransferase, RNA 5′-triphosphatase, mRNA guanylyltransferase, and capI-mRNA (nucloside-$O^{2'}$-)methyltransferase are detected in common with respect to their basic reaction nature, both in viral and cellular systems. How-

TABLE I
ISOLATED ENZYME ACTIVITIES INVOLVED IN CAP SYNTHESIS

Enzymes	Reactions catalyzed	Enzyme sources
1. mRNA guanylyltransferase	ppN- + GTP ⟶ GpppN- + PP_i	Vaccinia virus (*27, 75, 28*) Rat liver (*41, 42*) HeLa cells (*36–38*) Yeast (*50–53*) *A. salina* (*45*) Wheat germ (*44, 47*) Calf liver (*46*)
2. mRNA (guanine-7-)methyltransferase	GpppN(m)- + AdoMet ⟶ m^7GpppN(m)- + AdoHcy	Vaccinia virus (*27, 28, 55*) Rat liver (*40, 41*) HeLa cells (*33*) Yeast (*48, 50, 54*) *N. crassa* (*49*) Wheat germ (*48*)
3. CapI-RNA (nucleoside-$O^{2'}$-) methyltransferase	(m^7)GpppN- + AdoMet ⟶ (m^7)GpppNm- + AdoHcy	Vaccinia virus (*30*) HeLa cells (*39*)
4. CapII-RNA (nucleoside-$O^{2'}$-) methyltransferase	m^7GpppN^{1}m-N^2- + AdoMet ⟶ m^7GpppN^{1}m-N^{2}m- + AdoHcy	HeLa cells (*39*)
5. mRNA ($O^{2'}$-methyladenosine-N^6-)methyltransferase[a]	m^7GpppAm- + AdoMet ⟶ m^7Gpppm^{6}Am- + AdoHcy	HeLa cells (*35*)
6. RNA 5′-triphosphatase[b] (RNA 5′-terminal phosphohydrolase)	pppN- ⟶ ppN- + P_i	Rat liver (*43*) Yeast (*50, 51*) *A. salina* (*45*) Vaccinia virus (*28, 55*)

[a] In animal cells, Am in the N^1 position is frequently methylated further to give m^{6}Am.
[b] Associated with guanylyltransferase.

ever, the remaining two enzyme activities, capII-mRNA (nucleoside-$O^{2'}$-)methyltransferase and mRNA ($O^{2'}$-methyladenosine-N^6-)methyltransferase (EC 2.1.1.62) are detected only in cellular systems. In contrast to the vaccinia virus enzyme, guanylyltransferase and (guanine-7-)methyltransferase from cellular sources are readily resolved in the early stage of purification.

The purpose of the present review is to summarize the current knowledge of the reaction mechanism and the structure of the capping enzyme with an emphasis on our recent work.

I. Isolation of the Capping Enzyme System from Rat Liver Nuclei

Mizumoto and Lipmann (*41*) found that the extract of purified rat liver nuclei catalyzes the formation of the capI structure (m^7GpppNm-) on externally added RNA substrates. When the sonic extract of rat liver nuclei was incubated with (p)ppG-RNA and [α-^{32}P]GTP in the presence of AdoMet, four different cap structures, [^{32}P]m^7GpppGm, -m^7GpppG, -GpppGm, and -GpppG were detected after nuclease P_1 (EC 3.1.30.1) digestion of the RNA. This indicated that the nuclear extract contains at least three activities: mRNA (guanine-7-)methyltransferase, mRNA guanylyltransferase, and mRNA (nucleoside-$O^{2'}$-) methyltransferase. The former two activities were rather stable in the crude extract, but the (nucleoside-$O^{2'}$-)methyltransferase activity was quite unstable during storage.

Cell fractionation experiments indicated that both mRNA guanylyltransferase and mRNA (guanine-7-)methyltransferase activities were mainly (60%) localized in the nuclear fraction and this fraction also possessed the highest specific activities of both enzymes. The mRNA guanylyltransferase was purified about 700-fold starting from the sonic extract of rat liver nuclei through successive column chromatographies on Hypatite C, Sephadex G-150, CM-Sephadex, and Blue dextran-Sepharose (*42*). Guanylyltransferase and (guanine-7-)methyltransferase were readily resolved at the step of Sephadex G-150 column chromatography. The resolution of both activities was also observed with the enzymes from HeLa cells (*33, 36*), *A. salina* (*45*), wheat germ (*47*), and yeast (*50, 51, 53*). This is in contrast to the vaccinia virus capping enzyme complex in which guanylyltransferse and (guanine-7-)methyltransferase are tightly associated. Rat liver (guanine-7-)methyltransferase has not been purified beyond the step of Sephadex G-150 column chromatography because of its instability.

Partially purified (guanine-7-)methyltransferase had an M_r of about 130,000 and catalyzed the formation of m^7GpppN-RNA from capped but unmethylated reovirus or CPV mRNA. The cap core structures G(5′)pppN without the RNA chain also accepted a methyl group in position 7 with activities in the order of GpppG > GpppA > GpppC > GpppU. G(5′)ppppG and GTP were completely inert. The following points should be mentioned about the (guanine-7-)methylation. When $O^{2'}$-methylated compounds, such as GpppGm and GpppAm, were used as methyl acceptors, a 2- or 5-fold increase, respectively, in methyl transfer over the non-ribose-methylated counterparts was observed (Table II). With partially purified (guanine-7-)methyltransferase from rat liver nuclei, K_m values of 50 and 10 μM for GpppG and GpppGm, respectively, were obtained.

A similar stimulation by $O^{2'}$-methylation was observed with extracts of *A. salina* embryos, but not with wheat germ extracts. This is probably due to the fact that the ribose-2′-OH of the cap structure of plant cell mRNA is not methylated (*56*). On the contrary, an opposite effect was obtained with reovirus cores, that is, GpppGm was less effective than GpppG, and GpppA was inactive as a substrate (Table II). The substrate specificity of reovirus cores is consistent with the observation that all of the 10 reovirus mRNAs contain a common cap structure, m^7GpppGm-, and that (guanine-7-)methylation precedes (ribose-O^2-)methylation (*22*).

These results and the above observation that the nuclear extract of rat liver could form GpppGm- on (p)ppG-RNA indicated that (guanine-7-)methylation does not necessarily precede the (ribose-O^2-) methylation in animal cells. Langberg and Moss (*47*) observed that CapI-RNA (nucleoside-$O^{2'}$-)methyltransferase partially purified from HeLa cells can recognize GpppN- as efficiently as m^7GpppN-. It is also interesting that, in α and β globin mRNAs of nucleated erythroid cells of mice, the $O^{2'}$-methylation of cap structures occurs as early as (guanine-7-)methylation (*57*).

In the studies on the capping reaction catalyzed by mRNA guanylyltransferase partially purified from rat liver nuclei, the following features appear: (1) among α-, β-, or γ-^{32}P-labeled GTP used as substrates, only the radioactivity of the α-phosphate is incorporated into

TABLE II
EFFECT OF (RIBOSE-O^2-)METHYLATION ON THE (GUANINE-7-)METHYLTRANSFERASE ACTIVITY[a]

Substrate (100 μM)	[^{3}H]Methyl transfer to the guanine-7 position (cpm): Rat liver nuclear extract	Reovirus cores	*Artemia* extract	Wheat germ extract
GpppG	4,890	6,444	12,411	13,210
GpppGm	8,050	2,975	22,528	15,783
GpppA	305	<50	2,547	869
GpppAm	1,490	<50	7,935	1,020

[a] Methyl transfer from *S*-[methyl-^{3}H]adenosyl-L-methionine to the guanine-7 position of GpppN or GpppNm was measured as described (*41*). Crude extracts were prepared from rat liver nuclei (*41*), *A. salina* (*76*), and wheat germ (*77*) as reported. Reovirus cores were obtained by digestion of purified virions with chymotrypsin (*78*).

the cap structure to give [^{32}P]GpppN-, whether the N is a purine or a pyrimidine (*46*); (2) ppN- and pppN-RNA are active as cap acceptors, while pN- and $_{HO}$N-RNA are completely inert (*41*); (3) the reaction with unlabeled GTP and [β-^{32}P]ppN-RNA gives [^{32}P]GpppN-RNA (*41*); (4) cap formation is strongly inhibited by PP_i (*42*); (5) capping with [α-^{32}P]GTP is not affected by the addition of a 50-fold molar excess of unlabeled m^{7}GTP. Based on these observations, it is proposed that the capping reaction by rat liver capping enzyme system proceeds through the following steps:

$$\text{pppG} + \text{ppN-} \xrightarrow{\text{mRNA guanylyltransferase}} \text{GpppN} + \text{PP}_i \qquad (1)$$

$$\text{GpppN-} + \text{AdoMet} \xrightarrow[\text{capI-mRNA (nucleoside-}O^{2'}\text{-)methyltransferase}]{\text{mRNA (guanine-7-)methyltransferase}} \text{m}^7\text{GpppNm-} + \text{AdoHcy} \qquad (2)$$

That is, guanylyltransferase transfers the GMP moiety from GTP to the 5′-diphosphate terminus of an mRNA chain to form a 5′–5′-triphosphate bridge with a concomitant release of pyrophosphate. Formation of this structure is followed by methylation at the guanine-7 and ribose-2-OH positions of the penultimate nucleoside by two different methyltransferases. At present, no conclusion regarding the order of guanine-7-methylation and $O^{2'}$-methylation can be drawn from the above results. However, since the (guanine-7-)methyltransferase had higher affinity for GpppNm than for GpppN, it is possible that a preferred order of methylation reactions is (ribose-O^2-)methylation → (guanine-7-)methylation.

II. Isolation and Characterization of a Guanylylated Enzyme Intermediate

Mizumoto and Lipmann (*41*) found that guanylyltransferase partially purified from rat liver nuclei catalyzes, in addition to the overall capping reaction, a GTP–PP_i exchange in the absence of an acceptor RNA. This suggested that the capping reaction proceeds through a covalent enzyme–GMP intermediate.

The presence of such a complex was directly demonstrated by incubating the partially purified rat liver guanylyltransferase with [α-^{32}P]GTP and subjecting it to polyacrylamide gel electrophoresis in sodium dodecyl sulfate (*42*). A single radioactive band with M_r 69,000 was observed (Fig. 2B, lane 1). The activity to form the 69,000 complex always copurified with both the capping and the GTP–PP_i exchange activities (Fig. 2A). The complex formation required Mg^{2+} and

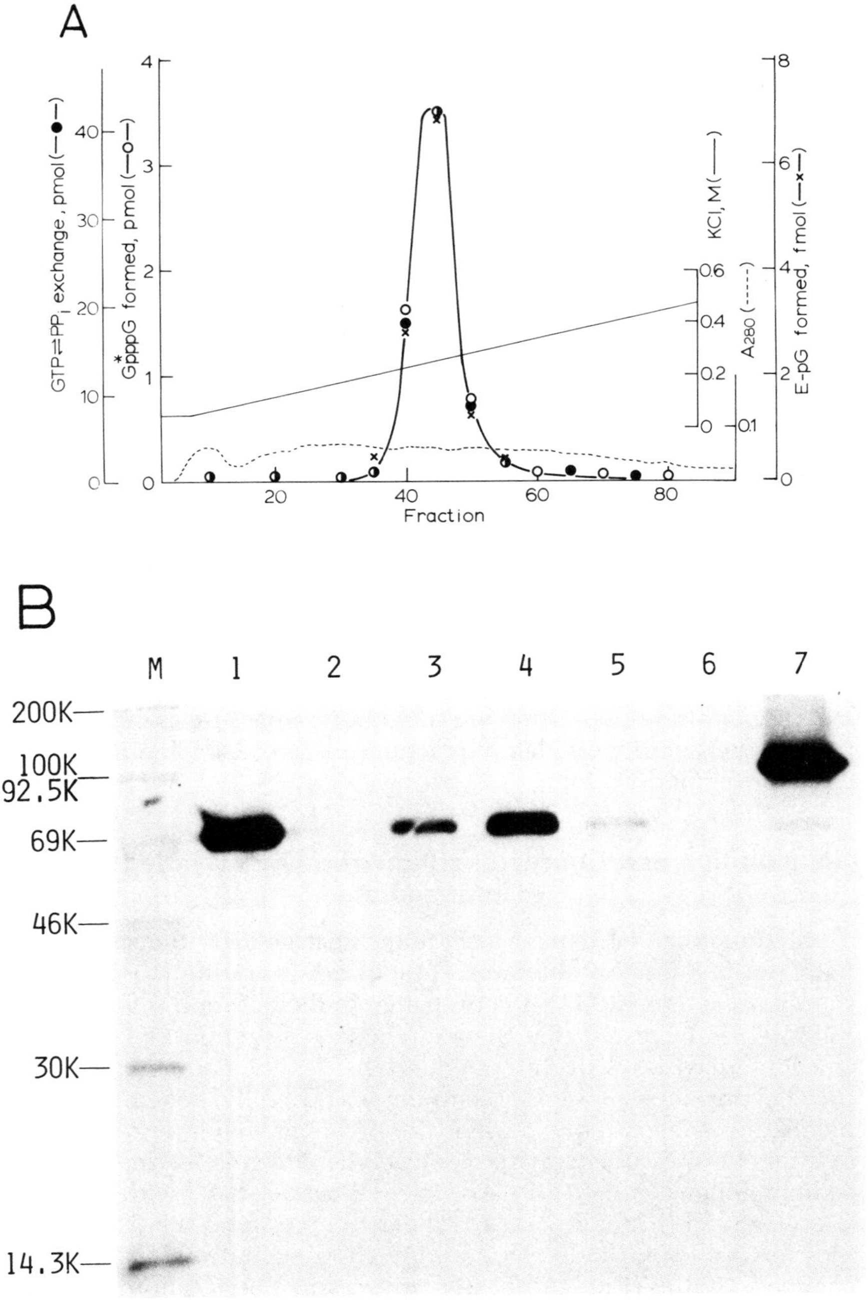
A
GTP⇌PPi exchange, pmol (—●—)
Gp*ppG formed, pmol (—○—)
KCl, M (——)
A280 (-----)
E-pG formed, fmol (—×—)
Fraction
B
M 1 2 3 4 5 6 7
200K
100K
92.5K
69K
46K
30K
14.3K

was strongly inhibited by low concentrations of PP_i but not with P_i. Formation of the 69,000 complex was specific to GTP and the incorporation of ^3H from [8-^3H]GTP was nearly stoichiometric to that of ^{32}P from [α-^{32}P]GTP. No radioactivity was incorporated when [α-^{32}P]GTP was replaced by [β-^{32}P]GTP (*42*), indicating that the GMP moiety of the GTP molecule is tranferred to the enzyme with the displacement of PP_i. These properties are consistent with the mechanism shown in Eq. (1).

The complex of enzyme and [^{32}P]GMP was isolated in the undenatured form to test whether it acts as an intermediate in transguanylylation. As shown in Fig. 3, when the isolated enzyme–GMP complex is incubated with ppG-RNA, the enzyme-bound GMP is quantitatively transferred to the RNA to yield GpppG-RNA. The reaction depends on the presence of an acceptor RNA and Mg^{2+}. On the other hand, when the complex is incubated with pyrophosphate, all the radioactivity is released as GTP. Thus, it is clearly demonstrated that the enzyme–GMP complex is a functional intermediate of the capping reaction. These results indicate that the reaction catalyzed by rat liver guanylyltransferase occurs through the following two partial steps:

$$\text{E} + \text{pppG} \longrightarrow \text{E-pG} + \text{PP}_i \qquad (3)$$

$$\text{E-pG} + \text{ppN-RNA} \longrightarrow \text{GpppN-RNA} + \text{E} \qquad (4)$$

First, the enzyme reacts with GTP to form a covalent enzyme–GMP intermediate with a displacement of PP_i. Then, the enzyme-bound GMP is transferred to diphosphate-terminated RNA to form a cap structure. The first reaction is freely reversible. Although we did not test the reversibility of the second reaction, Shuman (*38*), using

FIG. 2. Covalent enzyme–GMP complex formation with rat liver mRNA guanylyltransferase (A) Blue dextran-Sepharose column chromatography of partially purified guanylyltranferase. Approximately 0.2 mg of the enzyme from the CM-Sephadex column (*41*) was applied to a column of Blue dextran-Sepharose 4B (0.5 × 13 cm). The proteins were eluted with a 50–500 mM KCl gradient in a buffer solution containing 20 mM Tris–HCl (pH 7.9), 5 mM dithiothreitol, 0.1 mM EDTA, 20% glycerol. Fractions of 0.7 ml were collected, and 10 μL of each fraction was assayed for the capping activity (○), GTP–PP_i exchange (●), and formation of the protein–GMP complex (×). The amount of the complex was determined by measuring the radioactivity of the α-^{32}P in the M_r 69,000 bands in B. (B) Demonstration of the enzyme–GMP complex on dodecyl sulfate/polyacrylamide gel. Enzyme fractions were incubated with [α-^{32}P]GTP. Proteins were precipitated with 5% trichloroacetic acid and electrophoresed. Lanes: 1, sample before chromatography (0.6 μg); 2–6, aliquots (10 μl) from Blue dextran-Sepharose fractions 35, 40, 45, 50, and 55; 7, vaccinia virus capping enzyme. K = 10^3 Da. From Mizumoto *et al.* (*42*).

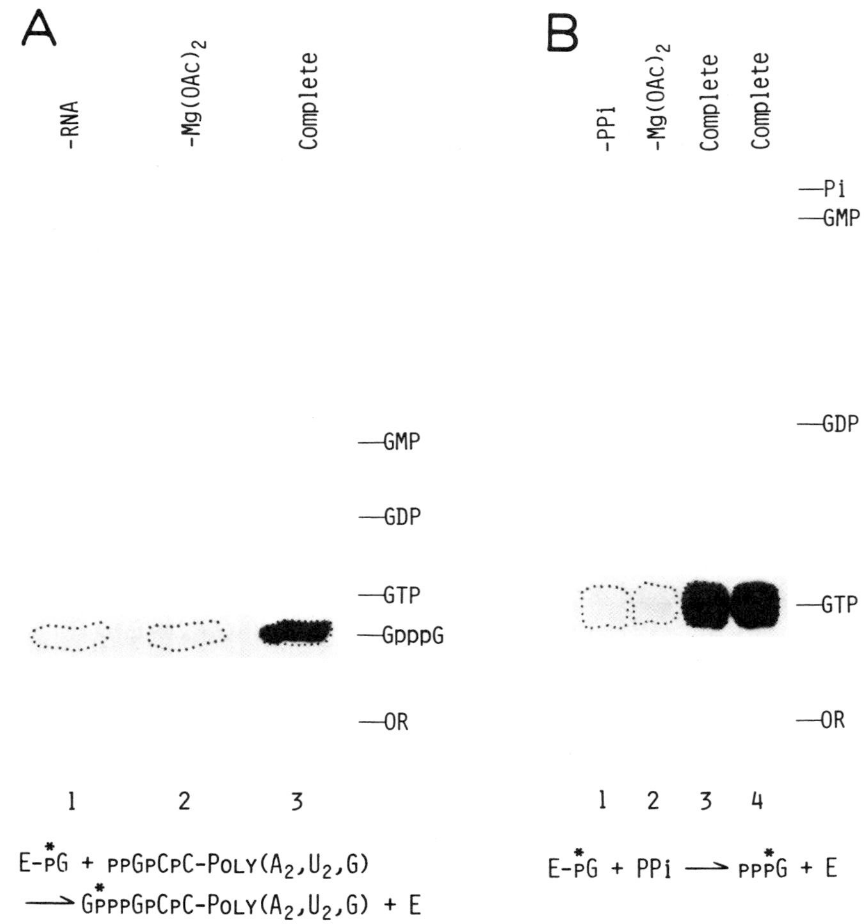

FIG. 3. Synthesis of capped RNA or GTP from the isolated enzyme–GMP intermediate. After incubation of the rat liver guanylyltransferse with [α-^{32}P]GTP, the enzyme–[^{32}P]GMP complex was isolated by gel filtration on a Sephadex G-50 column and used for the following experiments. (A) Forward reaction. The complete reaction mixture (50 μL) contained 40 mM Tris–HCl (pH 7.9), 10 mM dithiothreitol, 10 μg bovine serum albumin, 7000 cpm of the enzyme–[^{32}P]GMP complex, 6 mM Mg(OAc)$_2$, and 200 pmol of ppGCC(A$_2$, U$_2$, G)$_n$ as the cap acceptor. After incubation at 30°C for 10 minutes, RNA was digested with nuclease P$_1$ and subjected to paper electrophoresis on DEAE-cellulose paper. Lanes: 1, without RNA; 2, without Mg^{2+}; 3, complete. (B) Reverse reaction. The complete reaction mixture (50 μL) contained 7000 cpm of the enzyme–[^{32}P]GMP, 6 mM Mg(OAc)$_2$, and 120 μM PP$_i$ (lane 3) or 40 μM PP$_i$ (lane 4). After incubation for 10 minutes at 30°C, the nucleotides in the cold acid-soluble fraction were analyzed by thin-layer chromatography on polyethyleneimine-cellulose plates. The positions of authentic nucleotides (GpppG and GTP) are marked with dotted circles. From Mizumoto *et al.* (*42*).

HeLa cell guanylyltransferase, has shown that this reaction is also reversible.

Messenger RNA guanylyltransferase has been partially purified from various sources and tested for the ability to form an enzyme–GMP complex (*44*). Enzymes from HeLa cells, brine shrimp *Artemia salina,* wheat germ, and yeast each gave a single radioactive band on polyacrylamide gel electrophoresis with M_r values of 69,000, 73,000, 77,000, and 45,000, respectively (Fig. 4). The capping enzyme complex from vaccinia virus yielded a 95,000 complex confirming the results of Shuman and Hurwit (*29*). Guanylyltransferase–GMP inter-

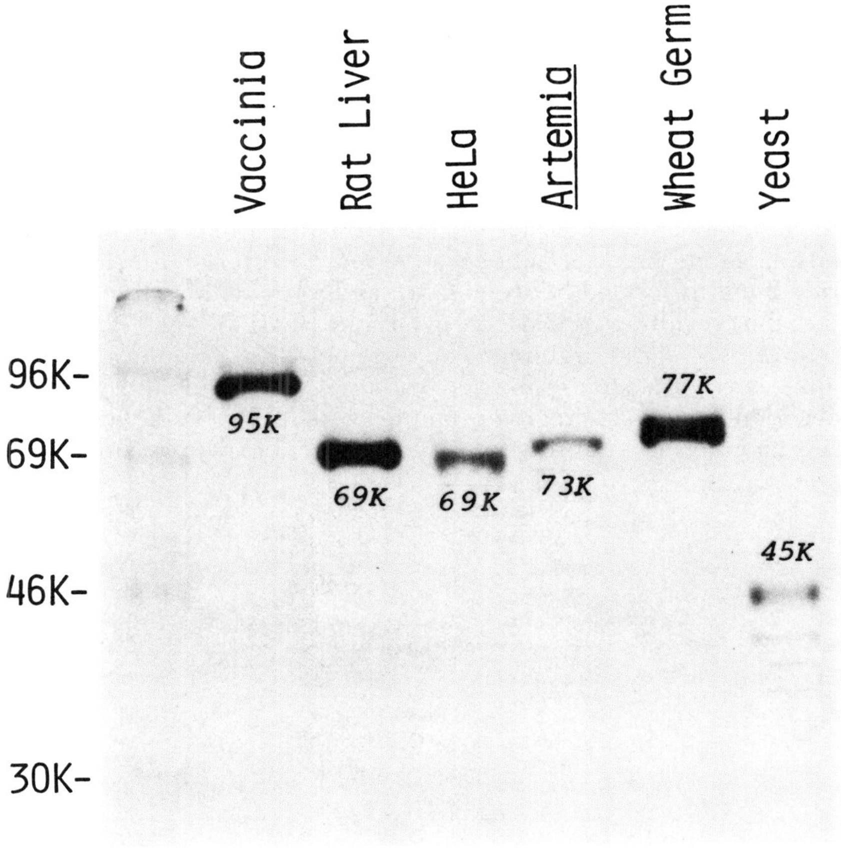

FIG. 4. Guanylylated capping enzymes from various sources. Guanylyltransferase was partially purified from various enzyme sources. Enzyme–[^{32}P]GMP was formed by incubating the enzyme with [α-^{32}P]GTP, acid-precipitated and electrophoresed on dodecyl sulfate/polyacrylamide gel. K = 10^3 Da. From Toyama *et al.* (*44*).

mediates of HeLa cells (*37, 38, 58*), calf thymus (*46*), and yeast (*53*) have also been demonstrated by others. Thus, the covalent catalysis through a guanylylated enzyme intermediate seems to be the universal mechanism of the mRNA capping reaction in eukaryotic cells.

III. Identification of the Amino Acid Residue to Which GMP Is Linked in the Guanylyltransferase–GMP Intermediate

The stability of the rat liver guanylyltransferase–GMP complex in alkaline or acidic solutions has been examined (*42*). More than 75% of the ^{32}P remains in the complex even after heating for 5 minutes at 95°C in 0.1 M NaOH. On the other hand, the complex is cleaved almost completely in 1 to 2 minutes by exposure to 0.1 M HCl at 95°C. The acid lability and the alkali stability of the bond suggested that GMP may be linked to the enzyme through a phosphoamide bond (*59*). In agreement with this is the finding that in acidic hydroxylamine (pH 4.75), the enzyme–GMP complex is rapidly degraded at 37°C, whereas the linkage is stable in neutral hydroxylamine or in acetate buffer at pH 4.75. These characteristics strongly suggest that a phosphoamide linkage, rather than a phosphoester or a mixed anhydride bond, is formed between GMP and a lysine, histidine, or arginine, thus resulting in GMP(5′→N^{ε})lysine, GMP(5′→N^{imid})histidine, or GMP(5′→N^{guanido})arginine. The enzyme–GMP complexes formed with guanylyltransferase from *A. salina*, wheat germ, and yeast behave identically to the rat liver complex with respect to the stability in acids and the susceptibility to cleavage by acidic hydroxylamine (*44*).

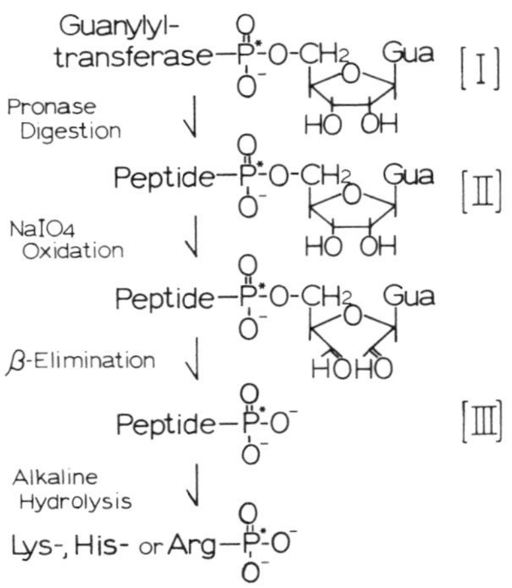

FIG. 5. Strategy for identification of the amino acid residue linked to GMP (*44*). The asterisks denote ^{32}P labels.

A similar covalent enzyme–AMP intermediate participates in the DNA ligase reaction (*60*). In DNA ligase from *E. coli* and T4 phage, the AMP is linked to the ε-amino group of a lysine residue of the enzyme (*61*).

In this work we identified the amino acid after the removal of the guanosine moiety from the enzyme–GMP complex obtained with rat liver guanylyltransferase (Fig. 5; *44*). The guanylyltransferase-[^{32}P]GMP (I) was digested with Pronase to a peptide-GMP (II) consisting of 6 to 8 amino-acid residues. This GMP-containing fragment (II) was then subjected to periodate oxidation followed by β-elimination to yield a phosphopeptide (III). The phosphopeptide (III) was hydrolyzed with potassium hydroxide and was analyzed for N-linked [^{32}P]phosphoamino acids. As shown in Fig. 6A, the major radioactive

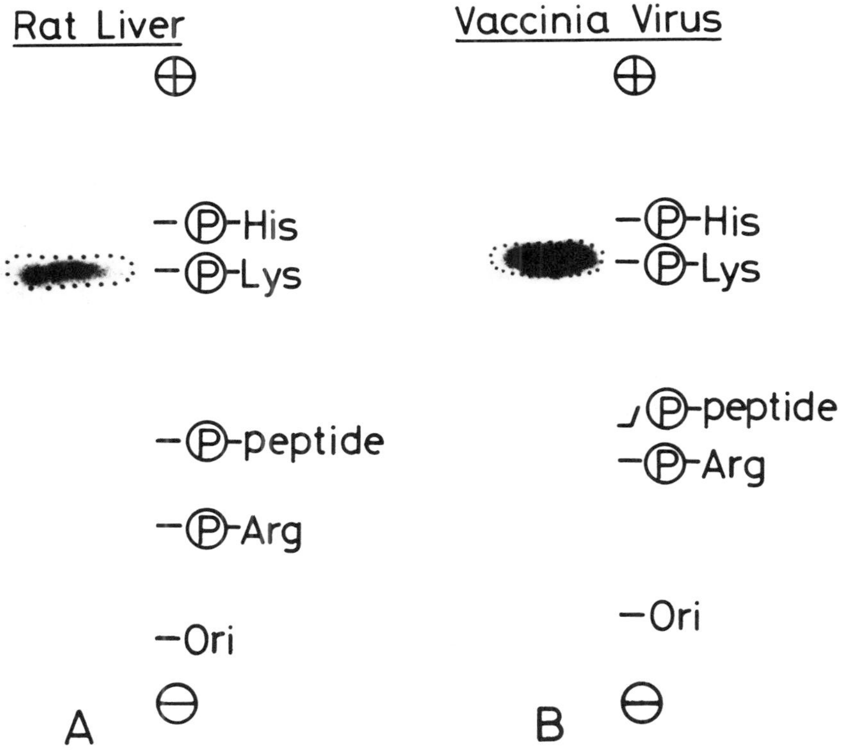

FIG. 6. Isolation of N^{ε}-phospholysine from guanylyltransferase–GMP reaction intermediates. The enzyme–[^{32}P]GMP complexes were prepared from rat liver (A) and vaccinia virus (B) capping enzyme. The complexes were subjected to Pronase digestion and β-elimination followed by alkaline hydrolysis as shown in Fig. 5. The hydrolysates were neutralized and electrophoresed on DEAE-cellulose paper.

product comigrated with the authentic N^{ε}-phospholysine when the alkaline hydrolysate was electrophoresed on a DEAE-cellulose paper. Neither ^{32}P-labeled N^{imid}-phosphohistidine nor N^{guanido}-phosphoarginine was detected. Therefore, it is concluded that the GMP is linked to the ε-amino group of a lysine residue of the enzyme.

N^{ε}-[^{32}P]Phospholysine was also isolated from vaccinia virus capping enzyme–[^{32}P]GMP by the same procedure as above (Fig. 6B; *44*). Recently, Roth and Hurwitz (*62*) also isolated N^{ε}-phospholysine from a vaccinia-virus capping-enzyme–GMP complex. These results suggest that an initial step of mRNA capping catalyzed by guanylyltransferase occurs by a nucleophilic attack of the ε-amino group of a lysine residue of the enzyme on the α-phosphate of GTP.

IV. Association of an RNA 5′-Triphosphatase Activity with mRNA Guanylyltransferase

As described above, the transguanylylation reaction requires the 5′-diphosphate end of the RNA chain. Since capping occurs generally on the initiating nucleotide of the nascent RNA chain (*63, 64*), the 5′-γ-phosphate of the primary transcript has to be removed before the transguanylylation reaction takes place. An RNA 5′-triphosphatase activity is associated with the purified capping enzyme complex from vaccinia virus (*28, 55*). Reovirus also possesses a nucleotide phosphohydrolase activity that removes the γ-phosphate group from the initiating dinucleotide pppGpC before capping (*21*). HeLa-cell extracts contain an RNA triphosphatase activity, which was separated from the guanylyltransferase activity during purification (*36*); however, it was not clear whether the triphosphatase activity is functionally related to the capping reaction or not.

By testing various fractions of rat liver nuclear extracts, we detected an activity that removes the γ-phosphate from [γ-^{32}P]-pppA(pA)$_n$. This activity could not be separated from the guanylyltransferase activity through successive chromatographies on Sephadex G-150, CM-Sephadex, and Blue dextran-Sepharose columns (*43*). Both activities remained physically associated during sedimentation through a glycerol gradient after high-salt treatment (Fig. 7). Since the molecular weight (65,000) of the undenatured enzyme (*41*) is close to that of the enzyme–GMP complex observed in dodecyl sulfate (Figs. 2 and 4), we concluded that rat liver guanylyltransferase is a bifunctional enzyme possessing both guanylyltransferase and RNA 5′-triphosphatase activities in a single polypeptide chain with an M_r of 69,000. The RNA 5′-triphosphatase specifically removed ^{32}P as inor-

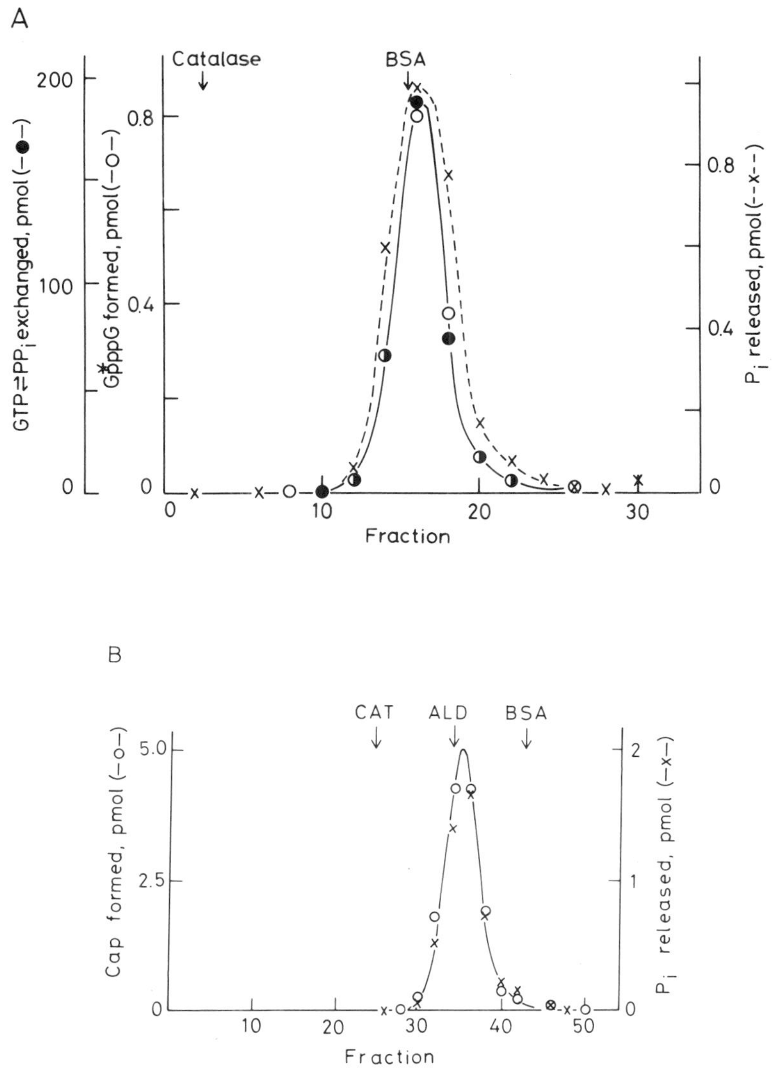

FIG. 7. Association of an RNA 5′-triphosphatase activity with mRNA guanylyltransferase. (A) Glycerol gradient centrifugation of rat liver guanylyltransferase purified through the CM-Sephadex column (*45*). Guanylyltransferase was assayed by cap formaguanylyltransferase purified through the poly(U)-Sepharose column (*51*). (C) DEAE-Sephadex column chromatography of *Artemia salina* guanylyltransferase purified through the CM-Sephadex column (*45*). Guanylyltransferase was assayed by cap formation (O), GTP–PP_i exchange (●), or enzyme–GMP covalent complex formation (△). RNA 5′-triphosphatase activity (×) was assayed by measuring the release of $[^{32}P]P_i$ from $[\gamma\text{-}^{32}P]pppA(pA)_n$ (*43*).

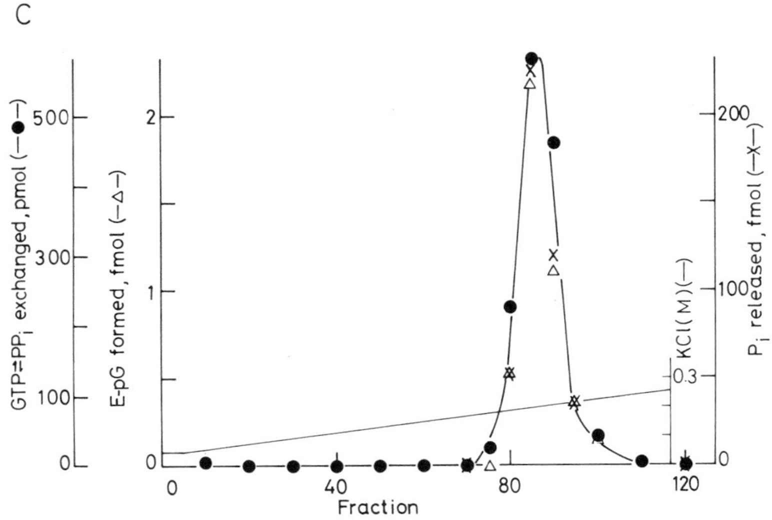

FIG. 7. (Continued)

ganic phosphate from [γ-^{32}P]pppA(pA)$_n$, but not from [β-^{32}P]-pppA(pA)$_n$ or from [γ-^{32}P]ATP (*43*). By testing the activity of the RNA molecules with various chain lengths as substrate, we found that at least one phosphodiester bond is required, and is probably sufficient, to be the substrate for the RNA 5′-triphosphatase activity.

The tight association of RNA 5′-triphosphatase and guanylyltransferase was also observed in the enzymes isolated from *A. salina* and *Saccharomyces cerevisiae* (Fig. 7). The close association of the two enzyme activities seems to be a general feature of the eukaryotic and viral mRNA capping systems.

V. Isolation of Domains for mRNA Guanylyltransferase and RNA 5′-Triphosphatase from *Artemia salina* Capping Enzyme

When the guanylyltransferase–[^{32}P]GMP complex from *A. salina* was digested partially with trypsin at 0°C, and subjected to gel electrophoresis in dodecyl sulfate, the 73-kDa enzyme–GMP complex was quantitatively converted to a 44-kDa ^{32}P-containing fragment (*45*; Fig. 10, lanes 1 and 2). To examine whether this 44-kDa fragment was catalytically active, we treated free guanylyltransferase, not complexed with GMP, with trypsin and then tested the ability to form a protein–[^{32}P]GMP complex. The 44-kDa fragment retained almost full ability to form such a complex upon incubation with [α-^{32}P]GTP.

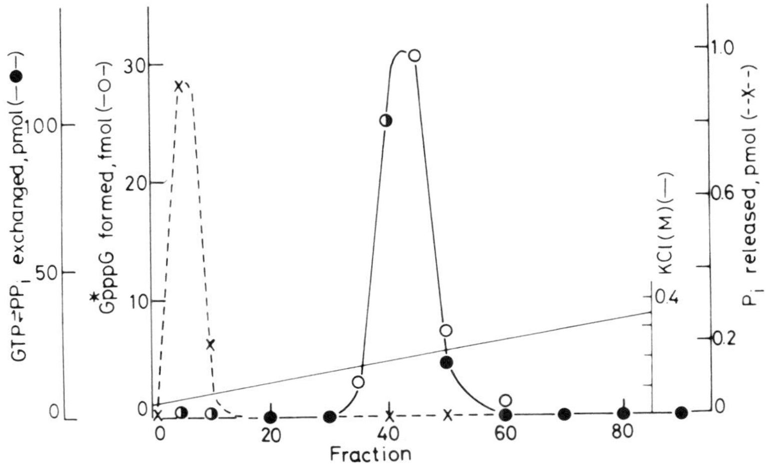

FIG. 8. Separation of domains for guanylyltransferase and RNA 5′-triphosphatase of the *Artemia* capping enzyme. The *A. salina* guanylyltransferase purified through the CM-Sephadex column (2.1 mg) was incubated with trypsin [protein/trypsin, 70 : 1 (w/w)] for 10 minutes at 0°C and trypsin inhibitor was added to terminate the reaction. The digest was applied to a CM-Sephadex column (0.7 × 13 cm) and eluted with 100 mL of a linear gradient of 50–400 mM KCl in a buffer solution containing 20 mM Tris–HCl (pH 7.9), 0.1 mM EDTA, 5 mM 2-mercaptoethanol, 0.5 mM $Mg(OAc)_2$, and 20% glycerol. Fractions (1 mL) were collected and assayed for GTP–PP_i exchange (●), capping (○), and RNA 5′-triphosphatase (×) activities. From Yagi *et al.* (*45*).

The triphosphatase activity was unimpaired, but rather stimulated about 1.5- to 3-fold under these conditions.

When the free enzyme was digested with trypsin and chromatographed on a Sephacryl S-200 column, the activities of guanylyltransferase and triphosphatase were eluted as two separate peaks at positions corresponding to 44 and 20 kDa, respectively (*45*). Complete separation of two domains was achieved by ion-exchange chromatography as shown in Fig. 8. The RNA 5′-triphosphatase domain released labeled inorganic phosphate from [γ-^{32}P]pppA(pA)$_n$ but not from [γ-^{32}P]ATP, indicating that the triphosphatase domain still retains the same specificity as the native enzyme.

The purified guanylyltransferase domain was active only with diphosphate-terminated RNA, and not with triphosphate-terminated RNA. This is in contrast to the native enzyme, which can transfer GMP to either ppN- or pppN-terminated RNA. Only in the presence of the triphosphatase domain could triphosphate-terminated RNA serve as a substrate for the purified 44-kDa guanylyltransferase domain (*45*). For the efficient transfer of GMP from GTP to form

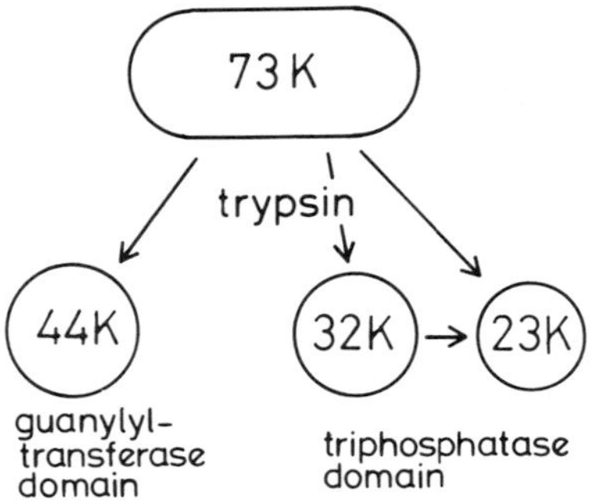

FIG. 9. Domain structure of *Artemia salina* capping enzyme. K = 10^3 Da.

GpppN-, at least one phosphodiester bond is required when either the native enzyme or the isolated guanylyltransferase domain was used.

Recently, we purified the guanylyltransferase of *A. salina* to an almost homogeneous state. Following tryptic digestion the two domains were separated by electrophoresis in dodecyl sulfate and the triphosphatase activity was assayed after *in situ* renaturation of proteins (for details, see Section VI) to estimate the exact size of the triphosphatase domain. Of three major polypeptides (44, 32, and 23 kDa) detected by silver staining, the 32- and 23-kDa fragments had triphosphatase activity. A summary of the separation of two catalytic domains of the capping enzyme from *A. salina* is presented in Fig. 9. Since a 43- to 45-kDa protein–[^{32}P]GMP fragment was also produced by partial digestion of the 73-kDa enzyme–[^{32}P]GMP complex with subtilisin, chymotrypsin, elastase, or papain, the *Artemia* capping enzyme seems to be composed of two domains connected by a protease-sensitive hinge region.

As shown in Fig. 10, a similar domain structure exists in the rat liver capping enzyme, since the 69-kDa enzyme–GMP complex was quantitatively converted to a 40-kDa complex by the trypsin digestion (lanes 3 and 4). In contrast, the size of the intact yeast guanylyltransferase–GMP complex was 45 kDa, a value similar to that of the guanylyltransferase domains of *A. salina* and rat liver, and was refractory to trypsin digestion (lanes 5 and 6). As described in the next section, the yeast enzyme had a subunit structure in which the guanylyltransferase and the triphosphatase were present in separate chains. From these observations, it was concluded that the cellular guanylyltransferase activity resides within a 40- to 45-kDa polypeptide fragment. Vaccinia virus guanylyltransferase–GMP (95 kDa) was converted to a

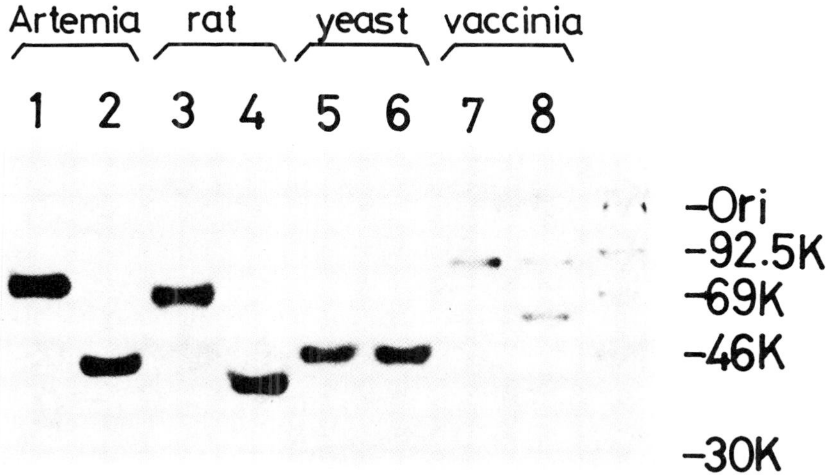

FIG. 10. Detection of the guanylyltransferase domain by limited proteolysis of guanylyltransferase–[^{32}P]GMP intermediate. The enzyme–[^{32}P]GMP complexes were formed by incubation of *A. salina* (lanes 1, 2), rat liver (lanes 3, 4), yeast (lanes 5, 6), and vaccinia virus (lanes 7, 8) guanylyltransferase with [α-^{32}P]GTP and treated for 30 minutes at 0°C without trypsin (lanes 1, 3, 5, 7) or with trypsin (protein/trypsin = 50/1, w/w) (lanes 2, 4, 6, 8). After incubation, proteins were precipitated with cold trichloroacetic acid and subjected to dodecyl sulfate/polyacrylamide gel electrophoresis. K = 10^3 Da.

60-kDa fragment by treatment with trypsin under similar conditions (lanes 7 and 8).

VI. Structure and Function of Yeast Capping Enzyme

The molecular weight of the guanylyltransferase–GMP complex from yeast on gel electrophoresis in dodecyl sulfate was considerably smaller than those from higher eukaryotes (see Figs. 4 and 11), while the size of the undenatured yeast enzyme was estimated to be about 140 kDa by gel filtration. This suggests that the yeast capping enzyme has a subunit structure.

We have purified the yeast capping enzyme to apparent homogeneity from the spheroplasts of a protease-deficient mutant *S. cerevisiae pep* 4 (*51*). The RNA 5′-triphosphatase activity hydrolyzing the γ-phosphate from pppN-RNA was copurified with guanylyltransferase activity through various ion-exchange and affinity column chromatographies, and by sedimentation through a glycerol gradient (Fig. 7). An $s_{20,w}$ value of 7.3 and an M_r of 140,000 were estimated from the

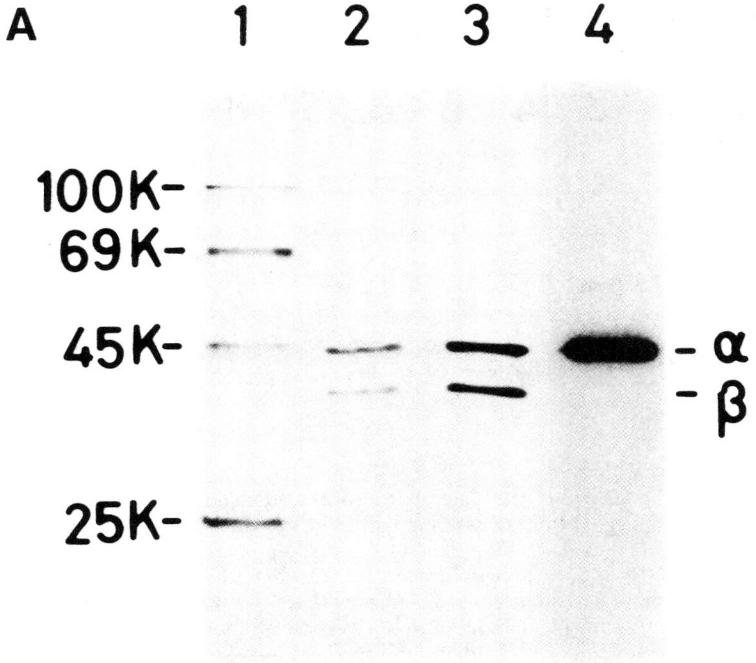

FIG. 11. Dodecyl sulfate/polyacrylamide gel electrophoresis of the purified yeast capping enzyme. (A) The guanylyltransferase purified through a poly(U)-Sepharose column (*51*) was analyzed on 10% polyacrylamide gels in the presence of dodecyl sulfate. The proteins were detected by silver staining (lanes, 1, 2, 3). 1, marker proteins; 2, 37 ng of the purified enzyme; 3, 74 ng of the purified enzyme which had been incubated with [α-^{32}P]GTP to form the E-[^{32}P]GMP; 4, autoradiogram of lane 3. From Itoh *et al.* (*51*). (B) *In situ* detection of guanylyltransferase and triphosphatase activities after gel electrophoresis. For RNA triphosphatase (lane 2), the capping enzyme was electrophoresed through a gel containing [γ-^{32}P]pppA(pA)$_n$; the proteins were then renatured and the gel was incubated under the conditions for the RNA 5′-triphosphatase reaction. For guanylyltransferase (lane 3), the capping enzyme was electrophoresed through a gel containing unlabeled ppA(pA)$_n$, proteins were renatured, and the gel was incubated with [α-^{32}P]GTP under the conditions for capping reaction. After the capping reaction, the gel was treated with cold trichloroacetic acid solution to remove the substrate [α-^{32}P]GTP and the [^{32}P]GMP bound to the enzyme. Lane 1 is the silver-stained protein bands of lane 2. From Itoh *et al.* (*79*).

glycerol gradient centrifugation analysis. Polyacrylamide gel electrophoresis of the purified enzyme gave two polypeptides of M_r 45,000 (α) and 39,000 (β) (Fig. 11A). Since their molar ratios were close to 1 : 1 when estimated by the relative density of silver staining, the

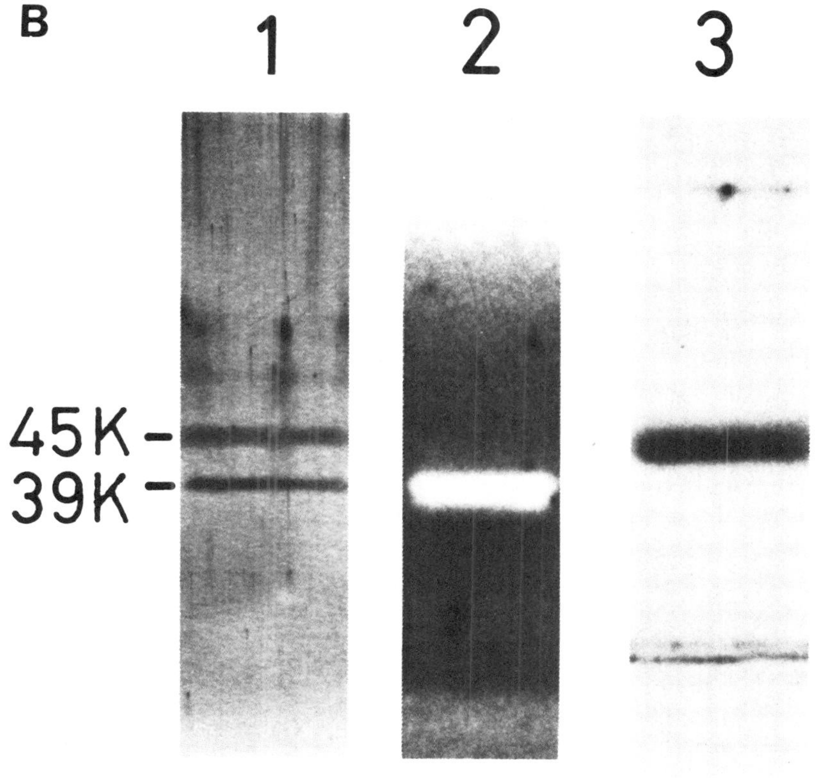

FIG. 11. (Continued)

yeast capping enzyme is an oligomeric protein that may consist of two α and two β chains ($\alpha_2\beta_2$).

When the enzyme was incubated with [α-^{32}P]GTP before dodecyl sulfate/polyacrylamide gel electrophoresis, [^{32}P]GMP was bound to the α subunit (Fig. 11A, lane 4) consistent with the results obtained with the partially purified enzyme (Figs. 4 and 10). This indicates that the catalytic center of guanylyltransferase is in the α-subunit. It is of interest to determine which subunit has the RNA triphosphatase activity. We tried to locate it (triphosphatase activity) on the gel containing [γ-^{32}P]ppp(pA)$_n$ by using the *in situ* assay. As shown in Fig. 11B, after renaturation of the separated proteins in the gel, the RNA triphosphatase activity was detected at a position corresponding to the β chain as a white band on a dark background in the autoradiogram (lane 2). The guanylyltransferase activity, on the other hand, assayed similarly on the gel containing unlabeled ppA(pA)$_n$ by incubation with [α-^{32}P]GTP to form [^{32}P]GpppA(pA)$_n$, was detected as a dark band on a white background corresponding to the α chain (lane 3). This is consistent with the fact that the α chain forms an enzyme–GMP complex. Thus it was clearly demonstrated that the α and the β subunits contain the guanylyltransferase and the triphosphatase, respectively, and that each subunit is active by itself in the absence of the other.

To minimize proteolytic artifacts, we have used a protease-deficient mutant, *S. cerevisiae pep* 4 (66) as the enzyme source, with phenylmethylsulfonyl (α-toluenesulfonyl) fluoride and pepstatin in all buffer solutions used for purification. Under these conditions, we have reproducibly observed only the 45-kDa enzyme–GMP intermediate even in the crude extracts. Furthermore, incubation of the purified rat liver guanylyltransferase with the yeast S-100 fraction caused no detectable fragmentation of the rat liver enzyme. However, these results do not rule out the possibility of proteolytic cleavage during the preparation of spheroplasts with Zymolyase,[2] which lyses yeast cell walls.

Recently, we developed a new procedure for large-scale purification of the capping enzyme from kilogram quantities of yeast cells (79). The procedure employs a glass-beads method to disrupt the cells instead of Zymolyase treatment. On purification of the capping enzyme by this new procedure, we found that the enzyme consists of two polypeptides of M_r 52,000 and 80,000. An $s_{20,w}$ of 9 and M_r of 180,000 were estimated for this preparation. The *in situ* assay after gel electrophoresis of the purified enzyme in dodecyl sulfate demonstrated that the guanylyltransferase activity resides in the 52-kDa polypeptide (α) while the triphosphatase activity is in the 80-kDa polypeptide (β). Probably, a more intact form of the yeast capping enzyme was obtained by the glass-bead disruption procedure. We detected a trace amount of the 45-kDa enzyme–GMP complex in addition to the 52-kDa complex, but no labeled polypeptides larger than 52 kDa appeared during purification. Recently, Wang and Shatkin (53) isolated the capping enzyme from *S. cerevisiae*. According to them, the molecular weight of the enzyme–GMP complex estimated by polyacrylamide gel electrophoresis in dodecyl sulfate was approximately 52,000, a value consistent with our results. However they claimed that their preparation contained no RNA 5′-triphosphatase activity.

[2] Endo-1,3(4)-β-glucanase (EC 3.2.1.6) and/or glucan endo-1,3-β-glucosidase (EC 3.2.1.39). [Eds.]

To see whether the α and β subunits are derived from a single polypeptide or are independent polypeptides encoded by two separate genes, we attempted to isolate the gene(s) coding for the yeast capping enzyme. A preliminary account of this study is given below. A yeast genomic DNA expression library in λgt11 (*67*; kindly supplied by R. W. Davis, Stanford University) was screened with the antibody raised against yeast capping enzyme. The antibody obtained from guinea pig recognized both 80- and 52-kDa subunits when analyzed by Western blotting[3] (*80*) of the 180-kDa yeast capping enzyme. Out of 2×10^6 recombinant phages screened, four plaques gave positive signals, one of which, λC3, contained a 3500-base pair insert of yeast DNA. From experiments based on the affinity selection of the antibody by antigens produced with λC3 in *E. coli*, this clone appears to contain the gene for guanylyltransferase. The insert contained an open reading frame encoding a polypeptide of 459 amino acids with a calculated M_r of 52,764, which corresponds to the size for the intact α subunit (52 kDA). The identity of the gene was further confirmed by expressing the gene in *E. coli* to give catalytically active yeast guanylyltransferase.

From Northern blot analysis[4] (*82*) of the yeast poly(A)$^+$-RNA using a fragment containing the open reading frame as a probe, the size of the guanylyltransferase mRNA was estimated to be about 1600 bases. Since this mRNA size corresponds to a 52-kDa polypeptide, the α and β chains of the yeast capping enzyme are probably encoded by two separate genes.

VII. Discussion

The sequence of reactions involved in the cellular mRNA cap formation is illustrated in Fig. 12. Following the initiation of RNA synthesis by RNA polymerase II, the γ-phosphate of the RNA chain is removed by RNA 5′-triphosphatase to generate the RNA with a 5′-

[3] In "Western blot analysis," electrophoretically separated polypeptides are transferred (by blotting) from the gel to a filter membrane; an antibody is used to form a complex on the membrane with a specific polypeptide, and the membrane-bound complex is detected by the addition of a second (radioactive) antibody directed against the first (*80*).

[4] In "Northern" analysis, poly(A)-containing RNA is electrophoresed through an agarose gel (containing formaldehyde) and transferred to a membrane. The membrane-bound, immobilized RNA is hybridized with radioactive DNA to reveal specific RNA species. (From Weissmann and Weber, in Vol. 33 of this Series, p. 254. Eds.)

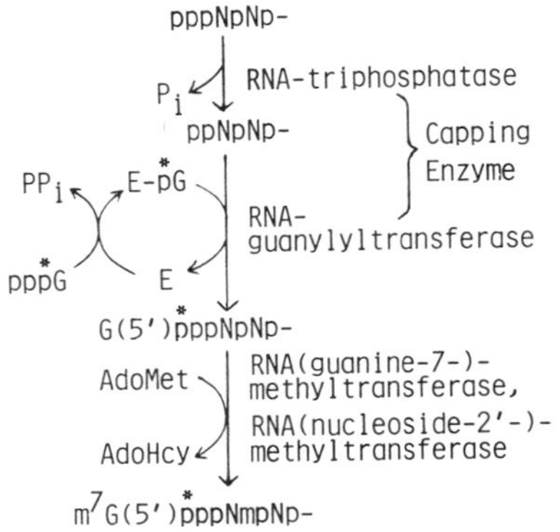

FIG. 12. Sequence of reactions involved in cellular mRNA cap synthesis.

diphosphate end that serves as the substrate for guanylyltransferase. Formation of the 5′-terminal blocked structure, GpppN-, is followed by methylation at the N-7 position of the blocking guanosine residue and in the ribose moiety of the penultimate nucleoside by two different methyltransferases. At present, we cannot predict, from data described here, which methylation step occurs first.

The characteristics of the capping enzymes from various sources are summarized in Table III. In all cases, transguanylylation proceeds through a covalent enzyme–GMP as a reaction intermediate. The vaccinia virus capping enzyme complex with an M_r of approximately 130,000 consists of two subunits and three enzymatic activities, i.e., mRNA guanylyltransferase, mRNA (guanine-7-)methyltransferase, and RNA 5′-triphosphatase. The 95-kDa subunit is known to be involved in the transguanylylation reaction. Cellular capping enzymes contain no (guanine-7-)methyltransferase activity, but guanylyltransferase and triphosphatase are associated physically. The association of triphosphatase and guanylyltransferase activities may be reasonable, since these two reactions could be a sequential process occurring at the very initial stage of mRNA cap formation.

The enzyme from mammalian and *A. salina* consists of a single polypeptide chain with an M_r of approximately 70,000, while the yeast enzyme is composed of two separate polypeptide chains. The 80- and 52-kDa subunits contain RNA 5′-triphosphatase and mRNA guanylyltransferase, respectively. The two subunits are apparently encoded by two separate genes in yeast (see Section VI). Comparison of the gene structures of the mRNA capping enzymes of yeast and animal cells

TABLE III
CHARACTERISTICS OF PURIFIED CAPPING ENZYME

Enzyme sources	MW (undenatured)	Subunit structure	MW of E-pG (SDS-PAGE)[a]	Associated activities[a]
Vaccinia virus	127K[b]	95K[b] 31K	95K[c]	GTase (95K subunit) MTase TPase
Rat liver nuclei	65K	No	69K	GTase (40K domain) TPase
Brine shrimp (*A. salina*)	73K	No	73K	GTase (44K domain) TPase (23K domain)
Yeast	140K (*180K*)	45K (*52K*) 39K (*80K*)	45K (*52K*)	GTase [45 (*52*)K subunit][d] TPase [39 (*80*)K subunit][d]

[a] K = 10^3. GTase, mRNA guanylyltransferase; MTase, mRNA (guanine-7-)methyltransferase; TPase, RNA 5′-triphosphatase.

[b] From ref. (*27*).

[c] From refs. (*29*) and (*58*).

[d] Values in italics are those obtained with the enzyme purified from extracts prepared by glass-bead disruption (*79*).

will be of interest in connection with the process of gene fusion that may have occurred during evolution.

Another important question we would like to address is why only the transcripts of RNA polymerase II are capped. Studies with the purified capping enzyme indicated no requirements for specific sequences of the RNA substrate, since synthetic homo- and heteropolymers were efficiently capped (*36, 41, 69*). No common 5′-terminal sequences were found in cellular mRNAs, although many mRNAs have A-Y-N-Y-Y-Y at their 5′-termini (*70*). Furthermore, primary transcripts of RNA polymerase I and III[1] can also be capped with the purified viral and cellular enzymes *in vitro* (*71–73*). Thus, it is unlikely that the 5′-terminal sequences of polymerase II transcripts are specifically recognized by capping enzyme.

By brief exposure of Chinese hamster ovary cells to methyl-labeled methionine, Salditt-Georgieff *et al.* (*13*) demonstrated that the addition of cap structure occurs at a very early stage of hnRNA synthesis. Also, analyses of *in vitro* transcription products from the adenovirus major late promoter indicate that capping seems to occur when RNA chains grow to between 20 and 50 nucleotides after transcription initiation (*15, 16*). AdoHcy specifically inhibits the *in vitro* transcription initiation by RNA polymerase II without affecting the transcription by RNA polymerase III (*10*), suggesting a close involvement of the cap methyltransferase(s) in the initiation by RNA polymerase II. These observations suggest that cap formation is a promoter-proximal event that occurs concomitantly with the synthesis of the nascent RNA chain. It is reasonable to imagine that the initiation complexes formed with RNA polymerase II may contain the capping enzyme system. We have isolated an RNA polymerase II transcription initiation complex using the HeLa whole cell extract and the DNA that contains the adenovirus major late promoter. The template DNA was incubated with the HeLa cell extract in the presence of ATP or dATP to form a transcription initiation complex (*74*). The initiation complex with an approximate size of 50 S could be isolated by glycerol gradient centrifugation free from the bulk of RNA polymerase II, mRNA guanylyltransferase, and mRNA (guanine-7-)methyltransferase. Specific transcription was detected with this complex when supplemented with the remaining nucleoside triphosphates. Analyses of the 5′-terminal structure of the transcript revealed the presence of a cap structure (m^7GpppA) in more than 50% of RNA chains. These results indicate that at least two cap-related enzymes, mRNA guanylyltransferase and mRNA (guanine-7-)methyltransferase, are specifically associated with the complex. The precise step of transcription during which capping takes place and the interactions between the capping enzyme system and transcription factors remain to be studied.

Acknowledgments

The authors acknowledge the contributions of N. Itoh, Y. Yagi, R. Toyama, H. Yamada, Z. Tsuchihashi, and Y. Shibagaki to the research from our laboratory.

References

1. A. J. Shatkin, *Cell* **9**, 645 (1976).
2. A. K. Banerjee, *Microbiol. Rev.* **44**, 175 (1980).
3. G. W. Both, Y. Furuichi, S. Muthukrishnan and A. J. Shatkin, *Cell* **6**, 185 (1975).
4. J. K. Rose, *JBC* **250**, 8098 (1975).
5. W. Filipowicz, *FEBS Lett.* **96**, 1 (1978).
6. Y. Furuichi, A. LaFiandra and A. J. Shatkin, *Nature* **266**, 235 (1977).
7. K. Shimotohno, Y. Kodama, J. Hashimoto and K. Miura, *PNAS* **74**, 2734 (1977).
8. Y. Furuichi, *JBC* **256**, 483 (1981).
9. A. K. Banerjee, G. Abraham and R. J. Colono, *J. Gen. Virol.* **34**, 1 (1977).
10. R. Jove and J. L. Manley, *PNAS* **79**, 5842 (1982).
11. A. R. Krainer, T. Maniatis, B. Ruskin and M. R. Green, *Cell* **36**, 993 (1984).
12. M. M. Konarska, R. A. Padgett and P. A. Sharp, *Cell* **38**, 731 (1984).

13. M. Salditt-Georgieff, M. Harpold, S. Chen-Kiang and J. E. Darnell, *Cell* **19**, 69 (1980).
14. J. L. Manley, P. A. Sharp and M. L. Gefter, *PNAS* **76**, 160 (1979).
15. J. A. Coppola, A. S. Field and D. S. Luse, *PNAS* **80**, 1251 (1983).
16. R. Jove and J. L. Manley, *JBC* **259**, 8513 (1984).
17. R. P. Perry and D. E. Kelley, *Cell* **8**, 433 (1976).
18. J. E. Darnell, Jr., this series **22**, 327 (1979).
19. Y. Furuichi and K. Miura, *Nature* **253**, 374 (1975).
20. K. Shimotohno and K. Miura, *FEBS Lett.* **64**, 204 (1976).
21. Y. Furuichi, S. Muthukrishnan, J. Tomasz and A. J. Shatkin, *JBC* **251**, 5043 (1976).
22. Y. Furuichi, S. Muthukrishnan, J. Tomasz and A. J. Shatkin, this series **19**, 3 (1976).
23. B. Moss, A. Gershowitz, C.-M. Wei and R. Boone, *Virology* **72**, 341 (1976).
24. B. Moss, S. A. Martin, R. F. Boone and C.-M. Wei, this series **19**, 63 (1976).
25. G. Abraham, D. P. Rhodes and A. K. Banerjee, *Cell* **5**, 51 (1975).
26. D. Testa and A. K. Banerjee, *J. Virol.* **24**, 786 (1977).
27. S. A. Martin, E. Paoletti and B. Moss, *JBC* **250**, 9322 (1975).
28. S. Shuman, M. Surks, H. Furneaux and J. Hurwitz, *JBC* **255**, 11588 (1980).
29. S. Shuman and J. Hurwitz, *PNAS* **78**, 187 (1981).
30. E. Barbosa and B. Moss, *JBC* **253**, 7698 (1978).
31. Y. Groner and J. Hurwitz, *PNAS* **72**, 2930 (1975).
32. I. Winicov and R. P. Perry, *Bchem.* **15**, 5039 (1976).
33. M. J. Ensinger and B. Moss, *JBC* **251**, 5283 (1976).
34. C.-M. Wei and B. Moss, *PNAS* **74**, 3758 (1977).
35. J. M. Keith, M. J. Ensinger and B. Moss, *JBC* **253**, 5033 (1978).
36. S. Venkatesan, A. Gershowitz and B. Moss, *JBC* **255**, 2829 (1980).
37. D. Wang, Y. Furuichi and A. J. Shatkin, *MCBiol.* **2**, 993 (1982).
38. S. Shuman, *JBC* **257**, 7237 (1982).
39. S. R. Langberg and B. Moss, *JBC* **256**, 10054 (1981).
40. K. Mizumoto and F. Lipmann, *FP* **37**, 1734 (1978).
41. K. Mizumoto and F. Lipmann, *PNAS* **76**, 4961 (1979).
42. K. Mizumoto, Y. Kaziro and F. Lipmann, *PNAS* **79**, 1693 (1982).
43. Y. Yagi, K. Mizumoto and Y. Kaziro, *EMBO J.* **2**, 611 (1983).
44. R. Toyama, K. Mizumoto and Y. Kaziro, *EMBO J.* **2**, 2195 (1983).
45. Y. Yagi, K. Mizumoto and Y. Kaziro, *JBC* **259**, 4695 (1984).
46. Y. Nishikawa and P. Chambon, *EMBO J.* **1**, 485 (1982).
47. J. M. Keith, S. Venkatesan, A. Gershowitz and B. Moss, *Bchem.* **21**, 327 (1982).
48. C. Locht and J. Delcour, *EJB* **152**, 247 (1985).
49. J. Germershausen, D. Goodman and E. W. Somberg, *BBRC* **82**, 871 (1978).
50. N. Itoh, K. Mizumoto and Y. Kaziro, *FEBS Lett.* **155**, 161 (1982).
51. N. Itoh, K. Mizumoto and Y. Kaziro, *JBC* **259**, 13923 (1984).
52. N. Itoh, K. Mizumoto and Y. Kaziro, *JBC* **259**, 13930 (1984).
53. D. Wang and A. J. Shatkin, *NARes* **12**, 2303 (1984).
54. C. Locht, J.-L. Beaudart and J. Delcour, *EJB* **134**, 117 (1983).
55. S. Venkatesan, A. Gershowitz and B. Moss, *JBC* **255**, 903 (1980).
56. M. H. Haffner, M. B. Chin and B. G. Lane, *Can. J. Biochem.* **56**, 729 (1978).
57. T. Cheng and H. H. Kazazian Jr., *JBC* **253**, 246 (1978).
58. S. Venkatesan and B. Moss, *PNAS* **79**, 340 (1982).
59. Z. A. Shabarova, this series **10**, 145 (1970).
60. I. R. Lehman, *Science* **186**, 790 (1974).
61. R. Gumport and I. R. Lehman, *PNAS* **68**, 2559 (1971).

62. M. J. Roth and J. Hurwitz, *JBC* **259,** 13488 (1984).
63. D. Gidoni, C. Kahana, D. Cananni and Y. Groner, *PNAS* **78,** 2174 (1981).
64. O. Hagenbüchle and U. Schibler, *PNAS* **78,** 2283 (1981).
65. K. C. Gupta and P. Roy, *J. Virol.* **33,** 292 (1980).
66. E. W. Jones, *Genetics* **85,** 23 (1977).
67. R. A. Young and R. W. Davis, *Science* **222,** 778 (1983).
68. J. R. Morgan, L. K. Cohen and B. E. Roberts, *J. Virol.* **52,** 206 (1984).
69. S. A. Martin and B. Moss, *JBC* **251,** 7313 (1976).
70. R. Breathnach and P. Chambon, *ARB* **50,** 349 (1981).
71. H. Saiga, K. Mizumoto, T. Matsui and T. Higashinakagawa, *NARes* **10,** 4223 (1982).
72. I. Financsek, K. Mizumoto and M. Muramatsu, *PNAS* **79,** 3092 (1982).
73. I. Financsek, K. Mizumoto and M. Muramatsu, *Gene* **18,** 115 (1982).
74. H. E. Tolunay, L. Yang, W. F. Anderson and B. Safer, *PNAS* **81,** 5916 (1984).
75. G. Monroy, E. Spencer and J. Hurwitz, *JBC* **253,** 4481 (1978).
76. M. Zasloff and S. Ochoa, *PNAS* **68,** 3059 (1971).
77. K. Marcu and B. Dudock, *NARes* **1,** 1385 (1974).
78. R. E. Smith, H. J. Zweerink and W. K. Joklik, *Virology* **39,** 791 (1969).
79. N. Itoh, H. Yamada, Y. Kaziro, and K. Mizumoto, *JBC* **262,** 1989 (1987).
80. W. N. Burnette, *Anal. Biochem.* **112,** 195 (1981).
81. H. Towbin, T. Staehelin and J. Gordan, *PNAS* **7,** 4350 (1979).
82. T. Maniatis, E. F. Fritsch and J. Sambrook, *in* "Molecular Cloning, A Laboratory Manual," p. 202. Cold Spring Harbor Laboratory, New York, 1982.

The Genome of *Mycoplasma capricolum*

Akira Muto,
Fumiaki Yamao, and
Syozo Osawa

Laboratory of Molecular Genetics
Department of Biology
Faculty of Science
Nagoya University
Chikusa, Nagoya, Japan

Mycoplasmas are the smallest and perhaps the simplest free-living organisms so far known. Their genome DNA size is about 5×10^8 Da (700 kilobase pairs, kbp), which is one-fourth to one-fifth of that of *Escherichia coli*. Thus, mycoplasmas have been regarded as the "minimum living system." In spite of this biological interest, the genetics of mycoplasmas has remained undeveloped until recently, mainly because of the inadequacy of the classical genetic techniques to the organisms. The recent development of recombinant DNA techniques has opened the way to analyze mycoplasma genes directly. Several specific genes have already been isolated from various species of mycoplasmas and characterized by DNA sequencing techniques. This article describes the organization and structure of the mycoplasma genes, especially of *Mycoplasma capricolum* ATCC27343 (KID), which our group has been studying for years. Several characteristic genetic features of this organism have been found, the most outstanding being the deviation of one of the codons from the "universal genetic code" (see Table I, on p. 45).

Progress in Nucleic Acid Research
and Molecular Biology, Vol. 34

I. Characteristics of Mycoplasmas and Their Genomes

The wall-less prokaryotes have been grouped in the class *Mollicutes* and referred to as mycoplasmas. The class *Mollicutes* includes five genera: *Mycoplasma, Ureaplasma, Spiroplasma, Acholeplasma,* and *Anaeroplasma.* They are widely distributed in nature as parasites of animals and plants. Some of them are significant pathogens for mammalian, green plant, and insect diseases.

The genome of *Mollicutes* is distinguished by its small size and low guanine-plus-cytosine content from most of other bacteria. The mycoplasma genomes fall into two classes in size. The masses of *Acholeplasma, Spiroplasma,* and *Anaeroplasma* genomes are about 1×10^9 Da; those of *Mycoplasma* and *Ureaplasma* genomes are about 5×10^8 Da, the smallest of all cellular genomes known. The G+C contents of all the mycoplasmas so far reported range between 25 and 41%. Many mycoplasmas, including *M. capricolum,* exhibit values as low as 25%, the lowest in all prokaryotes. Like other bacteria, the genome of mycoplasmas is a single circular double-stranded DNA (for reviews, see refs. *1–4*).

II. Phylogeny of Mycoplasmas

The origin and phylogenetic relationships of mycoplasmas have been of interest because of their small genome size and simple structure. These features led some workers to suggest that they might be descendants of a primitive type of organism that preceded the present-day eubacteria in the evolutionary process (*5*). However, a number of studies indicate that it is not. A clear view of mycoplasma evolution has come from comparative sequence analyses of ribosomal RNA (rRNA). Hori *et al.* (*6*) sequenced the 5-S rRNA of *M. capricolum* and demonstrated that the overall nucleotide sequence and the secondary structure of the RNA are closer to those of gram-positive than gram-negative eubacteria. Comparisons of mycoplasma tRNA and 16-S rRNA sequences with those of other bacteria also revealed the same relationship (*7–9*). Successively, the 5-S rRNA sequences from 10 species of the class *Mollicutes,* including *Acholeplasma, Spiroplasma, Ureaplasma,* and *Anaeroplasma,* were determined and used to construct a phylogenetic tree (*10, 11*) showing that all the species studied form a phylogenetically related group. It also showed that two clostridia, *Clostridium innocuum* and *C. ramosus,* which are gram-positive bacteria having low genomic G+C content, are related to the mycoplasma group (*10*). Figure 1 shows the phylogenetic tree of eu-

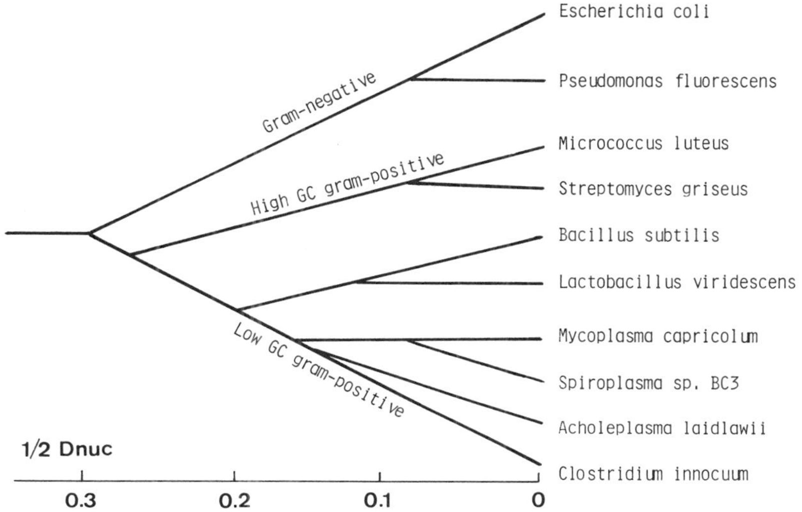

FIG. 1. Phylogenetic tree of eubacterial 5-S rRNA. *Dnuc* represents the rate of nucleotide substitutions, corrected by DNA base composition (*11*).

bacteria constructed from 5-S rRNA sequences and including some representative mycoplasma species (*11*). The mycoplasma and clostridia groups originated from the *Bacillus–Lactobacillus* branch of the gram-positive bacterial stem, followed by separation of the mycoplasma group from the clostridia group.

The tree provides two interesting aspects of mycoplasma evolution. First, mycoplasmas would not be primitive forms of eubacteria, but would have been brought about from certain gram-positive bacterial groups by degeneration. Therefore, the small size of their genome would be due to deletions of parts of the genome during evolution under the constraints of limited genome complexity. Second, the low G+C content of their genomic DNA is an evolutionary event. All species so far placed on the *Bacillus–Mycoplasma* branch of the gram-positive bacteria have low G+C contents in their genomes. They must have evolved with some constraints keeping the G+C content in their genomes low. We assume that some mutation pressure replacing G·C pairs by A·T pairs of DNA has been exerted on the genome during mycoplasma evolution. Our main interest is to see how these two evolutionary constraints have affected the organization and structure of mycoplasma genes.

III. Overall Composition of the Genome

The small size of the mycoplasma genome should reflect the number of encoded genes. Assuming that the average molecular weight of the mycoplasma proteins is 40,000, and that the structural genes for proteins are packed in the genome without spacers, 5×10^8 Da DNA is capable of coding about 650 different proteins (*12*). Since the bacterial genome usually contains spacers, including various signals, that comprise 20–30% of the total DNA, the actual number of genes would be around 500.

The simplest way to estimate the approximate number of genes encoded in the genome may be to count the number of the gene products (proteins and RNAs) in the cell. We analyzed by two-dimensional polyacrylamide gel electrophoresis (2D-PAGE) of O'Farrell *et al.* (*13, 14*) the entire protein content of *M. capricolum* (*15*). Figure 2 shows the 2D-PAGE patterns of the total proteins labeled with ^{14}C-containing amino acids. Acidic proteins were separated using isoelectrofocusing gels for the first dimension with a gradient of pH 5–7 (b), and basic proteins with a gradient of pH 7–9 (a). A total of about 350 spots, 200 in the acidic and 150 in the basic range, is detected. Under the same conditions, the numbers of the whole cell proteins of *E. coli* and *B. subtilis* are about 1,100 (*14, 15*). The results indicate that at least 350 genes for proteins are expressed in the growing *M. capricolum* cells. The average molecular weight of the proteins is about 40,000. The number of expressed genes would represent roughly the number of genes for proteins encoded in the genome. The small number of the protein spots is consistent with the view that mycoplasmas contain only a limited number of the genes necessary for growth (*12*), and that most, if not all, of them must be expressed in the growing cells.

A characteristic feature of the *M. capricolum* proteins is the relative abundance of basic proteins, especially large ones. The proportion of basic proteins in the total proteins of *M. capricolum* is about 40%, which is much higher than for *E. coli* and *B. subtilis* (about 15%).

The ribosomal proteins have been isolated from the *M. capricolum* ribosomes and separated by 2D-PAGE (*16*). The number of spots detected was about 30 for 50-S and 20 for 30-S subunits, indicating that the *M. capricolum* ribosome contains at least about 50 different proteins (*15*). The molecular weights of these proteins range from 9,000 to 40,000, averaging about 15,000. Thus, the number and size of the *M. capricolum* ribosomal proteins are not significantly different from

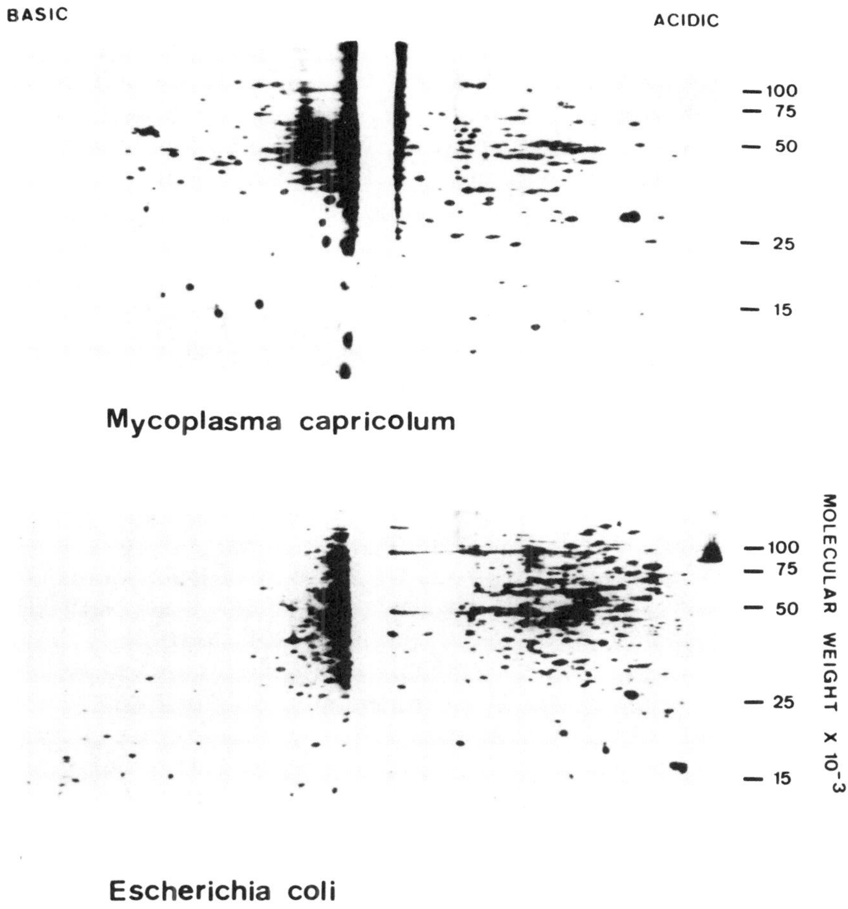

FIG. 2. Protein composition of *Mycoplasma capricolum*. 2D-PAGE patterns of whole cell proteins from *M. capricolum* and *E. coli* (15). Left: basic proteins (pH 7–9 at the first dimension). Right: acidic proteins (pH 5–7).

other eubacterial ribosomal proteins. This suggests that the genes indispensable for growth such as those for ribosomal proteins are well conserved in the mycoplasma genome, in spite of a great reduction of the total number of genes (see Section IV,C).

The number of tRNA genes in the *M. capricolum* genome has been estimated to be about 44 by hybridization and saturation experiments (*17*). The 2D-PAGE analysis of the *M. capricolum* total tRNA revealed about 35 spots (F. Yamao, unpublished data). Thus, the number of tRNA species may be around 40. There are three rRNA species,

16 S, 23 S and 5 S, in mycoplasmas; these are common in all prokaryotes. The *M. capricolum* genome contains two sets of genes for each rRNA species (see Section IV,A) (*18*). When the number of the genes for 350 protein species is added to about 50 genes for tRNA and rRNA species, a total of about 400 genes is expressed in *M. capricolum* cells. This is not far from the total number of the genes calculated from the size of the genome mentioned above (about 500).

What kind of protein genes are they? The mycoplasma cell has a simple structure; the only highly ordered architectures observable by electron microscopy are cell membranes and ribosomes. Mycoplasmas require highly complex media for growth. A partially defined culture medium for *M. capricolum* has been established (*19*); it contains 20 amino acids, 4 bases, 5 or 6 vitamins, cholesterol, and several fatty acids. These complex nutritional requirements suggest that most, if not all, of the genes for the metabolic pathway enzymes for producing building blocks of macromolecules are missing in the genome. Instead, membrane systems for transport of these building blocks and vitamins from the environment should exist in this organism.

An energy-generating system is also indispensable for autonomous growth of these cells. *M. capricolum* contains glycolysis and lactate fermentation pathways to support the ATP production, while the tricarboxylate cycle and cytochromes are not present. Thus, the number of the genes necessary for ATP production must be very small.

Protein-, DNA-, and RNA-synthesizing systems are also essential. On the *E. coli* genome, about 270 genes for the components of systems synthesizing proteins and nucleic acids (rRNAs, tRNAs, ribosomal proteins, aminoacyl-tRNA synthetases, RNA polymerase, DNA polymerase, transcription and translation factors) have been found and mapped on the chromosome. Most of these genes should be essential for growth and so must also be present on the mycoplasma genome. Well-conserved organization and structures of the genes for rRNA, tRNA, and ribosomal protein in *M. capricolum* (see Section IV) support this view. Since the total number of genes of this organism has been estimated to be 400–500, the major part (possibly more than a half) of the genome may be occupied by the genes for the components of systems synthesizing proteins and nucleic acids mentioned above. The rest of the genome may be composed of the genes for membrane proteins, the ATP-generating system, and some others.

A possible and perhaps the simplest way to know the precise number and kind of the genes encoded in the *M. capricolum* genome may be the complete sequencing of the total DNA. The task of sequencing a 700 kbp DNA is possible, since the 172 kbp genome of the Epstein–

Barr virus has been sequenced. Most of the protein genes of *M. capricolum* could be identified by comparing their sequences with those of other organisms, since many of the essential genes of *E. coli* and other bacteria have already been sequenced. Recently, Morowitz (*20*) proposed launching an international project for sequencing the entire mycoplasma genome. This project is naturally laborious, but involves no conceptual difficulties. If achieved, it would lead to the establishment of the genetic basis of cell proliferation, the most fundamental feature of living systems.

IV. Gene Organization and Structure

A library of the *M. capricolum* genomic clones was constructed by insertion of *Eco*RI-, *Hin*dIII-, or *Bgl*ll-digested fragments of the DNA into plasmid pBR322 or pBR325 using *E. coli* HB101 as a host. A total of about 2500 clones were isolated and used to select several representative genes of *M. capricolum.* We have so far isolated recombinant plasmids including rRNA genes, tRNA genes, and some ribosomal protein genes, respectively, and analyzed their gene organization and structures mainly by DNA sequencing techniques.

A. rRNA Genes

An early hybridization and saturation experiment (*17*) suggested that the *M. capricolum* genome carries only one set of rRNA genes. However, Southern hybridization[1] of restriction enzyme-digested DNA with ^{32}P-labeled 16-S, 23-S, and 5-S rRNAs, indicates the presence of two sets of genes for these three rRNA species (*18*). A recent survey for the number of rRNA gene sets reveals that all species of *Mollicutes* so far tested carry one or two sets of rRNA genes (*21–23*). The rRNA gene sets are 7 in *E. coli* (*24, 25*) and 10 in *B. subtilis* (*26*). Thus, the small size of the mycoplasma genome seems to reflect the small number of rRNA gene sets.

We have isolated the clones including the two rRNA gene clusters and analyzed their organization and structure by Southern hybridization and DNA sequencing (*27, 28*). Both of the clusters, referred to as

[1] In "Southern" analysis, restriction fragments are separated by electrophoresis on an agarose gel and transferred (by blotting) from the gel to a filter membrane. The membrane-bound, immobilized DNA is hybridized with a radioactive, specific DNA (for example, a cloned DNA) or RNA to identify the DNA fragments carrying cognate sequences. (From Weissman and Weber, in Vol. 33, of this Series, p. 254.) [Eds.]

rrnA and *rrnB*, have the same gene arrangement, (5′)-16S-23S-5S-(3′). The order is the same as those of other bacteria, such as *E. coli* (*29*), *B. subtilis* (*30*), and *Anacystis nidulans* (*31*). However, the spacers between 16-S and 23-S rRNA genes of both *rrnA* and *rrnB* do not include the tRNA gene (*28*). All the seven rRNA operons of *E. coli* (*32–34*), all the two rRNA gene clusters of *A. nidulans* (*31*), and 2 out of 10 rRNA gene clusters of *B. subtilis* (*35*) contain one or two tRNA genes in the spacers. Thus, the overall organization of *M. capricolum* rRNA gene clusters resembles those of the eight *B. subtilis* clusters that do not contain spacer tRNA gene.

To see the organization of the rRNA gene cluster in more detail, we determined the DNA sequence of most parts of the *rrnB* cluster, including the total 16-S rRNA gene, the spacers between 16-S and 23-S rRNA genes and between 23-S and 5-S rRNA genes, and about 1-kbp 5′- and 3′-flanking sequences of the structural genes (*27, 28*; unpublished data). The spacers and the 5′- and 3′-flanking sequences are extremely poor in G and C (20%) in contrast to a relatively high G+C content (48%) of the coding sequences of rRNAs. The G+C content of the spacers is even lower than that of the total genomic *M. capricolum* DNA (25%). There are two promoter-like sequences about 200 and 300 bp upstream from the 5′ end of the 16-S rRNA gene. The sequences are homologous to other typical bacterial promoter consensus sequences including −10 (Pribnow box) and −35 regions (see Section IV,D). The rRNA gene cluster is followed by a terminator-like sequence; a dyad symmetrical structure and a stretch of thymidine residues (see Fig. 3). The structure suggests that the three rRNA genes form a single transcription unit, an operon, as in the cases of other bacterial rRNA gene clusters. Long complementary stretches between the 5′- and 3′-flanking sequences of both 16-S and 23-S rRNA genes can be constructed, suggesting that the rRNA gene transcript can generate large loop-stem structures by pairing the flanking sequences (Fig. 3). These stem structures are also known for other bacterial rRNA genes, such as *E. coli* (*36*) and *B. subtilis* (*37, 38*), as possible substrates for processing enzymes during rRNA maturation (*36, 37*). The base-paired stretches of 12 bp with almost the same sequence exist in the root of the 16-S and 23-S stems (sequences boxed in Fig. 3). Interestingly, this structure with very similar sequences has also been observed in the homologous regions of *B. subtilis* (*37, 38*). These sequences do not exist in *E. coli*. These regions may be the target sites for a processing enzyme in *B. subtilis* (*37*). This would mean that the processing signals of the rRNA gene transcripts as well as the specificity of the processing enzymes are conserved between *M. capricolum*

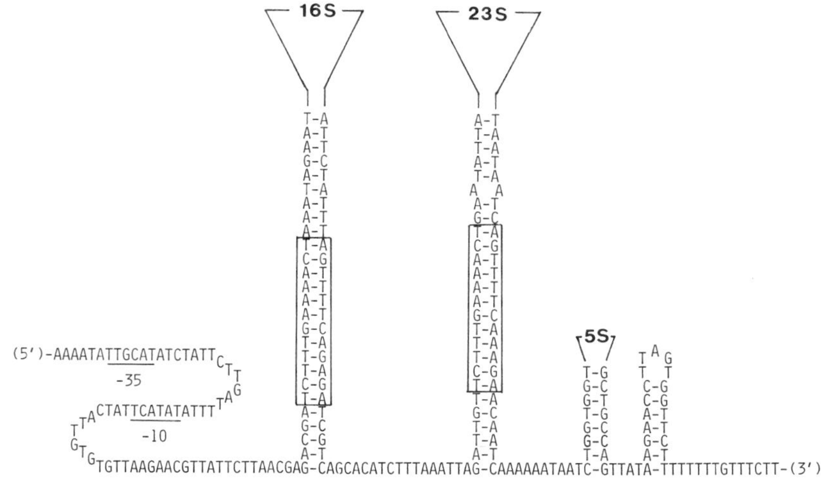

FIG. 3. Organization of rRNA genes in *M. capricolum*. A possible secondary structure for the transcript of *rrnB* gene cluster. The repeated stem structures are boxed. One of the promoter-like sequence (−10 and −35 regions: P2) and the terminator-like sequence ("hairpin" structure) are also indicated.

and *B. subtilis* but are different from *E. coli*, providing further evidence that mycoplasmas are related to gram-positive bacteria.

The total sequence of 16-S rRNA gene in *rrnB* of *M. capricolum* (*28*) is 1521 bp long, being 21 bp shorter than that of the *E. coli* 16-S rRNA gene (*39*), with a remarkable sequence similarity to those of *E. coli* (74%) and *A. nidulans* (75%) (*40*). The similarity to the *B. subtilis* sequence is much higher (82%) as expected (*41*). The 16-S rRNA gene of *Mycoplasma* strain PG50 is 99% identical with the *M. capricolum* sequence (*22*).

Secondary structure models of eubacterial 16-S rRNA have been proposed (*42–44*). Almost all the stem and loop structures in these models can be observed in *M. capricolum*, suggesting that the secondary structures of rRNAs are strongly conserved among eubacteria. The sequence conservation is greater in nonpaired (loop) regions than in paired (stem) regions. This means that a base change occurring in the stem regions has been cancelled by a compensatory change of the opposite base, so that the secondary structure can be maintained. However, there are several regions where the primary as well as the secondary structures are not conserved. Figure 4 shows the comparison of a part of the secondary structure model of the 16-S rRNAs between *M. capricolum* and *E. coli*; a stem of 8 bp between positions

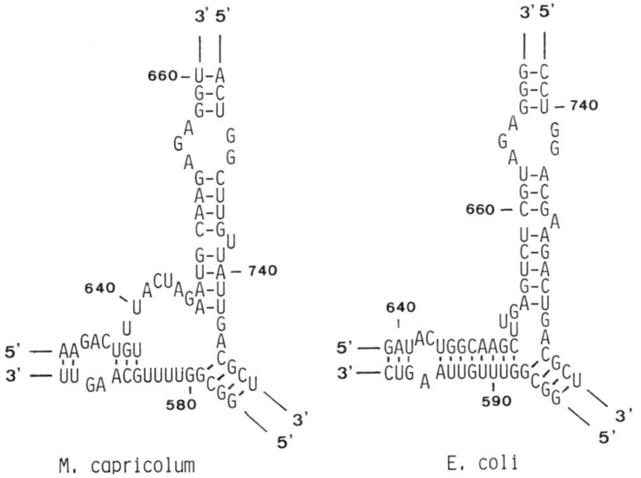

FIG. 4. Secondary structures of ribosomal protein S8 and S15 binding sites in 16-S rRNA. *M. capricolum* (left); *E. coli* (right).

588–595 and 644–651 in *E. coli* has been proposed as a binding site of a part of ribosomal proteins S8 and S15 (*45, 46*), while the corresponding regions in *M. capricolum* (positions 580–587 and 636–643) have only three base pairs, forming a fairly large loop with the rest of the bases in this region. Nevertheless, cross-binding experiments of *M. capricolum* 16-S rRNA with *E. coli* ribosomal proteins S8 and S15 (R. A. Zimmermann and A. Muto, unpublished data) showed that the *E. coli* proteins recognize and bind to exactly the same region of the *M. capricolum* rRNA as that of *E. coli*. The fact suggests that the structures essential for ribosomal architecture are well conserved between the two organisms despite some differences in their structures.

Reflecting the low G+C content of the genome, the G+C content of the 16-S rRNA gene of *M. capricolum* is significantly lower (48%) than those of other bacterial species (around 55%). The sequence of the *M. capricolum* 5-S rRNA gene in *rrnB* is 109 bp long, which is shorter than other bacterial 5-S rRNAs (116–120 bp), and its G+C content is the lowest (42%) of all the 5-S rRNAs so far assayed.

B. tRNA Genes

Early studies on the mycoplasma tRNAs suggested an apparent lack of many isoaccepting tRNA species (*8*). For example, *M. mycoides* subsp. *capri* was reported to contain only one tRNA species for Phe, Gly, Val, and Met. It is interesting to speculate that some evolutionary pressure decreasing their genome size has been exerted on DNA to minimize the genes for tRNA species. In fact, the total number of spots for tRNA of *M. capricolum* separated by 2D-PAGE is about 35, less than that of *E. coli* (F. Yamao, unpublished data).

We have isolated several clones having *M. capricolum* tRNA genes, and have characterized two of them by DNA sequencing. One of them, pMCH964, contains two $tRNA^{Trp}$ genes (*47*); an interesting feature of these genes is described in detail in Section V. The other clone, pMCH502, contains a cluster of five tRNA genes, the order of which is

(5′)-$tRNA^{Thr}$(AGU)-$tRNA^{Tyr}$(GUA)-$tRNA^{Gln}$(UUG)-$tRNA^{Lys}$(UUU)-$tRNA^{Leu}$(UAA)-(3′)

(M. Iwami *et al.*, unpublished data). The tRNA gene cluster is preceded by the putative promoter structure and followed by the probable terminator structure (see Fig. 6). These tRNA genes are divided with short (4–7 bp) spacers. Thus, these genes seem to consist of a single transcription unit. Clustering of tRNA genes is a common phenomenon in many organisms. The genes for these mycoplasma tRNAs are very similar (70–80%) to those for the corresponding *E. coli* and *B. subtilis* tRNAs. The spacers and the 5′- and 3′-flanking sequences of the cluster are extremely rich in A and T (about 80%).

A tRNA gene cluster of *Spiroplasma* sp. strain BC-3 (*48, 49*) consists of genes for

(5′)-$tRNA^{Cys}$(GCA)-$tRNA^{Arg}$(ACG)-$tRNA^{Pro}$(UGG)-$tRNA^{Ala}$(UGC)-$tRNA^{Met}$(CAU)-
$tRNA^{Ile}$(NAU)-$tRNA^{Ser}$(UGA)-$tRNA^{fMet}$(CAU)-$tRNA^{Asp}$(GUC)-$tRNA^{Phe}$(GAA)-(3′)

presumably constituting a single transcription unit. A tRNA gene cluster from *M. mycoides* also includes nine tRNA genes with the same gene order as that of the *Spiroplasma* BC-3, except that the $tRNA^{Cys}$(GCA) gene is lacking (*50*). Interestingly, the order of the above nine tRNA genes is identical to a portion of the tRNA gene cluster from *B. subtilis* (*51, 52*). In *B. subtilis,* these nine tRNA genes occur in the middle part of the 21 tRNA gene cluster that constitutes a single operon with one of the rRNA gene sets (*rrnE*). A comparison of individual tRNA genes between *M. mycoides* and *B. subtilis* shows a substantial similarity; the corresponding genes all have the same anticodons and the overall sequence identity is 85% on average. This is also consistent with the view that mycoplasmas are closely related to gram-positive bacteria. However, the *Spiroplasma* and *M. mycoides* tRNA gene clusters have no neighboring rRNA genes in contrast to *B. subtilis.* It should be recalled here that there are no tRNA genes in the spacer regions between the 16-S and 23-S rRNAs nor in the 3′-distal region of the *rrnB* cluster of *M. capricolum* (*27*), differing from the rRNA operons of *E. coli* and several other bacteria. A $tRNA^{Arg}$(UCU) gene also occurs in *M. mycoides* (*50*), which seems to be a single gene operon.

All the CCA-(3′) terminal sequences of tRNAs of mycoplasmas so far characterized are encoded in DNA. A very low G+C content (about 20%) in the spacers and flanking regions of tRNA genes is also the common feature in all the above tRNA gene clusters.

C. Protein Genes

As described in Section III, the *M. capricolum* cell contains at least 350 protein species (*15*). Among them, ribosomal proteins comprise about 50, about one-seventh of the total proteins (*15*). The number and size of the ribosomal proteins of *M. capricolum* are similar to those of other bacteria, suggesting that the organization and structures of ribosomal protein genes are conserved throughout bacteria. We attempted to isolate the ribosomal protein genes from *M. capricolum* (*53*). For screening ribosomal protein genes from the DNA library, we used the maxi-cell method (*54*), assuming that the genes of *M. capricolum* can be expressed in *E. coli* cells. As we will see in Section V, this assumption is true only when the protein does not include Trp coded by the UGA (opal) codon, because UGA is a Trp codon in *M. capricolum* in contrast to its chain termination property in all organisms except for mitochondria (*47*). Fortunately, the Trp content in the ribosomal proteins is low, and we observed the expression in *E. coli* of several *M. capricolum* ribosomal proteins having no Trp (*53*).

The recombinant plasmids including relatively large DNA fragments of *M. capricolum* in pBR322 were randomly introduced into *E. coli* CSR603, a DNA repair deficient mutant. The plasmid-coded proteins were labeled with ^{14}C-labelled amino acids after pretreatment of the cells by UV irradiation to abolish the host–chromosome-dependent protein synthesis. Since most of the ribosomal proteins are basic and small in size, their identification is relatively easy compared with that of other proteins. The labeled proteins were separated by 2D-PAGE with added nonlabeled *M. capricolum* ribosomal proteins to see whether the labeled spots overlap with authentic ribosomal proteins. Among about 50 plasmids tested, one of the recombinant plasmids included a DNA fragment coding for at least eight ribosomal proteins (*53*). The recombinant plasmid, pMCB1088, includes a 9 kbp *Bgl*II fragment of *M. capricolum* DNA in the *Bam*HI site of pBR322. The restriction map of the plasmid pMCB1088 is shown in Fig. 5. We

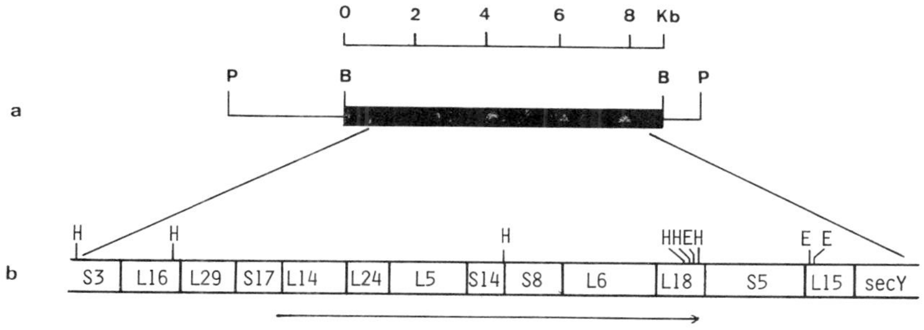

FIG. 5. Organization of *M. capricolum* ribosomal protein genes in pMCB1088. (a) Structure of pMCB1088. The thick line shows the inserted *M. capricolum* DNA, and the thin line represents the vector pBR322 DNA. (b) Ribosomal protein gene arrangement. The horizontal arrow shows the direction of transcription. P (*Pst*I), E (*Eco*RI), H (*Hind*III), and B (*Bam*HI/*Bgl*II).

have determined the DNA sequence of a part (about 6 kbp) of the inserted fragment, and have identified 13 ribosomal protein genes in the order S3, L16, L29, S17, L14, L24, L5, S14, S8, L6, L18, S5, L15, and a secretion protein (*secY*), based on the deduced amino-acid sequence homologies with the corresponding *E. coli* proteins (*55, 47*; unpublished data).

The arrangement of the genes in the insert of pMCB1088 is shown in Fig. 5b. All are aligned in the same strand (direction) with short spacers, suggesting that these are a part of the operon. The order of the genes in this fragment is essentially the same as the gene order in the 72 minute (*spc*) region of *E. coli*, where the genes for about 30 ribosomal proteins are clustered (*56*). However, the gene for ribosomal protein L30, located between the S5 and L15 genes in the *E. coli* chromosome, is absent in *M. capricolum*. The other difference exists in the spacer between the S17 and L14 genes. In *E. coli*, the spacer of about 200 bp between the two genes includes transcriptional termination and initiation (promoter) sequences, indicating that a new operon (*spc* operon) starts from the L14 gene, and thus the genes for S3, L16, L29, and S17 belong to the preceding operon (*S10* operon) (*57*). On the other hand, the spacer between the S17 and L14 genes in *M. capricolum* is only 15 bp and does not include transcription signals. It thus appears that in *M. capricolum* the two operons are fused together to constitute one operon. Nevertheless, the similarity of the gene order in the two organisms is remarkable, suggesting that the basic organization of ribosomal protein gene clusters had been established before the separation of *M. capricolum* and *E. coli* during evolution.

The deduced amino-acid sequences of the *M. capricolum* riboso-

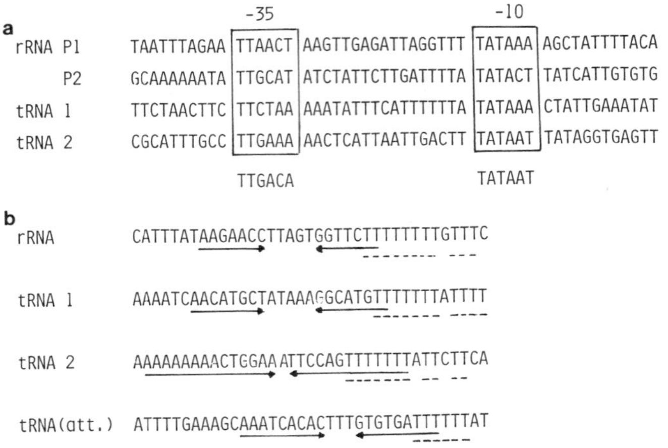

FIG. 6. Transcription signals. (a) Promoter-like sequences. The −35 and −10 regions are boxed. (b) Terminator-like sequences. Arrows indicate dyad symmetrical structures, and dotted lines represent stretches of thymidine residues. rRNA, *rrnB*; tRNA1, gene cluster in pMCH502; tRNA2, gene cluster in pMCH964; tRNA(att.), attenuator-like sequence between the tRNA genes in pMCH964 (see Fig. 9).

mal proteins are also highly homologous (40–60% identity) with those of *E. coli*, showing that the structures of the individual genes are also conserved between the two organisms. Among 13 ribosomal proteins so far identified, 6 proteins (S3, S5, S8, S14, L16, and L24) include Trp coded by UGA (S8 and S14 contain Trp at the C-terminus). Hence, the translation of mRNA for these *M. capricolum* proteins must be terminated by the codons in *E. coli* cells in the maxi-cell experiments. The rest of the proteins (S17, L5, L6, L14, L15, L18, and L29) do not contain Trp.

An extremely biased codon usage for the ribosomal proteins and the genetic code change in *M. capricolum* is discussed in Section V.

D. Transcription and Translation Signals

The rRNA gene cluster (*rrnB*) and the two tRNA gene clusters described above constitute one transcription unit (operon), respectively; the 5′-upstream regions of these clusters contain promoter-like sequences, and the 3′-downstream regions contain terminator-like sequences. The promoter-like sequences are similar to the −10 (Pribnow box) consensus sequence (TATAAT) and the −35 consensus sequence (TTGACA), separated by a 17 bp spacer (Fig. 6a), just like the *E. coli* RNA polymerase sigma-dependent transcription–initiation signal. The clusters are followed by terminator-like structures (Fig. 6b):

dyad symmetrical structures and stretches of thymidine residues, that resemble the *E. coli* rho-independent transcription termination signal. Similar promoter and terminator structures before and after the tRNA gene clusters are also reported in *Spiroplasma* and *M. mycoides* (*48–50*). These facts suggest that the mycoplasma RNA polymerase recognizes transcription signals similar to those of *E. coli*. Several genes that have been cloned from mycoplasmas can be expressed in *E. coli* cells (*58–60*), suggesting that the *E. coli* RNA polymerase recognizes transcription signals of mycoplasmas.

The *M. capricolum* 16-S rRNA contains the mRNA binding sequence (5′)-ACCUCC-(3′) at the 3′-end (*28*). In Fig. 7 are listed the 5′ termini and their upstream sequences of all the genes belonging to the ribosomal protein operon (pMCB1088). All the genes start with ATG fMet codons (boxed in the figure). The 3- to 7-base stretch of probable ribosome binding sequence, complementary to the mRNA binding sequence, exists at about 10 bp upstream from the initiation codons (underlined sequence), suggesting that the structure is actually required for the initiation of translation.

AAAGGAGGTAAAAGATAATT	ATG	TTA	—— (L16) ——
GTTAAAAGAGGTGAAAATTA	ATG	GCT	—— (L29) ——
ACTAAAGGAGAAACTAAATA	ATG	CAA	—— (S17) ——
TATAATAGGAGATAAAAAAT	ATG	ATT	—— (L14) ——
ATTATAGGAGGAACCACATT	ATG	GCA	—— (L24) ——
AAGAAAGGAAATAAATTAGT	ATG	AAA	—— (L5) ——
AGAACAAGGGAGATATTATA	ATG	GCA	—— (S14) ——
AGAGAGAAGATTCAAAAAGT	ATG	ACA	—— (S8) ——
TTGATAATAGGAGTTTAAAT	ATG	TCT	—— (L6) ——
GAGCTTAGGGATTACTAAGT	ATG	AAA	—— (L18) ——
AAGAAAGGGTAGAAGAGATT	ATG	ACT	—— (S5) ——
GTAGAAAGGAGTCATAATTA	ATG	AAA	—— (L15) ——
GGAAAAGTAGAGGTGATTTA	ATG	GTT	—— (secY) ——

FIG. 7. Translation signals. Probable ribosome binding sequences of 5′-upstream regions of ribosomal protein genes in pMCB1088 are underlined. The initiation codons (ATG) of the protein genes are boxed.

There is no spacer between the genes or L16 and L29, L29 and S17, and L15 and secY, where the third letter of the termination codon (TAA) of the preceding gene is overlapped with the first letter of the initiation codon (ATG) of the following gene (---TAATG---).

V. Codon Usage

A. Preferential Use of A- and U-Biased Codons

The sequence of pMCB1088 includes the most of the ribosomal protein genes in the *spc* operon. Since the complete nucleotide sequence of the *E. coli spc* operon has been determined (57), the codons in the ribosomal protein genes between the two bacteria can be compared directly. The G+C content of the coding regions in *M. capricolum* is about 30%, which is much lower than that in the *E. coli* genes (51%), showing that *M. capricolum* uses A- and U-biased codons more frequently than does *E. coli*. Since the G+C content of *E. coli* genome is 50%, the level of G+C in the protein genes is roughly proportional to that of the genomic DNA. In Table I the codon usages in the ribosomal protein genes in the *spc* operon between *M. capricolum* and *E. coli* are compared. The choice of synonymous codons in *M. capricolum* clearly differs from that in *E. coli* and is strongly biased in favor of A and U. For example, UUA (Leu), AUU (Ile), and AGA (Arg) are the most frequently used among synonymous codons in *M. capricolum*, while they are rarely used in *E. coli*, where CUG (Leu), AUC (Ile), and CGU (Arg) are predominant.

The most outstanding feature of codon usage can be seen in the very high frequency of codons ending in either A or U in *M. capricolum*. More than 90% of the *M. capricolum* codons have A or U at the third position, in contrast to only 49% of the *E. coli* codons. On the other hand, codons ending in G or C are rarely used except for the AUG codon for Met. Fifteen codons (CUC, CUG, GUC, UCC, UCG, CCC, ACC, ACG, GCC, GCG, CAG, CGG, UGG, AGG, and GGC), all of which end in G or C, have not yet been found. Thus, *M. capricolum* preferentially uses codons ending in A or U among synonymous codons, since most of synonymous codons differ at the third positions. The preferential use of A- and U-rich codons by this organism is also seen at the first and/or second positions among synonymous codons for Leu, Arg, and Ser, which have six codon degeneracies. For example, among six Leu codons, UUA, UUG, CUU, CUC, CUA, and CUG, 86% of the codons used in *M. capricolum* are UUA. The G+C contents of the codon first and second positions in *M. capricolum* are 43

TABLE I

CODON USAGE FOR RIBOSOMAL PROTEINS IN *M. capricolum* AND *E. coli*[a]

	Second												
	U			C			A			G			
First		M.c.	E.c.		M.c.	E.c.		M.c.	E.c.		M.c.	E.c.	Third
U	Phe	37	11	Ser	20	27	Tyr	15	6	Cys	4	4	U
	Phe	7	28	Ser	0	17	Tyr	7	16	Cys	2	2	C
	Leu	78	3	Ser	41	2	Ocher	4	9	Opal	4[b]	0	A
	Leu	2	6	Ser	0	1	Amber	5	1	Trp	0	3	G
C	Leu	3	6	Pro	12	13	His	12	9	Arg	17	68	U
	Leu	0	8	Pro	0	2	His	6	8	Arg	1	35	C
	Leu	8	1	Pro	27	5	Gln	44	13	Arg	2	3	A
	Leu	0	65	Pro	2	24	Gln	0	29	Arg	0	1	G
A	Ile	84	28	Thr	52	30	Asn	53	13	Ser	12	2	U
	Ile	9	58	Thr	0	26	Asn	9	34	Ser	3	8	C
	Ile	14	0	Thr	29	3	Lys	162	79	Arg	48	1	A
	Met	27	35	Thr	0	6	Lys	14	42	Arg	0	0	G
G	Val	63	53	Ala	56	53	Asp	33	27	Gly	49	82	U
	Val	0	22	Ala	0	20	Asp	6	27	Gly	0	41	C
	Val	39	31	Ala	38	37	Glu	77	54	Gly	53	0	A
	Val	4	14	Ala	0	29	Glu	5	21	Gly	5	7	G

[a] The numbers indicate the number of times a codon appeared in the ribosomal proteins analyzed. M.c., *Mycoplasma capricolum* data (1295 codons) for 9 ribosomal proteins in *spc* operon (55; unpublished). E.c., *Escherichia coli* data (1299 codons) for 10 ribosomal proteins in *spc* operon, taken from Cerretti *et al.* (57).

[b] Opal (UGA) is a Trp codon in *M. capricolum*.

and 37%, respectively, which are also significantly lower than those in *E. coli* (62 and 43%, respectively).

To see further the preferential use of A- and U-biased codons in *M. capricolum*, individual codons for ribosomal protein genes of the two organisms may be compared. There exist 360 synonymous codon replacements at the homologous positions between *M. capricolum* and *E. coli* ribosomal proteins. In Table II representative synonymous codon substitutions observed between the two organisms are listed. Seventy-three percent (264/360) of the codon changes occur so that *M. capricolum* contains much more A and U than does *E. coli*. A striking example is that 21 AAG codons for Lys in *E. coli* are replaced by AAA in *M. capricolum*, showing clearly that the choice of synonymous codons in *M. capricolum* is strongly biased toward A and U.

TABLE II
SYNONYMOUS CODON SUBSTITUTIONS[a]

	M. capricolum	*E. coli*	(A+U gain)[b]	Occurrence[c]
Phe	UUU	UUC	(+1)	12
	UUC	UUU	(−1)	1
Lys	AAA	AAG	(+1)	21
	AAG	AAA	(−1)	2
Arg	AGA	CGU	(+1)	20
	AGA	CGA	(+1)	1
	AGA	CGC	(+2)	9
	CGU	CGC	(+1)	1
	CGA	CGC	(+1)	2
	CGC	CGU	(−1)	1
Leu	UUA	CUG	(+2)	25
	UUA	CUC	(+2)	2
	UUA	CUU	(+1)	2
	UUA	UUG	(+1)	2
	CUU	CUG	(+1)	1
	CUA	CUG	(+1)	5
Gln	CAA	CAG	(+1)	9
		Total	360 codons	
		A+U gain (+)	264	
		(0)	87	
		(−)	14	

[a] Synonymous codon substitutions between ribosomal protein genes in *spc* operon of *M. capricolum* and *E. coli*. Upper column shows, as examples, all the codon substitutions for Phe, Lys, Arg, Leu, and Gln in a total of 360 substitutions.

[b] A+U gains in *M. capricolum* as compared with *E. coli*.

[c] Number of occurrence.

A total of 170 conservative amino-acid replacements (i.e., Lys/Arg, Ser/Thr, Leu/Ile/Val, etc.) exists at the corresponding positions of the ribosomal proteins between the two organisms (Table III). Sixty-seven percent (114/170) of these substitutions take place to choose amino acids having higher A and U content in *M. capricolum*. For example, 28 Arg coded by CGU, CGA, CGC, CGG, and CGA in *E. coli* are replaced by Lys coded by AAA or AAG in *M. capricolum*; no reversed cases (replacement of Lys in *E. coli* by Arg in *M. capricolum*) have been found so far. Accordingly, the Lys content of the *M. capricolum* ribosomal proteins is 14%, which is significantly higher than that of the *E. coli* proteins (9%). Thus, the choice of codons in *M. capricolum* seems here again to discriminate against G and C and to use A and U whenever possible.

The above results suggest that some mutation pressure replacing G·C pairs by A·T pairs has been exerted to account for the A- and U-biased codon usage in this bacterium during evolution. It should be

TABLE III
CONSERVATIVE AMINO-ACID SUBSTITUTIONS[a]

Amino acids	*M. capricolum*	*E. coli*	(A+U gain)[b]	Occurrence[c]
Lys/Arg	Lys(AAA)	Arg(CGU)	(+2)	16
	Lys(AAA)	Arg(CGA)	(+2)	1
	Lys(AAA)	Arg(CGC)	(+3)	6
	Lys(AAA)	Arg(CGG)	(+3)	1
	Lys(AAG)	Arg(CGU)	(+1)	1
	Lys(AAG)	Arg(CGA)	(+1)	1
	Lys(AAG)	Arg(CGC)	(+2)	2
Leu/Val	Leu(UUA)	Val(GUU)	(+1)	4
	Leu(UUA)	Val(GUA)	(+1)	4
	Leu(UUA)	Val(GUG)	(+2)	2
	Leu(UUA)	Val(GUC)	(+2)	3
	Val(GUU)	Leu(CUC)	(+1)	1
	Val(GUU)	Leu(CUG)	(+1)	4
	Val(GUA)	Leu(CUG)	(+1)	2
	Val(GUG)	Leu(CUG)	(0)	1
	Total	170 codons		
	A+U gain (+)	114		
	(0)	50		
	(−)	6		

[a] Conservative amino-acid substitutions between ribosomal protein genes in *spc* operon of *M. capricolum* and *E. coli*. Upper column shows, as examples, all the substitutions for Lys/Arg and Leu/Val in a total of 170 substitutions.

[b] A+U gains in *M. capricolum* as compared with *E. coli*.

[c] Number of occurrence.

noted that the pressure is exerted not only on the choice of synonymous codons but also on the amino-acid composition of its proteins.

B. UGA (opal) Is a Tryptophan Codon

The most outstanding finding in the *M. capricolum* genes is that one of the codons deviates from the "universal code" (*61, 47*). Deviant codons were also discovered in mitochondria and some ciliates (*62–66*). Thus the genetic code is not universal even in the nuclear genes of both prokaryotes and eukaryotes, and was changed at least in some spheres during evolution.

So far, eight UGA (opal) codons appear in the reading frames of the *M. capricolum* genes for ribosomal proteins S3, S5, S8, S14, L16, and L24. UGA is a chain termination codon throughout prokaryotes and eukaryotes. In the present case, however, it is unlikely that these UGAs are termination codons, because they occur in the regions having extensive sequence homology with the corresponding *E. coli* proteins. More importantly, several UGA codons appear at positions corresponding to Trp in the *E. coli* proteins, suggesting strongly that UGA is a "sense" codon for Trp in *M. capricolum.* Figure 8 shows examples of the appearance of UGA codons in the parts of ribosomal protein S3 and L16 genes of *M. capricolum* aligned with the amino-acid sequences of the corresponding regions of *E. coli* S3 and L16. In the universal code, Trp is assigned by a single codon, UGG. The UGG codon also appears but to a much lesser extent than UGA (unpublished data), suggesting that UGA may be the major codon for Trp in this organism.

If UGA is read as Trp in *M. capricolum*, one would expect the

```
(a) S3
                                                 *
      ACA TAT GGT CAA ATT GGA GTT AAA GTA TGA ATT AAT CAT GGT GAA GTA TTT
M.c.  Thr Tyr Gly Gln Ile Gly Val Lys Val(Trp)Ile Asn His Gly Glu Val Phe
E.C.  Thr Tyr Gly Val Ile Gly Val Lys Val Trp Ile Phe Lys Gly Glu Ile Leu

(b) L16
                                                         *
      GTT CGT ATG GGT TCA GGA AAA GGA AAT CCT GAA AAA TGA GTA GCA GTA GTT
M.c.  Val Arg Met Gly Ser Gly Lys Gly Asn Pro Glu Lys(Trp)Val Ala Val Val
E.c.  Val Arg Met Gly Lys Gly Lys Gly Asn Val Glu Tyr Trp Val Ala Leu Ile
```

FIG. 8. DNA sequence of a portion of ribosomal protein genes in pMCB1088. The DNA sequences of a part of mRNA-like strand, together with the predicted amino-acid sequences (M.c.) of portions of *M. capricolum* S3 (a) and L16 (b) genes, are aligned with the corresponding *E. coli* protein sequences. Sites of identical amino-acid sequences are boxed. TGA triplet is marked with an asterisk (*). The data are taken from F. Yamao *et al.* (*47*).

occurrence of an opal tRNA that can accept Trp and decode the UGA codon. The plasmid pMCH964, having a 2 kbp *Hin*dIII fragment in pBR322, was isolated from the DNA library as one of the clones that hybridizes with unfractionated *M. capricolum* tRNAs (*47*). By restriction mapping of this fragment followed by hybridization with ^{32}P-labeled tRNAs, the tRNA genes were localized within a 600 bp *Alu*I fragment derived from the middle part of the *Hin*dIII fragment. The DNA sequence of this region revealed the presence of a pair of tRNA genes with a 40 bp spacer between them. The coding region for these two tRNAs is preceded by a putative promoter structure and is followed by a probable termination signal, suggesting that the genes are arranged in a single operon (Fig. 9). In the spacer between the two tRNA genes, there is a terminator signal-like sequence: a dyad symmetry and a stretch of thymidine residues. The tRNA encoded by the first gene has an anticodon sequence (5′)-UCA-(3′) that can decode both opal codon UGA and universal Trp codon UGG by wobble pairings, whereas the second one has an anticodon sequence (5′)-CCA-(3′) for only the universal UGG codon of Trp. The tRNA(UCA) gene could have emerged by duplication of the tRNA(CCA) gene, since the tRNA genes are tandemly arranged and have a high sequence homology (78%).

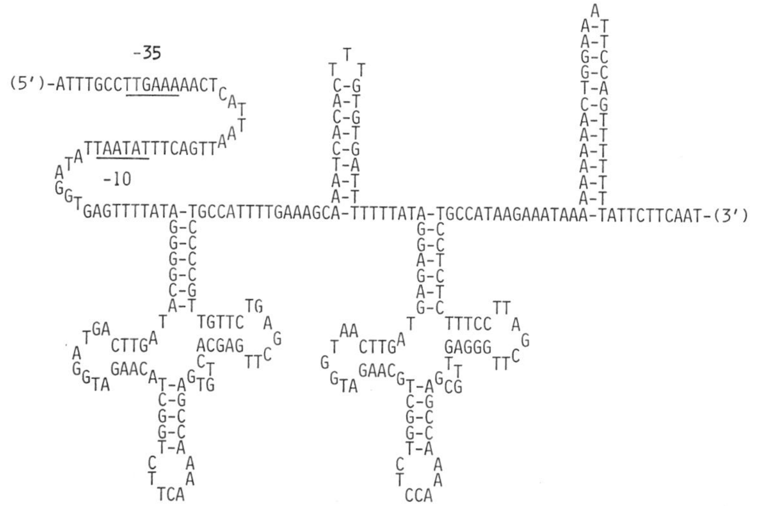

FIG. 9. Organization of tRNA^{Trp} genes in pMCH964. The tRNAs are shown as clover-leaf structures. Probable promoter- (−35 and −10 regions), terminator-, and attenuator-like sequences (hairpin structures) are indicated. The data are taken from F. Yamao *et al.* (*47*).

To see whether the two tRNA genes described above are expressed *in vivo,* tRNAs that hybridize with the DNA fragment containing these genes have been purified. Since tRNA(UCA) is one base longer than tRNA(CCA) as deduced from their DNA sequences, two tRNA bands could easily be distinguished by polyacryamide gel electrophoresis (*47*). Subsequent sequence analyses of the tRNAs isolated from the gel reveals that the first migrating one agrees with the DNA sequence for the tRNA(UCA) gene, and the other with that for the tRNA(CCA) gene. Thus, in the cells, both tRNA genes are transcribed and processed. Both of the two tRNAs accept Trp *in vitro* (*47*).

The presence of $tRNA^{Trp}$(UCA) in *M. capricolum* strongly supports the idea that the UGA codon is translated as Trp by this tRNA. In other organisms, a single Trp codon UGG is decoded by a tRNA with the anticodon CCA. In mitochondria, not only UGG but also UGA is used as a Trp codon (*67–69*); both of these are translated by a single tRNA with the anticodon UCA (*70–72*). Thus, the discovery of two tRNA species having anticodon sequences of CCA and UCA, respectively, in one genetic system contrasts with previous observations in other systems. Recent analysis revealed that the level of the $tRNA^{Trp}$(UCA) gene expression *in vivo* is about 10 times higher than that of the $tRNA^{Trp}$(CCA) (S. Iwagami *el al.*, unpublished data), suggesting that the terminator signal-like sequence between the two tRNA genes operates as an attenuator. The level of Trp charging to $tRNA^{Trp}$(UCA) *in vivo* is also higher than that to $tRNA^{Trp}$(CCA) (*47*). These findings also support the view that UGA is used much more than UGG as a Trp codon in *M. capricolum.*

C. Evolutionary Aspects

It has been proposed that the genetic code was fixed by accident prior to the divergence of all life forms (*74*). As any mutation that leads to change the codon assignment would cause many "mistakes" in the proteins, it seems reasonable to assume that such a mutation would be deleterious or lethal. The use of UGA as a Trp codon in *M. capricolum,* however, indicates that the genetic code can change during evolution. Since *B. subtilis* uses UGA as a chain termination codon, UGA as a Trp codon in *M. capricolum* must have appeared after the separation of the two organisms in the "low G+C" gram-positive bacterial stem (see Fig. 1). The separation time is estimated to be roughly 0.8 billion years ago. In this respect, it is of interest to see whether the use of UGA as a Trp codon is observed in other species belonging to the class *Mollicutes* and *Clostridium* species. The gene for spiralin, the major membrane protein of *Spiroplasma citri,* can be correctly transcribed and translated in *E. coli* cells (*60*). This would mean that spiralin does not contain Trp coded by UGA, or that *Spiroplasma* does not use UGA as a Trp codon, since UGA is a chain termination codon in *E. coli*. If the latter is the case, the change of the Trp UGA codon should have appeared after the separation of *Mycoplasma* and *Spiroplasma*.

As shown in Table I, the codon usage of *M. capricolum* is strongly biased toward A and U, suggesting that some mutation pressure replacing G·C pairs by A·T pairs on DNA has been exerted on the genome during mycoplasma evolution. We suppose that the occurrence of the Trp UGA in this bacterium has also been brought about by mutation pressure. Jukes (*75*) has proposed a possible pathway whereby the UGA Trp codon evolved in *M. capricolum* without deleterious or lethal changes. The genetic code change might have occurred in sequential steps as follows. First, the pressure to replace G by A could lead to replacements of all UGA stop codons by UAA stop codons. This step would have been selectively neutral. The next step could be a duplication of the gene for $tRNA^{Trp}$(CCA), followed by a mutation of one of them to $tRNA^{Trp}$(UCA) (replacement of C by T). The tandem arrangement of the two $tRNA^{Trp}$ genes (see Fig. 9) in pMCH964 supports this view. If all the UGA stop codons had been replaced by UAA (the first step), the emergence of $tRNA^{Trp}$(UCA) would have been not deleterious. The $tRNA^{Trp}$(UCA) can also decode the universal Trp codon UGG by wobble pairing. The last stage would be the replacement of UGG Trp codons by UGA engendered by the pressure for substituting A for G. The UGA so produced would then become translatable by $tRNA^{Trp}$(UCA). Since $tRNA^{Trp}$(UCA) can translate both UGG and UGA as seen in mitochondria (*70–72*), $tRNA^{Trp}$(CCA) would no longer be needed. Interestingly enough, $tRNA^{Trp}$(CCA) is present in a less amount and is less charged with Trp than $tRNA^{Trp}$(UCA) in the cell (*47*). Perhaps, $tRNA^{Trp}$(CCA) persists in *M. capricolum* as a vestigial remnant. The series of evolutionary steps outlined above could have taken place by silent and possibly neutral nucleotide substitutions from G·C pairs to A·T pairs in DNA.

Thus, the above explanation provides an example of an evolutionary pathway leading to the development of codons differing from the universal code without deleterious or lethal effect. In a similar way, it is possible that other genetic code changes could have taken place in organisms having biased DNA nucleotide compositions. In fact, certain ciliates having extremely low G+C content (about 30%) use UAA and UAG (chain termination codons in the universal code) as Gln codons (*62–66*).

It has been suggested that mitochondria evolved from certain prokaryotes by endosymbiosis (*76, 77*). The use of UGA as a Trp codon in *M. capricolum* as in mitochondria (*67–72*) raises an interesting possibility of phylogenetic relationship between mycoplasmas and mitochondria. Mycoplasmas have a small genome and are parasitic in eukaryotes. The richness in A and T of their DNAs resembles that of mitochondria of lower eukaryotes. Furthermore, the *M. capricolum* $tRNA^{Trp}$(UCA) sequence is more similar to yeast mitochondrial $tRNA^{Trp}$(UCA) (*70*) (66% identity) than to cytoplasmic $tRNA^{Trp}$(CCA) (*78*) (55% identity). Thus, the mycoplasma-like organisms might have played some role in the evolution of mitochondria.

VI. Biased Mutation Pressure

The mean G+C content of bacterial genomic DNA varies from approximately 25 to 75% (*79, 80*). Recent phylogenetic analyses of eubacterial 5-S rRNA have clearly indicated that bacterial phylogenetic relationships are closely reflected in the genomic G+C content (*11*). For example, all the low-G+C gram-positive bacteria so far known, such as *B. subtilis* (genomic G+C: 42%), *Lactobacillus viridescens* (40%), *Staphylococcus aureus* (33%), *Clostridium perfringens* (38%), and *M. capricolum* (25%), are phylogenetically related to one another, while the high-G+C gram-positive bacteria, such as *Micrococcus luteus* (75%), *Steptomyces griseus* (73%), and *Mycobacterium tuberculosis* (67%), that comprise one phylogenetic group, separated long ago from the low G+C group on the phylogenetic tree. *E. coli* (50%), *Proteus vulgaris* (40%), *Salmonella typhimurium* (51%), and *Pseudomonus fluorescens* (60%) are shown to have been derived from a common gram-negative bacterial branch, and most of them have intermediate amounts of G+C in their genomes. The close relation between bacterial phylogeny and the G+C content can best be explained by an assumption that some mutation pressure has been exerted on the entire genomic DNA to maintain more or less a given G+C level during evolution. Such biased mutation pressure seems to have been exerted on the DNA of individual phylogenetic branches at different rates. For example, on the low-G+C gram-positive bacterial branch including all the species of *Mollicutes*, a relatively strong mutation pressure replacing G or C by A or T seems to have been exerted. The mutations caused by this pressure are then subjected to selective constraints, usually operating in the form of negative selection to eliminate functionally deleterious changes. A certain fraction of the selected mutations is then fixed in the population due to random genetic drift. Thus, functionally less important parts in the genome evolve faster than more important one (*81–83*).

The biased pressure that has operated in the process of mycoplasma evolution would affect various parts of the mycoplasma genome all positively, but differentially depending on their functional importance. In other words, the G+C level of functionally less important parts of the *M. capricolum* genome would have become lower than that of important parts. The G+C levels of various parts of the *M. capricolum* genome so far characterized clearly demonstrate that this is the case. In Fig. 10 the G+C levels of different parts of the *M. capricolum* genome that we have cloned and sequenced are illustrated. The spacer regions between the genes are the lowest with respect to the G+C content (about 20%). The second lowest are the protein genes (about 30%). The G+C contents of the rRNA and

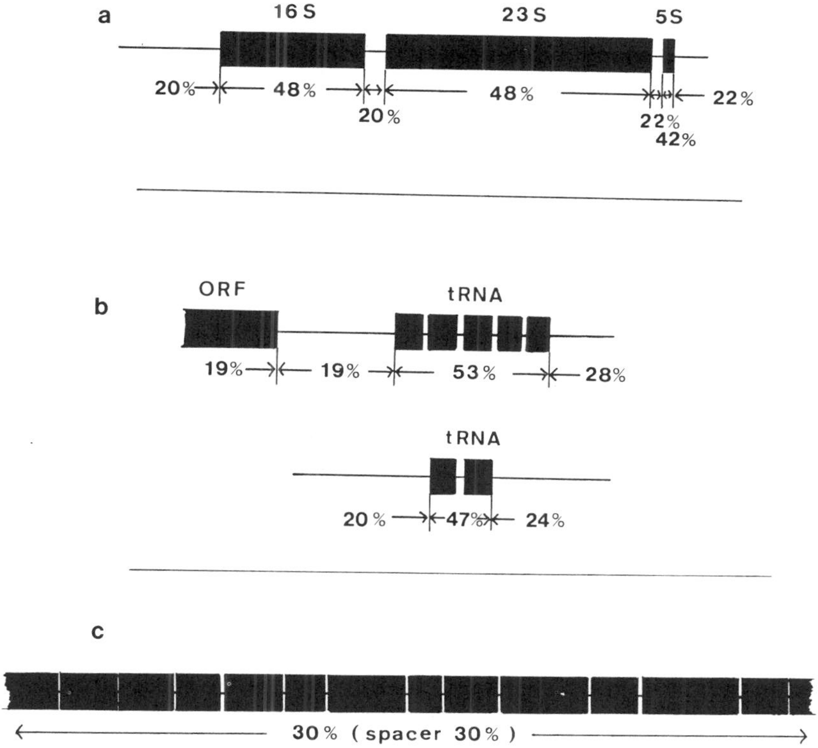

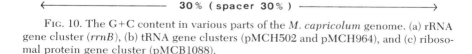

FIG. 10. The G+C content in various parts of the *M. capricolum* genome. (a) rRNA gene cluster (*rrnB*), (b) tRNA gene clusters (pMCH502 and pMCH964), and (c) ribosomal protein gene cluster (pMCB1088).

tRNA coding regions are relatively high (47–54%) compared with those of other parts. This indicates that all the different parts of the *M. capricolum* genome contribute positively to the low G+C content of the genome.

The results strongly support the idea that the biased mutation pressure has actually been exerted uniformly on the entire DNA during mycoplasma evolution. The different G+C levels in the different components of the genome must be a consequence of negative selections that have operated to eliminate deleterious mutations. Since spacers are functionally the least important parts of the genome, the bulk of replacements of G·C pairs by A·T pairs caused by the pressure is selectively neutral with an evolutionary rate faster than other parts. On the contrary, the rRNA and tRNA genes are the most invariable, because their transcripts are nontranslatable, and their sequences are directly concerned with biological function. Protein genes are more variable than the stable RNA genes, because synonymous codon changes and conservative amino-acid changes are thought to be selectively nearly neutral.

The effects of negative selection can also be seen in the codon usage of this organism. As shown in Table I, the G+C level of the third positions in the codons of *M. capricolum* is extremely low (less than 10%). This suggests that the pressure is exerted most effectively on the "silent" codon substitutions at the third positions. This is clearly seen in comparisons of individual codons in the ribosomal protein genes between *E. coli* and *M. capricolum* (Table II). The pressure is also exerted on conservative amino-acid substitutions.

As already mentioned, there are several conservative amino-acid changes between the *M. capricolum* and *E. coli* ribosomal protein genes; most of them occur so that *M. capricolum* uses more amino acids having A- and U-rich codons than does *E. coli* (Table III), reflecting the higher G+C levels not only at the third positions but also at the first and the second positions of the *M. capricolum* codons. The first position is more variable than the second one, because many conservative amino-acid changes occur by replacement of the first codon position (i.e., Leu/Ile/Val, Ser/Thr, etc.). These facts suggest that the biased mutation pressure changing G·C pairs to A·T pairs has been exerted equally on all positions of the codons, and then selected by constraints according to the functional importance.

Thus, we suppose that mutation pressure is the major determinant of the biased codon usage pattern in *M. capricolum*. Ikemura (*84*, *85*) demonstrated a positive correlation between codon usage and tRNA abundance in *E. coli* and yeast, claiming that tRNA population is the major factor acting as a selective constraint to determine the codon usage pattern. However, the results in *M. capricolum* suggest that the biased mutation pressure much predominates over such selective constraints for determining nonrandom codon usage in this organism. As discussed in Section V,D, the genetic code change found in *M. capricolum* can also be explained as the consequences of successive replacements of G·C pairs by A·T pairs on DNA by the biased mutation pressure.

The nature of the biased mutation pressure is unknown at present. One possibility might be that mutational modifications of some components in the DNA synthesis bias the mutation rate toward A·T pairs (*86*). The other possibility might be in a DNA base-modification system. Methylation and deamination of bases cause directional changes that tend to accumulate A·T pairs in DNA (*87, 88*). Thus, an abundance of modification enzymes or their repair system could be candidates for the mutation pressure.

VII. Conclusion

The gene analyses described in Sections III to V reveal several characteristic features of the mycoplasma genome. First, the genes encoded in the genome are few in number. Mycoplasmas can self-replicate with products of 400–500 genes under very complex nutritional conditions. This suggests that the mycoplasma genome is composed of only a limited number of genes indispensable for growth. Since mycoplasmas seem to be the degenerated descendants of gram-positive bacteria, the small number of the genes must be a consequence of deletions of nonessential genes. Parasitic behavior might have caused the reduction of the DNA. In eukaryotic organs and tissues, the environmental conditions are relatively constant and nutritionally very rich; amino acids, bases, fatty acids, cholesterol, and vitamins are constantly supplied. Thus, the metabolic pathways for production of these materials would become unnecessary, and the genes related to these processes could have been eliminated. On the other hand, the genes indispensable for autonomous growth would be conserved in the genome. The genes for the components of protein- and nucleic acid-synthesizing systems seem to occupy the major part of the mycoplasma genome.

Second, the organization and structure of the essential genes are remarkably conserved. In spite of the great reduction in the number of total genes in *M. capricolum*, the genes retained in the genome, such as for rRNAs, tRNAs, and ribosomal proteins, have essentially the same structures as in *E. coli*, *B. subtilis*, and other eubacteria, not only in their arrangement in operons but also in the sizes and structures of individual genes. This is also the case for the transcription and translation signals as well as for possible processing signals for rRNA maturation. There is much evidence that the organization and structures of mycoplasma genes have a higher homology with those of *B. subtilis* than with those of *E. coli*, supporting the view that mycoplasmas are related to the gram-positive bacteria.

Third, the mycoplasma genes are poor in G and C. All the parts of the genome—protein genes, stable RNA genes, and spacers—contribute positively to the low G+C content of the total genomic DNA. This would be due to some evolutionary mutation pressure that has been exerted on the entire genomic DNA to replace G·C pairs by A·T pairs (see Section VI).

In *M. capricolum*, the codon UGA (opal) is read as Trp. This finding has revised the general concept that the genetic code is unchangeable except for mitochondria. It remains to be seen if all species in the class *Mollicutes* use the deviant genetic code.

Acknowledgments

We thank our co-workers, M. Sawada, Y. Kawauchi, M. Iwami, S. Iwagami, Y. Azumi, S. Ohkubo, and E. Kenmotsu for their active contributions to this study. We are also grateful to Dr. H. Hori of this laboratory for many discussions. This work was supported by grants from the Ministry of Education, Science and Culture of Japan and the Naito Scientific Foundation Research Grant.

References

1. J. Maniloff and H. J. Morowitz, *Bacteriol. Rev.* **36**, 263 (1972).
2. S. Razin, *Microbiol. Rev.* **42**, 414 (1978).
3. E. J. Stanbridge and M. E. Reff, *in* "The Mycoplasmas" (M. F. Barile and S. Razin, eds.), p. 157. Academic Press, New York, 1979.
4. S. Razin, *Microbiol. Rev.* **49**, 419 (1985).
5. H. J. Morowitz and D. C. Wallace, *Ann. N.Y. Acad. Sci.* **225**, 63 (1973).
6. H. Hori, M. Sawada, S. Osawa, K. Murao and H. Ishikura, *NARes* **9**, 5407 (1981).
7. S. H. Chang, C. K. Brum, M. Silberkland, U. L. RajBhandary, L. I. Hecht and W. E. Barnett, *Cell* **9**, 717 (1976).
8. M. W. Kilpatrick and R. T. Walker, *NARes* **8**, 2783 (1980).
9. C. R. Woese, J. Maniloff and L. B. Zablen, *PNAS* **77**, 494 (1980).
10. M. J. Rogers, J. Simmons, R. T. Walker, W. G. Weisburg, C. R. Woese, R. S. Tanner, I. M. Robinson, D. A. Stahl, G. Olsen, R. H. Leach and J. Maniloff, *PNAS* **82**, 1160 (1985).
11. H. Hori and S. Osawa, *Biosystems* **19**, 163 (1986).
12. H. J. Morowitz, *Prog. Theor. Biol.* **1**, 35 (1967).

13. P. H. O'Farrell, *JBC* **250**, 4007 (1975).
14. R. Z. O'Farrell, H. M. Goodman and P. H. O'Farrell, *Cell* **12**, 1133 (1977).
15. Y. Kawauchi, A. Muto and S. Osawa, *MGG* **188**, 7 (1982).
16. E. Kaltschmidt and H. G. Wittmann, *Anal. Biochem.* **36**, 402 (1970).
17. J. H. Ryan and H. J. Morowitz, *PNAS* **63**, 1282 (1969).
18. M. Sawada, S. Osawa, H. Kobayashi, H. Hori and A. Muto, *MGG* **181**, 176 (1981).
19. A. W. Rodwell, *in* "Methods in Mycoplasmology" (S. Razin and J. G. Tully, eds.), Vol. 1, p. 163. Academic Press, New York, 1983.
20. H. J. Morowitz, *Isr. J. Med. Sci.* **20**, 750 (1984).
21. D. Amikam, G. Glaser and S. Razin, *J. Bact.* **158**, 376 (1984).
22. J. Frydenberg and C. Christiansen, *DNA* **4**, 127 (1985).
23. U. Gobel, G. H. Batler and E. J. Stanbridge, *Isr. J. Med. Sci.* **20**, 762 (1984).
24. M. E. Kenerley, E. A. Morgan, L. Post, L. Lindahl and M. Nomura, *J. Bact.* **132**, 931 (1977).
25. A. Kiss, B. Sain and P. Venetianer, *FEBS Lett.* **79**, 77 (1977).
26. H. Kobayashi and S. Osawa, *FEBS Lett.* **141**, 161 (1982).
27. M. Sawada, A. Muto, M. Iwami, F. Yamao and S. Osawa, *MGG* **196**, 311 (1984).
28. M. Iwami, A. Muto, F. Yamao and S. Osawa, *MGG* **196**, 317 (1984).
29. C. Vola, B. Jarry and R. Rosset, *MGG* **153**, 337 (1972).
30. G. C. Stewart, E. E. Wilson and K. Bott, *Gene* **19**, 153 (1982).
31. N. Tomioka, K. Shinozaki and M. Sugiura, *MGG* **184**, 359 (1981).
32. E. Lund, J. E. Dahlberg, L. Lindahl, S. R. Jaskunas, P. P. Dennis and M. Nomura, *Cell* **7**, 165 (1976).
33. T. Ikemura and M. Nomura, *Cell* **11**, 779 (1977).
34. E. A. Morgan, T. Ikemura and M. Nomura, *PNAS* **74**, 2710 (1977).
35. K. Loughney, E. Lund and J. E. Dahlberg, *NARes* **10**, 1607 (1982).
36. R. A. Young and J. A. Steitz, *PNAS* **75**, 3593 (1978).
37. N. Ogasawara, S. Moriya and H. Yoshikawa, *NARes* **11**, 6301 (1983).
38. G. C. Stewart and K. Bott, *NARes* **11**, 6289 (1983).
39. J. Broius, M. L. Palmer, P. J. Kennedy and H. F. Noller, *PNAS* **75**, 4801 (1978).
40. N. Tomioka and M. Sugiura, *MGG* **191**, 46 (1983).
41. C. J. Green, G. C. Stewart, M. A. Hollis, B. S. Vold and K. Bott, *Gene* **37**, 261 (1985).
42. C. R. Woese, R. Gutell, R. Gupta and H. Noller, *Microbiol. Rev.* **47**, 621 (1983).
43. P. Maly and R. Brimacombe, *NARes* **11**, 7263 (1983).
44. P. Stiegler, P. Carbon, M. Zuker, J. Ebel and C. Ehresmann, *NARes* **9**, 2153 (1981).
45. E. Ungewickell, R. Garrett, C. Ehresmann, P. Stiegler and P. Fellner, *EJB* **51**, 165 (1975).
46. R. A. Zimmermann, G. Makie, A. Muto, R. Garrett, E. Ungewickell, C. Ehresmann, P. Stiegler, J. Ebel and P. Fellner, *NARes* **2**, 279 (1975).
47. F. Yamao, A. Muto, Y. Kawauchi, M. Iwami, S. Iwagami, Y. Azumi and S. Osawa, *PNAS* **82**, 23061 (1985).
48. M. J. Rogers, A. A. Steinmetz and R. T. Walker, *Isr. J. Med. Sci.* **20**, 768 (1984).
49. M. J. Rogers, A. A. Steinmetz and R. T. Walker, *NARes* **14**, 3145 (1986).
50. T. Samuelsson, P. Elias, F. Lustig and Y. S. Guindy, *BJ* **232**, 223 (1985).
51. C. J. Green and B. S. Bold, *NARes* **11**, 5763 (1983).
52. E. F. Wawrousek, N. Narasimhan and J. N. Hansen, *JBC* **259**, 3694 (1984).
53. Y. Kawauchi, A. Muto, F. Yamao and S. Osawa, *MGG* **196**, 521 (1984).
54. A. Sancer, A. M. Hack and W. D. Rupp, *J. Bact.* **137**, 692 (1979).
55. A. Muto, Y. Kawauchi, F. Yamao and S. Osawa, *NARes* **12**, 8209 (1984).
56. M. Nomura and E. A. Morgan, *Annu. Rev. Genet.* **11**, 297 (1977).

57. D. P. Cerretti, D. Dean, G. R. Davis, D. M. Bedwell and M. Nomura, *NARes* **11,** 2599 (1983).
58. M. A. Taylor, M. A. McIntosh, J. Robbins and K. S. Wise, *PNAS* **80,** 4154 (1983).
59. C. Mouches, G. Barroso and J. M. Bove, *J. Bact.* **156,** 952 (1983).
60. C. Mouches, T. Candresse, G. Barroso, C. Saillard, H. Wroblewsky and J. M. Bove, *J. Bact.* **164,** *1094 (1985).*
61. A. Muto, F. Yamao, Y. Kawauchi and S. Osawa, *Proc. Jpn Acad.* **61B,** 12 (1985).
62. F. Caron and E. Meyer, *Nature* **314,** 185 (1985).
63. J. R. Preer, Jr., L. B. Preer, B. M. Rudman and A. J. Barnett, *Nature* **314,** 188 (1985).
64. E. Helftenbein, *NARes* **13,** 415 (1985).
65. S. Horowitz and M. A. Gorovsky, *PNAS* **82,** 2452 (1985).
66. Y. Kuchino, N. Hanyu, F. Tashiro and S. Nishimura, *PNAS* **82,** 1499 (1985).
67. B. G. Barrell, A. T. Bankier and J. Drouin, Nature **282,** 189 (1979).
68. G. Macino, G. Coruzzi, F. G. Nobrega, M. Li and A. Tzagoloff, *PNAS* **76,** 3784 (1979).
69. T. D. Fox, *PNAS* **76,** 6534 (1979).
70. N. C. Martin, H. D. Pham, K. Underbrink-Lyon, O. L. Miller and J. E. Donelson, *Nature* **285,** 579 (1980).
71. B. G. Barrell, S. Anderson, A. T. Bankier, M. H. C. DeBruijin, E. Chen, A. R. Coulson, J. Drouin, I. C. Eperon, D. P. Nierlich, B. A. Roe, F. Sanger, P. H. Schreier, A. J. H. Smith, R. Staden and I. G. Young, *PNAS* **77,** 3164 (1980).
72. J. E. Heckman, J. Sarnoff, B. Alzner-DeWeerd, S. Yin and U. L. RajBhandary, *PNAS* **77,** 3159 (1980).
73. F. H. C. Crick, *JMB* **19,** 548 (1966).
74. F. H. C. Crick, *JMB* **38,** 367 (1968).
75. T. H. Jukes, *J. Mol. Evol.* **22,** 361 (1985).
76. L. Margulis, "Symbiosis in Cell Evolution." Freeman, San Francisco, 1981.
77. H. Kuntzel and H. G. Kochel, *Nature* **293,** 751 (1981).
78. G. Keith, A. Roy, J. Ebel and G. Dirheimer, *Biochimie* **54,** 1405 (1972).
79. K. Y. Lee, R. Wahl and E. Barbu, *Ann. Inst. Pasteur* **91,** 212 (1956).
80. N. Sueoka, *JMB* **3,** 31 (1961).
81. M. Kimura, *Nature* **217,** 624 (1969).
82. J. L. King and T. H. Jukes, *Science* **164,** 788 (1969).
83. M. Kimura, "The Neutral Theory of Molecular Evolution." Cambridge Univ. Press, Cambridge, UK, 1983.
84. T. Ikemura, *JMB* **151,** 389 (1981).
85. T. Ikemura, *JMB* **158,** 573 (1982).
86. J. F. Speyer, *PNAS* **47,** 1141 (1980).
87. A. P. Bird, *NARes* **8,** 1499 (1980).
88. B. Dujon, *in* "Molecular Biology of the Yeast Saccharomyces " (J. N. Strathern, E. W. Jones and J. R. Broach, eds.), p. 505. Cold Spring Harbor, New York, 1981.

Regulation of Gene Transcription by Multiple Hormones: Organization of Regulatory Elements

ANTHONY WYNSHAW-BORIS,
J. M. SHORT,[1] AND
RICHARD W. HANSON

Department of Biochemistry
Case Western Reserve University
School of Medicine
Cleveland, Ohio 44106

Recombinant DNA technology has led to an increased understanding of the way in which hormones regulate the transcription of specific genes (*1–7*). Much of the progress in this area arose from the availability of simple techniques to isolate and manipulate eukaryotic genes, as well as the ability to introduce such genes into cells. In hormonally regulated genes, these methods have led to the definition of those sequences required for hormonal control. Genes regulated by steroid hormones, such as those for mouse mammary tumor virus, have served as models for understanding the sequence requirements for the hormonal regulation of gene transcription (for review, see *8*). The steroid hormone–receptor complex binds to hormone regulatory elements,

[1] Present address: Stratagene Cloning Systems, 3770 Tansy St., San Diego, California 92121.

Progress in Nucleic Acid Research
and Molecular Biology, Vol. 34

which are sequences usually located within the 5′-flanking region of hormonally regulated genes (*5, 7, 9–13*). The result of hormone–receptor complex binding to the hormone regulatory element is a change in chromatin conformation of the regulated gene, which leads to a stimulation (or repression) of transcriptional efficiency. Thus, the regulatory element is the functional unit that confers hormonal sensitivity to specific genes.

The examination of hormone regulatory elements in genes regulated by single hormones provides insight into the mechanism of hormonal regulation of transcription. Although a number of genes transcriptionally regulated by single hormones have been isolated and analyzed, few that respond to multiple hormones have been cloned. For even fewer genes has the location of each of the various hormone regulatory elements been found. The interactions of multiple regulatory elements in a single gene, studied by the techniques of gene transfer and mutagenesis, help us to understand the organization of complex promoter-regulatory regions of eukaryotic genes, and provide clues for the design of promoter-regulatory regions regulated by various hormones. Also, in conjunction with studies identifying protein factors that bind to regulatory elements, the interactions between different factors in genes regulated by multiple hormones can be studied.

The gene for cytosolic phospho*enol*pyruvate carboxykinase (GTP) (PEPCK) (EC 4.1.1.32), from the rat, has been analyzed in our laboratory, and provides a model for the study of the organization of genes that respond transcriptionally to a number of hormones (*3, 4, 14–16*; for review, see *17, 18*). The transcription of this gene is stimulated by cAMP, glucocorticoids, and thyroid hormone, and is inhibited by insulin. Other genes in which multiple hormone regulatory elements have been described include chicken ovalbumin, regulated by glucocorticoids, estrogens, androgens, and progestins (1, 19); chicken lysozyme, regulated by glucocorticoids, estrogens, androgens, and progestins (20); human proenkephalin, which responds to cAMP and phorbol esters (*68*); and bovine and rat prolactin, regulated by, among other hormones, glucocorticoids, estrogens, cAMP, and calcium (*21*).

This review focuses on the organization of regulatory elements in genes regulated by multiple hormones. The techniques used to identify regulatory elements are discussed, since they are essential to the identification of these elements. As examples of the way in which DNA sequences containing hormone regulatory elements are identified and organized in genes, glucocorticoid and cAMP regulatory elements are discussed, as are negative regulatory elements, which depress gene expression in the absence of hormones. The positional flexibility of positive and negative elements are described. Finally, we examine the organization of these elements in genes regulated by multiple hormones, whether the multiple elements are independent or interdependent, and whether multiple hormones can act through the same element. As the number of genes in which multiple hormone regulatory elements have been identified is small this discussion is necessarily speculative.

I. Techniques for the Identification of Hormone Regulatory Elements

We have recently reviewed the methodology used for determining the sequence requirements for the hormonal regulation of gene expression (22). Because these techniques are the basis for the identification of regulatory elements in genes, they are briefly described here. There are basically two approaches, both of which depend upon having the gene of interest isolated and characterized. The gene or a fragment of the gene can then be introduced into cells, and its regulation by the appropriate hormones measured. This provides a functional assessment of whether the gene of gene fragment contains regulatory elements. Using this procedure, specific sequences can be identified precisely by deletion, insertion, or substitution mutagenesis. These sequences then can be linked to a heterologous gene in order to test for hormone responsiveness, and the binding of proteins required for the regulation of gene transcription can be determined. Important sequences can be identified by mutagenesis, as described above, or by DNA "footprinting" (see Section I,C), in which the specific binding of regulatory proteins can be determined directly. If regulatory elements for the same hormone are identified in a number of genes, they can be compared in order to determine minimum sequence requirements for hormonal regulation.

A. Gene Transfer

To determine whether a cloned gene contains hormone regulatory elements, the entire gene or derivatives are introduced into cells, and the hormonal regulation of the gene is determined. Genes can be introduced into cells by transfection, particle-mediated gene transfer, microinjection, electroporation, or viral infection. It is important, of course, that the recipient cell be responsive to the hormones that regulate the transcription of the transferred gene, especially as many of hormonally regulated genes are expressed and regulated in a highly tissue-specific manner. For example, the rat PEPCK gene is induced by glucocorticoids in liver and kidney, but is repressed by the same hormone in adipose tissue, and is inactive in many tissues (for review, see *17*).

It is useful to use a recipient cell line derived from the tissues in which the gene being studied is normally expressed. If the cell line is well-differentiated, it may provide tissue-specific factors necessary for the expression of the transferred gene. Also, a well-differentiated cell line may express the endogeneous chromosomal gene, regulated by the proper hormones, providing an internal control for the hormonal regulation of the transfected gene. For many tissues, no stable cell lines have been established in long-term culture. In such cases, genes can be transferred into primary cell lines, if a suitable and efficient method of transfer can be employed.

An intact gene may also be transferred into cells without modification. However, the expression of the transfected copies of the intact gene cannot be differentiated from the products of the chromosomal gene, if they are expressed in the recipient cell line. In such cases, the intact gene can be modified by the removal of exon sequences, to create a "minigene." Alternatively, an exon can be tagged with an inserted piece of DNA, which is used as a marker. Either of these modifications will allow the differentiation of the product of the transferred gene from that of the endogenous gene. However, any modification in the coding sequence of the gene may render the resultant mRNA unstable, or interfere with translation of the mRNA.

Portions of the hormonally regulated gene can be fused to the transcriptional unit of a gene that is not hormonally regulated to form a chimeric, or fusion gene. Generally, chimeric genes contain the 5′-flanking region and the start site of transcription of the hormonally regulated gene, which is fused to the 5′ untranslated region of the marker gene. This type of chimeric gene produces an mRNA and protein from the marker gene that are under the control of the promoter-regulatory region of the hormonally regulated gene. The marker gene can code for a protein required for survival of the cell, or it may simply be an easily detectable and/or assayable gene product. Another useful type of chimeric gene consists of an isolated sequence from a hormonally regulated gene linked to a heterologous gene with its own promoter. Any putative regulatory sequence can be tested in this manner, as the promoter elements are provided by the gene. Also, as sequences placed 5′ to the heterologous gene are not included in mRNA, hormonal regulation of this type of chimeric gene indicates that the sequences confer transcriptional regulation to the gene.

When a cloned gene is introduced into cells, the transferred DNA is assembled into large molecules containing multiple copies of the transfected DNA. These transgenomes are extrachromosomal, generally unstable, and are actively transcribed in a small percentage of cells (the percentage is dependent on the cell line, gene, vector, and method of introduction). In such cells, the expression increases for 24–72 hours after transfection, and then declines as the genes are gradually lost from the cells. The expression and regulation of these transiently expressed genes can be assayed after 1–3 days. A small percentage of cells contain transgenomes that become stably integrated into chromosomal DNA. These cells can be isolated if the transfected DNA codes for a selectable marker required for the survival of cells in selective media. The selectable marker gene can be cotransfected with, or directly linked to, the gene being studied, or a chimeric gene that codes for a selectable marker can be used. For example, in thymidine kinase-deficient cells, the integration and expression of the thymidine kinase (TK) gene of the Herpes simplex virus protect cells from the lethal effects of HAT[2] medium (*23, 24*). Several genes coding for selectable markers can be introduced into cells; among these are the bacterial *neo* gene, which confers resistance to the antibiotic G418 (*25, 26*); the bacterial *gpt* gene, which confers resistance to mycophenolic acid (*27, 28*); and dihydrofolate reductase, which protects cells from methotrexate (*29, 30*).

There are advantages and disadvantages to the use of either transient or stable expression when examining hormone regulatory elements in genes. The *advantages of transient* expression are (1) a number of constructs can be analyzed rapidly, (2) all genes have the same extrachromosomal environment, and their function is not affected by the site of integration into the host genome, and (3) primary cell lines can be analyzed shortly after their establishment in culture. The *disadvantages of transient* expression are (1) no stable cell lines are produced, so that different populations of cells are used for different hormonal treatments, and (2) not all genes are active in transfected cells, so that changes in gene expression from hormonal treatment may arise from changes in the number of active genes. This mechanism is different from what is assumed to occur in chromosomal genes, where the activity of individual genes are modulated.

The major *advantages of stable expression* are (1) stable cell lines are produced, and (2) the transfected genes are integrated into the

[2] HAT medium is 10^{-4} M in hypoxanthine, 4×10^{-7} M in aminopterin, and 1.6×10^{-5} M in thymidine.

chromosomes, which is the usual location of cellular genes. The *disadvantages of stable expression* are (1) several weeks are required to isolate stable cell lines, (2) the site of integration of the gene may affect its expression, and (3) it may not be possible to maintain primary cell cultures long enough to isolate stable cell lines. It is possible that these methods for analyzing gene expression may not be equivalent, but instead may measure different properties of hormonal regulation. The identification of hormone regulatory elements by transient expression has not been compared systematically with stable expression using the same cell line and the same hormonally regulated gene, so the relative merits of the two systems have yet to be assessed.

The effect of hormones on the levels of protein or mRNA for the transferred gene can be measured, to infer indirectly changes in the rate of transcription of the gene. However, there are many steps in the production of mRNA and protein that are potential sites or hormonal control, and one cannot assume that the hormonal regulation of protein or mRNA levels is transcriptional. The rate of transcription of the transfected gene can be determined directly in the presence and absence of hormones by nuclear "run-on" transcription (*1*). Alternatively, the transcriptional regulation of the transfected gene can be determined by placing putative hormone regulatory elements in front of intact genes and measuring the production of mRNA initiating from the promoter of the marker gene. Since no sequences in the marker gene are modified, the regulation of expression of the gene must be at the level of transcription. This assumes that the unlinked marker gene is not itself regulated by the hormone being studied. Posttranscriptional regulation of the transfected gene for metallothionein I from the mouse by glucocorticoids has been demonstrated (*31*), so that it is important to establish that a specific hormone is acting to alter the transcription of the transfected gene.

B. Mutagenesis

The location of hormone regulatory elements in a gene fragment can be determined precisely by specifically modifying the fragment containing the elements and analyzing the hormonal regulation of the modified genes after their introduction into cells. The sequences of importance in the hormone regulatory elements can be identified by generating a series of graded deletions through the regions of the gene suspected to contain the elements. Deletions are made using specific nucleases in a manner allowing a controlled digestion of the DNA. The location of the elements is tentatively established when hormonal responsiveness of the gene is lost. Each hormone regulatory element can be further located by independently determining the 5′ and 3′ boundaries using 5′-to-3′ and 3′-to-5′ deletions, respectively. Alternatively, the boundaries can be established using 5′ deletion mutants only, assuming that the regulatory element occurs between the last mutant that was hormonally responsive and the first deletion where regulation was lost.

There are limitations to using direct deletion mutagenesis to identify regulatory elements. The construction of deletion mutants moves the remaining sequences in the gene closer together, or closer to sequences in the vector. This may interfere with hormonal regulation of the transfected gene and influence its transcription. If a gene has multiple regulatory elements, each capable of acting independently to modulate transcription, then deletion mutagenesis will identify only the last remaining element. Similarly, if the gene contains multiple, dependent elements required for hormonal responsiveness, then deletion analysis will identify only the first element deleted.

Linker-scanning mutagenesis (66) was developed as a means of specifically substituting a small cluster of nucleotides throughout a fragment of a gene, while maintaining the spatial arrangements of all other sequences. Nucleotides are replaced, without causing the addition or deletion of other sequences, and the functional significance of elements can be assessed more confidently than by direct deletion mutagenesis. However, as was the case with standard deletions, multiple elements cannot be identified by linker scanning mutagenesis.

Rather than deleting sequences, small DNA fragments can be inserted randomly into regions of a gene thought to contain regulatory elements. *Insertion mutagenesis* will disrupt the normal sequence of the gene, and if the insertion occurs in a hormone regulatory element, then the mutant will not be responsive to hormones after introduction into cells. Again, it is impossible, using this method, to identify multiple elements.

C. Binding of Regulatory Proteins

The binding of regulatory proteins to a hormonally responsive gene provides an independent assay for the sequences in the gene that confer hormonal stimulation, and is complementary to the gene transfer techniques outlined above. Qualitative methods for the determination of binding of regulatory proteins to DNA sequences depend upon physical changes in the properties of the DNA–protein complex upon binding, which can then be measured. For example, "naked" DNA will not bind to nitrocellulose filters, but protein or protein–nucleic acid complexes will. Therefore, the retention of a specifically labeled fragment of DNA to nitrocellulose filters after incubation with a trans-action protein factor indicates that sequences in the fragment are involved in binding. As an alternate approach, a mixture of labeled DNA fragments can be incubated with a regulatory protein. The fragments that bind the protein from the mixture can be isolated on nitrocellulose filters and identified by electrophoresis on agarose gels after the protein·DNA complex has been eluted from the filter. The specific sequences involved in binding can be identified by deletion, insertion, and substitution mutagenesis, as described above. These methods complement gene transfer studies, as both methods can use the same DNA. However, for this approach to provide unequivocal results, the regulatory protein factors should be relatively pure, since many proteins bind nonspecifically to DNA. Consequently, this approach has been used only for the identification of steroid hormone regulatory elements (*33–38*). Also, the amount of nonspecific binding to pure factors may be relatively high.

Competitive binding assays (*39*) have also been used to determine specific DNA binding to proteins. The protein can be labeled and bound to DNA-cellulose made from nonspecific DNA. The protein is then displaced from the DNA cellulose by soluble DNA, and the amount of factor remaining on the DNA-cellulose can be easily determined, since it is insoluble. Soluble DNA containing sequences that bind the factor specifically will displace the factor from DNA-cellulose more efficiently than nonspecific DNA. Competitive binding assays can be used in a manner similar to nitrocellulose binding assays.

DNA "footprinting" (*34, 38–41*) can yield specific information about sequences in a DNA fragment that bind a regulatory protein, as the binding of this factor to DNA fragments protects the DNA from enzymatic digestion or chemical modification. In practice, a cloned DNA fragment is incubated with a regulatory factor, and digested enzymatically or chemically. The resulting fragments are separated electrophoretically on high resolution sequencing gels. When the digestion products of the binding reaction are compared with the digestion pattern of naked DNA, undigested areas of the DNA are located by the absence of DNA bands in specific regions of the gel. This defines areas in the DNA that bind the regulatory protein. The precise area of binding can be determined by comparing the digestion reactions with sequencing reactions of the same DNA. DNA "footprinting" is currently the most precise method available for directly determining sequences that bind regulatory proteins.

D. Sequence Comparisons

The sequences contained in the hormone regulatory elements, identified as described above, can be compared with sequences of regulatory elements, or with genes regulated by the same hormones. There are several computer programs available that allow the comparison of sequences by both personal and mainframe computers. Such an analysis may provide minimum sequence requirements for hormonal regulation, or identify core elements that define each class of regulatory element. We will examine core sequence elements in more detail below.

II. Specific Regulatory and Inhibitory Elements

The sequence requirements for the hormonal regulation of gene transcription have been determined for a number of genes by the methods described above. Using both gene transfection and DNA–protein binding studies, the interaction of specific DNA sequences in genes regulated by steroid hormones with steroid hormone receptors has provided a paradigm for understanding the interaction of cis- and trans-acting factors to modulate gene transcription. Therefore, the identification of glucocorticoid regulatory elements will be examined and will provide a framework for a discussion of other regulatory elements in hormonally controlled genes. The number of genes that contain individual regulatory elements is constantly increasing, as is the number of hormones for which regulatory elements are being described. Consequently, it is not possible for us to present a current catalogue of these elements. Rather, we will examine specific examples to illustrate the use of the techniques described above for the identification of hormone regulatory elements.

A. Glucocorticoid Regulatory Elements

The sequence requirements for the regulation of gene transcription by glucocorticoid hormones have been determined for a number of genes. A recent review (*8*), discussing the regulation of transcription by steroid hormones, contains a detailed treatment of this subject.

Glucocorticoid regulatory elements in mouse mammary tumor virus (MMTV), the human metallothionein II_A gene (hMT-II_A), and the chicken lysozyme gene have been studied by both transfection and receptor binding (Fig. 1). With these genes, it is possible to discuss the relevance of specific binding sites for glucocorticoid–receptor

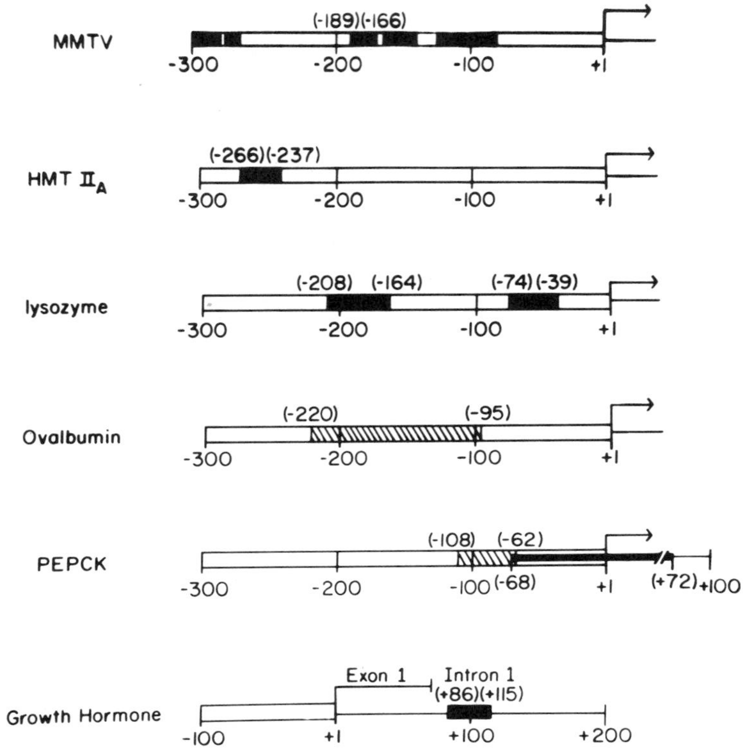

FIG. 1. The location of glucocorticoid regulatory elements in selected regulated genes. MMTV, mouse mammary tumor virus; HMT-II_A, human metallothionein II_A; PEPCK, phospho*enol*pyruvate carboxykinase. The filled boxes in the MMTV, HMT II_A, lysozyme, and growth hormone genes signify the location of glucocorticoid regulatory elements that have been identified by both transfection and binding experiments. The hatched boxes in the ovalbumin and PEPCK genes signify glucocorticoid regulatory elements identified by transfection and by linkage to a heterologous promoter. The solid line in the PEPCK gene refers to a glucocorticoid regulatory element identified by transfection, but not by linkage to a heterologous promoter. The numbers below the genes refer to the nucleotide relative to the major transcriptional start site, and the arrow above each gene indicates the transcriptional start site (+1) and the direction of transcription. Above the genes, in parentheses, are the nucleotide borders of the regulatory elements.

complexes, since the sequence known to be functional *in vivo* have also been determined.

A glucocorticoid regulatory element was identified in the 5′ long terminal repeat of MMTV using DNA transfection techniques (*42–46*). Deletion mutagenesis of this region demonstrated that the sequences responsible for glucocorticoid regulation are located between −190 and −140 relative to the start site of gene transcription (*47*). Of the nine glucocorticoid binding sites present in the MMTV genome (*40, 48*), three were mapped to this region using "footprinting" analysis (Fig. 1). Yamamoto and co-workers (*8, 40*) noted that these sites contain a consensus octanucleotide AGAWCAGW.[3] Mutations that disrupt this sequence decrease the binding of receptors to the altered sequence domain, and mutants that alter sequences outside of the octanucleotide have little effect on nuclear protection (*8, 49*). However, the eight-nucleotide motif is also present in bacteriophage λ, where it does not bind the glucocorticoid receptor (*8*), indicating that other sequences must be involved in receptor binding.

Linker-scanning mutants were made throughout a region of MMTV between −305 and −84, including five glucocorticoid receptor binding domains (Fig. 1). Mutants demonstrating reduced receptor binding *in vitro* were also less responsive to hormones after transfection into cells. Mutations in only one of the five binding sites reduced glucocorticoid responsiveness, implying that each binding domain contributes to hormonal regulation (*8, 49*).

Karin *et al.* (*7, 34*) demonstrated, by stable transfection of a chimeric gene, and receptor binding (nuclease protection and methylation protection), that the hMT-II_A gene contains a glucocorticoid regulatory element between −268 and −237 (*7, 34*) (Fig. 1). A 16-nucleotide sequence was identified as being critical for glucocorticoid regulation of the chimeric gene and for receptor binding. When this 16-nucleotide segment of DNA was computed with sequences in other genes regulated by glucocorticoids, a consensus glucocorticoid regulatory element was postulated: YGGTNWCAMNTGTYCT.[3] The octanucleotide sequence noted above is contained in this consensus sequence, and is the complement of the last nucleotides in the longer consensus sequence (*7, 34*). This suggests that there may be significance in short stretches of homology between elements.

The chicken lysozyme gene is regulated by several different steroid hormones, including glucocorticoids and progestins. Studies of the promoter-regulatory region of this gene for steroid hormone regulatory elements, by DNA transfection and protein binding experiments (*37, 41, 50, 51*), revealed that two steroid regulatory elements are located in the 5′-flanking region of lysozyme (Fig. 1) and empha-

[3] NC-IUB (Nomenclature Committee of the International Union of Biochemistry) recommends [see *Eur. J. Biochem.* **150,** (1985)] Y for "T or C," N for "G or A or T or C," W for "A or T." These are used in the sequences in this section. [Eds.]

sized the importance of the hexanucleotide TGTTCT for steroid-induced regulation of gene expression. This sequence is the complement of the octanucleotide noted in the glucocorticoid regulatory elements of MMTV (*48*) and is present at the end of the 16-nucleotide segment of DNA identified as part of the glucocorticoid regulatory element of the hMT-II_A gene (*7, 34*). Thus, a six- to eight-nucleotide sequence appears to be an important part of the glucocorticoid regulatory element, but is not sufficient for glucocorticoid binding.

Other genes containing glucocorticoid regulatory elements include the human growth hormone gene, and the long terminal repeats of Moloney murine sarcoma virus. The element in the human growth hormone gene is interesting since it is located in the first intron. This element, when placed 5′ to the thymidine kinase gene, renders the latter responsive to stimulation by glucocorticoids (*53*). In addition, a growth-hormone gene devoid of all 5′-flanking sequences is regulated by glucocorticoids (*54*). The glucocorticoid regulatory elements of the Moloney murine sarcoma virus are part of the 74/73-bp repeats in the long terminal repeat, which are transcriptional enhancers (*55, 56*).

The rabbit uteroglobin gene contains three binding sites for the glucocorticoid–receptor complex (*57*). The elements in all of these genes contain the hexanucleotide TGTYCT. A glucocorticoid regulatory element, which reduces transcription after hormonal treatment, is present in the bovine prolactin gene (*21*). While this element has not yet been mapped by deletion mutagenesis or protein binding, it will be interesting to compare its structure with those that positively regulate gene transcription.

There are glucocorticoid regulatory elements in the gene for cytosolic PEPCK from the rat (*13, 58*). A chimeric gene was constructed that contained the PEPCK transcriptional start site and 548 nucleotides of 5′-flanking sequence fused to the structural gene for TK. After transfection into FTO-2B rat hepatoma cells (*59, 60*), the transcription of the PEPCK-TK chimeric gene was regulated by dexamethasone (*13*), indicating that the glucocorticoid regulatory elements are contained within a 621-nucleotide fragment at the 5′ end of the gene. Systematic deletion of the 621-nucleotide fragment of the PEPCK gene from the 5′ end showed that the PEPCK gene contained a glucocorticoid regulatory element 3′ to −68 (relative to the start site of transcription, which is +1), as a deletion mutant containing just 68 bp of 5′-flanking sequence was regulated by dexamethasone (*58*). However, a chimeric gene consisting of −109 to −62 of the PEPCK gene fused to a TK gene with its own transcriptional start site and 109 bp of 5′-flanking sequence (TK-109; *4*) was also regulated by dexamethasone, even though the TK gene in TK-109 is insensitive to glucocorticoids (5, 58).

For all of the chimeric genes described, mRNA was produced from the predicted start site of transcription, ensuring that the fidelity of transcription was not affected.

The PEPCK gene contains two glucocorticoid regulatory elements, one between −109 and −62, and the other 3′ to −68 (Fig. 1). There appear to be no glucocorticoid regulatory elements between −548 and −109, since a chimeric gene containing this fragment fused to TK-109 is not regulated by dexamethasone (58). These elements act independently in the chimeric genes described above, and only one element is required to confer hormonal responsiveness to a heterologous gene. Other glucocorticoid regulatory elements may be present in the PEPCK gene either 5′ to −548 or 3′ to +72, but these sequences have not been analyzed to date.

Within the two regions identified, there are areas of sequence homology with the glucocorticoid elements described above. For example, the core sequence TGTCCT occurs between −48 and −42 in the PEPCK gene. However, between −109 and −68, there are no sequences related to the core sequences identified in other glucocorticoid-regulated genes. Although the identification of the precise sequences involved in the glucocorticoid regulation of transcription of the PEPCK gene awaits site-specific mutagenesis and/or nuclease "footprinting," these results support the idea that there is some flexibility in sequences that respond to glucocorticoid stimulation (8).

B. cAMP Regulatory Elements

The mechanism by which cAMP regulates gene transcription is not known at this time. The phosphorylation of chromosomal proteins may be altered by the catalytic subunit of the cAMP-dependent protein kinase, similar to the manner in which it is thought to regulate all cAMP-dependent cellular processes, or cAMP may bind to a receptor protein, and the cAMP–receptor complex binds to DNA in a similar to the steroid hormone receptor. In the absence of a well-characterized regulatory protein, the sequence requirements for the regulation of transcription by cAMP have been studied only by transfection of specific DNA fragments into various cell lines.

The presence of a cAMP regulatory element was first demonstrated in the gene for cytosolic PEPCK from the rat (Fig. 2). The PEPCK-TK chimeric gene described above, containing the transcriptional start site of the PEPCK gene as well as 548 bp of 5′-flanking sequences, fused to the structural gene for TK, was stably transfected

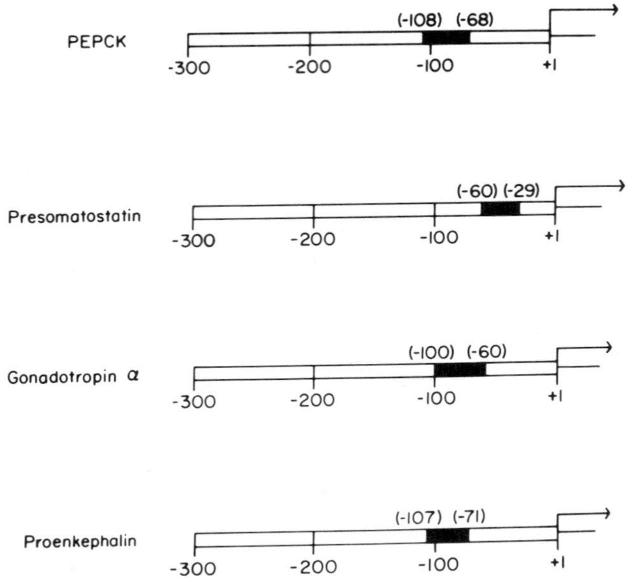

FIG. 2. The location of cAMP regulatory elements in the genes for phospho*enol*pyruvate carboxykinase (PEPCK), presomatostatin, gonadotropin α, and proenkephalin. The genes are represented as in Fig. 1, except that the filled boxes represent the location of the cAMP regulatory elements identified in each gene by transfection only, and the regulatory element indicated by the box in the proenkephalin gene is responsive to cAMP and phorbol esters.

into FTO-2B cells. In clonal cell lines or mass populations of cells containing this gene, TK enzyme activity and PEPCK-TK mRNA were stimulated by treatment of cells with Bt_2cAMP and theophylline (59). Deletions from the 5′ end of PEPCK demonstrated that the cAMP regulatory element is located between −109 and −68 relative to the PEPCK transcriptional start site (58). A chimeric gene consisting of 47 bp of the PEPCK gene between −109 and −62, fused to the TK gene with its own promoter and 109 nucleotides of 5′ flanking sequence (TK-109), as described above, was also regulated by cyclic nucleotides. TK-109 is insensitive to cAMP treatment, as are chimeric genes containing PEPCK sequences other than those between −109 and −62. Thus, it appears that the PEPCK gene contains a single cAMP regulatory element, located between −109 and −62 (58, 59).

The cAMP regulatory element in the PEPCK gene contains a 12-nucleotide sequence that is homologous to sequences present in other cAMP-regulated genes (Fig. 3). The sequence, CTTACGTCAGAG, is

-91	CTTACGTCAGAG	-80	Rat PEPCK
-49	CTGACGTCAGAG	-38	Rat Preprosomatostatin
-62	ACTTA·ATCAGAG	-51	Chicken PEPCK
-118	TCCGA·GTCAGAG	-107	Porcine Plasminogen Activator
-73	CGTCTTTCAGAG	-62	Human Vasoactive Intestinal Polypeptide

FIG. 3. Sequence homologies in the 5′-flanking regions of several genes regulated by cAMP. Homology (greater than 90%) between the regions in the rat PEPCK and rat presomatostatin genes that contain cAMP regulatory elements, identified by transfection, was used as a guide for comparing sequences in other cAMP-regulated genes.

located between −91 and −80 and is virtually identical to a similar sequence between −49 and −38 in the rat presomatostatin gene (*62*). It is also similar to sequences in the 5′-flanking regions of the genes for chicken cytosolic PEPCK (*63*), porcine plasminogen activator (*64*), and human vasoactive intestinal polypeptide (*65*). In all of these genes, the hexanucleotide TCAGAG is absolutely conserved. Promoter-deletion experiments with rat presomatostatin (*66*) and human vasoactive intestinal polypeptide (M. Montminy and R. Goodman, personal communication), transiently expressed in PC12 pheochromocytoma cells, have demonstrated that the cAMP regulatory elements in these genes are at regions containing these homologies. There is a palindromic octamer TGACGTCA in the same areas of these genes (*66*). The 3′ TCA in this octamer overlaps the 5′ TCA of the core sequences described above. The significance of these homologies will await binding studies with purified regulatory factors, once they have been identified, or fine deletion studies.

The area of the PEPCK gene containing the cAMP regulatory element has significant secondary structure. The element is part of the double-stranded region of dyad symmetry, with the potential of forming a stem-loop structure, which has a $\Delta G = -14.2$ kcal (Fig. 4). The significance of such structures in cells and whether they exist *in vivo* are unknown. However, a region of dyad symmetry in the 5′-flanking region of the c-*fos* gene is important for serum stimulation of transcription (*67*). It is tempting to speculate that a steady state equilibrium may exist between classically double-stranded DNA and stem-loop structures. The binding of proteins in native chromatin may favor one form in the absence of hormones, and this form would be unable to bind factors required for transcriptional initiation. Hormonal stimulation, on the other hand, increases the activity or amount of a factor

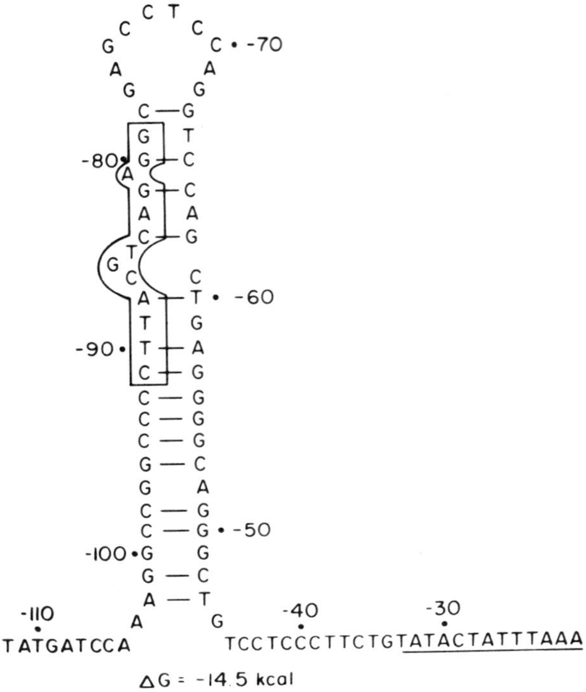

FIG. 4. A stable stem-loop structure in the area of the PEPCK gene which contains the cAMP regulatory element for this gene. This structure is quite stable, with a ΔG of -14.5 kcal. See text (Section II,B) for discussion.

that binds to this critical area of DNA, and shifts the equilibrium toward the other form of secondary structure at these sequences. This secondary structure of the gene would be accessible to factors necessary for transcriptional initiation, and the gene would be activated.

We are testing the significance of this secondary structure to the hormonal regulation of the PEPCK gene. Mutants containing this stem-loop structure, or that have this structure disrupted, have been constructed, and the hormonal regulation of each mutant will be tested. It must be noted, however, that the sequences between -109 and -62 in the PEPCK gene confer cAMP responsiveness to the TK gene, and that these sequences would not form a stem-loop structure (see Fig. 4).

The gene for proenkephalin from the human contains a cAMP regulatory element between 107 and 71 bp 5′ to the start of transcription of this gene, demonstrated by transfection, using transient expression (*68*). This region of the gene contains a sequence similar to, but not identical with, the sequences identified in the genes described above: CTGCGTCAGCTGCAG.

The cAMP regulatory element in the human gene for the α subunit of the gonadotropins has been identified by transfection, using transient expression (*69*). The element is located between −100 and −60 in this gene. This region of the gene does not contain the sequence homology described above for other cAMP-regulated genes. However, there is a sequence, TCATTGGA, in the subunit gene that is similar to the hexanucleotide TCAGAG in the PEPCK gene. The significance of these sequence homologies awaits fine deletion experiments, or binding studies once the trans acting factors responsible for cAMP regulation are identified.

C. Other Hormone Regulatory Elements

In addition to glucocorticoid regulatory elements, other steroid regulatory elements have been identified. There are estrogen receptor-binding domains in the rat prolactin (*70*) and chicken vitellogenin (*71*) genes, and estrogen regulatory elements in the chicken ovalbumin (*19*), the chicken transferrin (*72*), and the chicken vitellogenin A2 (*72a*) genes. An androgen regulatory element was found by transfection in the gene for prostatic-binding protein C3 from the rat (*73*). Finally, progesterone regulatory elements were demonstrated by receptor binding to the rabbit uteroglobin gene (*74*), and in the chicken lysozyme (*37, 41, 50, 51*) and chicken ovalbumin (*6, 71, 75*) genes by receptor binding and gene transfer studies. We discuss the latter two genes in more detail, since they are regulated by several hormones.

A number of genes regulated by calcium (as a second messenger for a variety of hormones) have been cloned. Cellular responses to calcium are mediated by two distinct, but related, second-messenger systems: the calcium–calmodulin pathway, involved in the activation of calcium-dependent protein kinases, or the phosphatidylinositol pathway, involved in the activation of calcium- and phospholipid-dependent protein kinases.

Calcium regulatory elements have been identified in some of these genes. For example, epidermal growth factor and/or thyrotropin-releasing hormone stimulate the transcription of the rat and bovine prolactin genes via calcium. After stable transfection of a prolactin–growth hormone fusion gene into human A431 cells, transcription of gene expression from the rat prolactin promoter is stimulated by epidermal growth factor (via calcium), and by phorbol esters, which activate the calcium-dependent protein kinase C (*76*). The transcription of the bovine prolactin gene is stimulated by epidermal growth factor or thyrotropin releasing hormone (via calcium) after transient expression in GH3 cells (*21*). The protein GRP78, which responds to glucose starvation, is also responsive to the calcium ionophore A23187[4] (*77*). The sequences required for this stimulation have been identified by DNA transfection experiments (*77, 78*). The gene from proenkephalin from the human contains sequences that respond to phorbol esters after transient expression in CV-1 cells (*68*). These sequences are in the same region of the gene as the cAMP regulatory element described above.

Similar to the second-messenger cAMP, much is known about the mechanisms by which calcium controls cytosolic processes, but little about the mechanism by which calcium regulates gene transcription. Further progress in this area awaits the isolation and characterization of specific proteins mediating the transcriptional effects of calcium.

Thyroid hormone regulates gene transcription by binding to a receptor that is thought to bind to DNA in a manner similar to the steroid hormone–receptor complexes. However, this receptor has been difficult to purify, so that the action of thyroid hormone on gene transcription is not as well understood as that of steroid hormones. Thyroid hormone regulatory elements have been identified in the rat (*79*) and human (*80*) growth hormone genes by DNA transfection and viral transfer techniques. The specific sequences required for hormonal regulation have not been precisely mapped by deletion mutagenesis, nor have binding studies been performed.

D. Negative Regulatory Elements

Stimulation of gene transcription by hormones may occur by increasing the activity or amount of a positive control factor. The effect of hormonal induction increases the concentration of active proteins required for transcription of the gene. Alternatively, it is possible that the activity of hormonally regulated genes is repressed by negative control factors, which are inactivated by hormonal stimulation. It is also possible that regulatory elements are under both positive and negative control. A number of inducible genes, including those for β-interferon (*82, 83*), cytochrome *P*-450 (*84*), the human gene for hypoxanthine phosphoribosyltransferase (EC 2.4.2.8) (*85*), and PEPCK (*58*), have been shown to contain such negative regulatory elements.

[4] A23187 is calcimycin, a highly substituted benzoxazolecarboxylic acid (C.A. number 52665.69.7) obtained from *Streptomyces chartreusensis* and also synthetically (*J. Org. Chem.* **45**, 3537, 1980). [Eds.]

They were identified during deletion mapping experiments of each of these genes. We examine briefly the negative regulatory elements in the genes for β-interferon and PEPCK as examples of this type of element.

Transcription of the β-interferon gene is induced by viruses or double-stranded RNA. The DNA sequences responsible for this regulation lie between −77 and −36, using deletion mutagenesis (9), and comprise the inducible regulatory element of the gene. Sequence deletion from the 3′ end of this element increased the basal level of transcription of an attached gene, while the inducibility of the gene by double-stranded RNA was diminished (82). There was a strong transcriptional enhancer, located between −77 and −55, which acted independently on a heterologous promoter. However, when the intact unit was placed in front of a heterologous promoter, transcription was strongly depressed, unless the cells were also treated with the inducing agent (double-stranded RNA), which caused a large increase in transcription of the marker gene. Deletion of the segment between −36 and −41 from the 3′ end of the element resulted in a large increase in the basal level of transcription of the marker gene, but it was not fully derepressed until sequences up to −55 were removed. The excision of these sequences also resulted in a loss of inducibility by double-stranded RNA (82). A protein bound to the area between −68 and −38 in the absence of inducing agent. Mapping of DNA from induced cells showed that this protein was no longer bound, but a protein was now bound to the region between −77 and −64 (83).

From these two studies, a model for negative control was developed in which the β-interferon gene, containing a negative regulatory element, binds a transcription factor in the absence of inducer (82, 83); in turn, this depresses the activity of the adjacent transcriptional enhancer. The inducing agent acts by removing the transcriptional factor bound to the negative regulatory element, relieving the inhibition of the transcriptional enhancer.

The rat PEPCK gene also contains a negative regulatory element. When the PEPCK promoter-regulatory region is attached to the TK structural gene, there is a reduction in the number of HAT-resistant colonies produced after transfection (59), when compared with the number of colonies produced after transfection with the intact TK gene. This reduction occurs in a variety of cell types, including FTO-2B rat hepatoma cells, mouse L cells, or NIH/3T3 cells (unpublished observations). The area of the PEPCK gene containing the negative regulatory element was determined by deletion mutagenesis from the 5′ end of the gene. When the sequence between −266 and −134 was deleted, a 10-fold increase in the transcription of the linked gene was observed, which was 3-fold higher than was noted for chimeric genes containing the undeleted promoter (*58*). The deletion mutant, containing 134 bp of the 5′-flanking sequence, was also insensitive to stimulation by glucocorticoid hormones, even though no glucocorticoid regulatory element could be demonstrated within this region. The basal level of expression decreased as additional transcriptional elements were deleted, and glucocorticoid inducibility was restored (*58*). The deletion of the area of the PEPCK gene between −266 and −134 had no effect on cAMP regulation. Thus, the PEPCK gene contains a putative negative regulatory element which depresses the transcription of the PEPCK gene. This element may be involved in glucocorticoid regulation of gene expression. Additional experiments, such as genomic DNase mapping of the PEPCK gene before and after glucocorticoid stimulation, are required to demonstrate unequivocally the existence of this regulatory element.

E. Positional Flexibility and Enhancer Elements

DNA sequence elements affecting a diverse range of genetic and cellular events have been identified in species from bacteria to humans. Many of these elements share the property that they can act from a variety of locations relative to a heterologous gene and yet retain their activity. In bacteria, for instance, the inversion of DNA sequences by site-specific recombination is stimulated by elements that act in an orientation- and distance-independent manner (*86, 87*). "Silencer" sequences in yeast decrease the expression of linked genes from a variety of locations (*88*). The first elements shown to act in this manner are the "enhancer" elements in the small DNA viruses, such as SV40 (for review, see *89*). These viral enhancer elements are strong transcriptional activators of linked genes, and the activation is independent of orientation and relatively independent of the distance from the promoter. Such elements have been found in a variety of viruses, as well as in cellular genes, such as the immunoglobulin genes (*90–93*).

Chandler *et al.* (*5*) made the original observation that the glucocorticoid regulatory element in mouse mammary tumor virus is active on a heterologous gene, in a manner independent of orientation and distance. Since this observation, a number of other inducible genes have been shown to contain regulatory elements that act independently of their orientation and distance from the start site of transcription. These include human β-interferon (*9*), human metallothionein II_A (*7*), murine metallothionein I (*11, 12*), rat PEPCK (*13*), Moloney murine sarcoma virus (*55, 56*), murine cytochrome *P*-450 (*94*), murine GRP78 (*78*), and human proenkephalin (*68*).

The hormone regulatory elements in the PEPCK gene are active from a variety of locations and in either orientation relative to gene transcription (*13*). The entire PEPCK promoter-regulatory region, or smaller fragments that contain the cAMP and glucocorticoid regulatory elements of the gene, were placed either 650 or 109 nucleotides 5′ from the transcriptional start site of the TK gene, or 3′ to this gene, 1800 bp 3′ from the start site of transcription, in both orientations. After transfection into FTO-2B cells and selection in HAT medium, the expression of these genes was shown to be regulated by cAMP and glucocorticoids. Thus, the regulatory elements in the PEPCK gene demonstrate positional flexibility.

Hormone regulatory elements and viral enhancer elements share a common property of positional flexibility, and the activity of both types of elements is dependent upon the binding of trans-acting factors to DNA sequences. These common properties imply a similar mechanism of action between regulatory elements of inducible genes and enhancer elements (*5, 9, 10*). In fact, hormone regulatory elements which have positional flexibility have been called hormonally regulated enhancers. Whether hormone regulatory elements and enhancer elements act via the same mechanism requires further study.

III. Genes Regulated by Several Hormones

A. Rat PEPCK

As described above, we demonstrated that the PEPCK gene contains cAMP and glucocorticoid regulatory elements, as well as a negative regulatory element (see Fig. 5). Each of the regulatory elements, with the exception of the negative regulatory element, can be isolated from the rest of the PEPCK gene sequences and still retain their activity. However, there is indirect evidence that these elements are part of a complex, interdependent promoter-regulatory region. Either the cAMP regulatory element or the glucocorticoid regulatory elements function most efficiently when they are in their normal location in the PEPCK promoter-regulatory region, even when they are acting on a heterologous promoter. For example, the sequences between −109 and −62 in the PEPCK gene contain cAMP and glucocorticoid regulatory elements. When this sequence is placed 109 bp 5′ from the transcriptional start site of the TK gene, the transcription of the resultant chimeric gene is approximately doubled by either hormone. If the

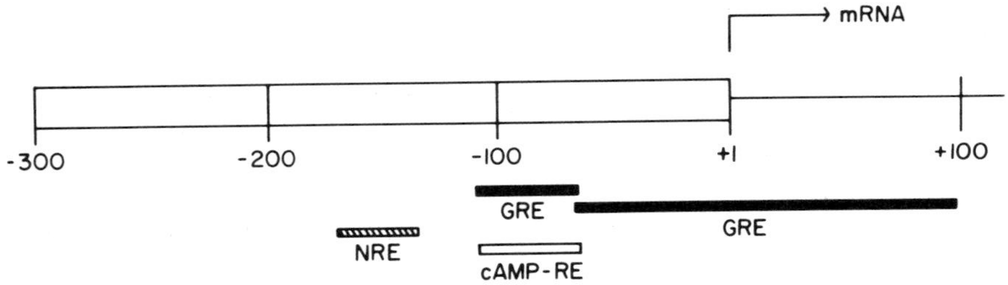

FIG. 5. The location of the regulatory elements in the PEPCK gene identified by transfection. The 5′-flanking sequence is indicated by an open bar, and the transcribed sequences are indicated by a straight line. The numbers below the gene refer to the location of nucleotides relative to the transcriptional start site (+1), which is also indicated by the arrow above the gene. The regulatory elements identified are indicated below the gene, in the positions in which they occur. NRE, negative regulatory element (hatched box); GRE, glucocorticoid regulatory elements (filled boxes); cAMP-RE, cAMP regulatory element (open box).

entire 621 bp promoter-regulatory region of the PEPCK gene is placed in the same position, the resulting chimeric gene is stimulated 3- to 5-fold by the same hormones (58). It is important to point out that this situation does not apply to all cAMP-regulated genes, since a 19-nucleotide sequence of the cAMP regulatory element in the human gene for the α subunit of the gonadotropins is fully as active with respect to cAMP inducibility as is the intact promoter (B. Silver and J. Nilson, personal communication).

It is apparent that the PEPCK promoter-regulatory region operates as a unit and that the multiple elements identified in this region appear to interact with one another. Linker-scanning deletions throughout this promoter-regulatory region are being used to determine which sequences are required for multihormonal responsiveness of the PEPCK gene. In this way, it may be possible to locate sequences modulating the activity of the regulatory elements.

B. Chicken Lysozyme

Regulatory elements for glucocorticoids and progestins were mapped in the chicken lysozyme gene by both gene-transfer and receptor-binding experiments. Mutant genes were microinjected into primary cultures of chicken oviduct, and their hormonal regulation was assayed by determining the number of cells expressing the injected genes in the presence and absence of hormones. Deletion of all sequences 5′ to −60 in the lysozyme gene caused the simultaneous loss of both glucocorticoid and progestin inducibility (*37*). Binding of both glucocorticoid and progestin receptors occurred between −74 and −39, although the glucocorticoid receptor had a much higher affinity for this site. Both receptors also bind to an upstream site, between −208 and −164, with the opposite relative receptor-binding affinities. At both binding sites, the progesterone receptor protected more nucleotides than the glucocorticoid receptor, but binding occurred at the same positions in both sites (*37, 51*). It appears that the chicken lysozyme gene contains a hormone regulatory element that responds to both glucocorticoid and progestin stimulation by binding the receptors for each of these hormones at the same site.

C. Chicken Ovalbumin

The ovalbumin gene is similar to the lysozyme gene as it is regulated by all classes of steroid hormones in the chicken oviduct. Regulatory elements for glucocorticoids, estrogens, and progestins in the ovalbumin gene have been located within the first 750 bp 5′ from the start site of transcription (*6, 71*). Using deletion mutants, the estrogen and progestin regulatory elements were located between −197 and −95 in the promoter-regulatory region. Surprisingly, deletion of sequences from −197 to −95 resulted in a gradual loss of responsiveness to both of these hormones, and not the abrupt loss of hormonal regulation that occur with other genes. It is probable that there are multiple independent binding sites for these hormones between −197 and −95, and that their effects on the transcripton of the ovalbumin gene are additive. The estrogen and progestin responsiveness of the deletion mutants were lost simultaneously, which was also observed with the glucocorticoid and progestin regulatory elements in the lysozyme gene. It seems that the steroid hormones act through different receptors on the same or similar elements in the two genes in which their effect has been studied.

D. Human Proenkephalin

This gene contains regulatory elements for both cAMP and phorbol ester stimulation (*68*). Both of these elements are located in the same 31 bp sequence 5′ from the transcriptional start site for this gene, between −107 and −71. The ability of this element to respond to cAMP and phorbol esters is stimulated by the presence of sequences between −133 and −102, which alone are not responsive to either agent. The effects of cAMP and phorbol esters on cellular processes are mediated via protein kinases. cAMP activates a cAMP-dependent protein kinase, while phorbol esters activate protein kinase C.

Although the mechanisms by which these agents activate transcription are unknown, it is tempting to speculate that they activate transcription by phosphorylating DNA-binding proteins via their respective protein kinases. If this is true, then the different types of kinases may phosphorylate the same DNA-binding protein, or different proteins that bind to one regulatory element, explaining why the cAMP and phorbol ester regulatory elements are located in the same sequences. Alternatively, phorbol esters may simply increase levels of cAMP, since the effects of phorbol esters on the expression of genes linked to the element in this gene depend on the presence of phosphodiesterase inhibitors (*68*). Thus, similar to the steroid hormone regulatory elements in the chicken lysozyme and ovalbumin genes, the human proenkephalin gene contains a common cAMP and phorbol ester regulatory element.

E. Bovine and Rat Prolactin

The prolactin gene is regulated by a number of factors, such as peptide hormones (via cAMP and calcium), steroids, and thyroid hormone, and is an excellent model for the regulation of gene transcription by multiple hormones. However, the regulatory elements in this gene have not been precisely located and await further study.

IV. Multihormonal Regulation: Conclusions

In this review, we have examined the organization of genes regulated by several hormones. The methods used to identify hormone regulatory elements, as well as examples of the identification of specific elements, provide the foundation on which the organization of multiple elements in a single gene can be studied.

It is obvious that hormonally regulated promoters consist of several different types of elements. Promoter elements necessary for polymerase binding and initiation of transcription are part of all eukaryotic genes, and seem to be similarly placed in hormonally regulated and constitutively expressed genes. These general promoter elements include the "TATA box" and the sequences surrounding the start site of transcription, which have already been reviewed in detail (*96*). Other upstream elements appear to be important for the expression of constitutively expressed and hormonally regulated genes. In constitutively expressed genes, such as the β-globin gene, an upstream element is required for maximal promoter activity (*96, 97*), by binding to transcriptional factors that are always present and active in cells.

In contrast, hormonally regulated genes contain elements binding to factors induced or activated by hormonal stimulation. The mechanism of binding and activation of gene transcription may be identical for upstream elements and hormone regulatory elements, but differs in the nature of the binding proteins: constitutively expressed genes bind constitutively expressed transcription factors, while hormonally regulated genes bind hormonally activated proteins. Other hormonally regulated genes may have repressors bound to negative regulatory elements, to depress transcription. Hormonal stimulation relieves the binding of the repressor, allowing the gene to be transcribed. Thus, hormonal regulation of transcription can involve the binding of activators, repressors, or both. Additionally, transcriptional enhancers, involved in either tissue specificity of gene expression or in maintaining basal levels of gene expression, may be associated with hormonally regulated genes. Other DNA sequences, yet to be identified, may also be essential or facilitating elements in hormonal regulation.

Promoter-regulatory regions of genes regulated by multiple hormones may contain elements that act as independent "cassettes," as interdependent elements, or both. For instance, the PEPCK gene contains cAMP and glucocorticoid regulatory elements, which can be individually linked to a heterologous gene and retain their activities (*13, 58*). Other genes contain elements with similar properties, as discussed above. However, the intact PEPCK promoter-regulatory region is more efficient in activating a heterologous gene than are the individual elements. This may be an artifact of plasmid constructions, but it may also indicate that the PEPCK gene is an integrated unit that requires interaction with other auxiliary DNA sequences for maximal activity. Also, the activity of the cAMP and phorbol ester regulatory element in the proenkephalin gene is increased by adjacent sequences in the 5′-flanking region of this gene (*68*). Other elements, such as the glucocorticoid regulatory element in mouse mammary tumor virus, can act as an independent cassette, but this gene is regulated by a single hormone.

Multiple hormones may regulate a single element, as appears to be the case for steroid hormones and the chicken ovalbumin and lysozyme genes (*37, 41, 50, 51*). For these genes, the same sequences appear to bind each of the steroid hormone–receptor complexes. Also, cAMP and phorbol esters regulate the transcription of the human proenkephalin gene via a single regulatory element (*68*). This raises the intriguing possibility that genes regulated by a particular class of hormone, such as steroid hormones, or hormones that stimulate protein kinases, will respond to all members of that class The specificity of regulation is determined by which receptors are present in a cell type. Whether all classes of hormones regulate gene transcription through the same regulatory element, or whether all regulatory elements respond to multiple hormones, awaits further experiments. Finally, in these genes, the relative ability of each hormone to stimulate transcription may be a function of the strength of binding of each receptor complex to the element. This may provide a means of "fine-tuning" the transcriptional response of a particular gene to various metabolic needs.

Many hormone regulatory elements have been identified in a variety of genes, and the list of elements is rapidly increasing. Progress in this area is related directly to the ease with which genes can be manipulated and introduced into cells. If a cloned gene is available, it is a relatively straightforward matter to introduce it into cells and to perform mutagenesis of sequences to be tested for functional control of gene expression. One limiting factor in such an analysis for hormonally regulated genes is to find the proper recipient cell lines for the modified genes. To date, virtually all progress made in this field has been in the identification of hormone regulatory elements. The next step in understanding hormonal regulation of gene expression will involve identifying the proteins that bind to regulated genes. Progress has been made in this area for a few genes, mostly viral genes that are constitutively expressed (for review, see *99*). The availability of an *in vitro* transcription system regulated by added factors will be required to understand the mechanisms of hormonal regulation of gene expression. It is probable that the development of this type of assay system will determine the rate of progress of this field.

Acknowledgments

We wish to thank Dr. Yaacov Hod for his useful comments on the manuscript. This work was supported in part by Grants AM-21859 and AM-24451 from the National Institutes of Health. A.W.-B. was a trainee on the Metabolism Training Program AM-07319 from the National Institutes of Health.

References

1. G. S. McKnight and R. D. Palmiter, *JBC* **254**, 9050 (1979).
2. W. A. Guyette, R. J. Matusiak and J. M. Rosen, *Cell* **17**, 1013 (1979).
3. W. H. Lamers, R. W. Hanson and H. M. Meisner, *PNAS* **79**, 5137 (1982).
4. D. Granner, T. Andreone, K. Sasaki and E. Beale, *Nature* **305**, 549–551 (1983).
5. V. L. Chandler, B. A. Maler and K. R. Yamomoto, *Cell* **33**, 489 (1983).
6. D. C. Dean, B. J. Knoll, M. E. Riser and B. W. O'Malley, *Nature* **395**, 551 (1983).
7. M. Karin, A. Haslinger, H. Holtgreve, G. Cathala, E. Slater and J. D. Baxter, *Cell* **36**, 371 (1984).

8. K. R. Yamamoto, *ARGen* **19,** 209 (1985).
9. S. Goodbourn, K. Zinn and T. Maniatis, *Cell* **41,** 509 (1985).
10. K. Ponta, N. Kennedy, P. Skroch, N. Hynes and B. Groner, *PNAS* **82,** 1020 (1985).
11. G. W. Stuart, P. F. Searle, H. Y. Chen, R. L. Brinster and R. D. Palmiter, *PNAS* **81,** 7318 (1984).
12. P. F. Searle, G. W. Stuart and R. D. Palmiter, *MCBiol* **5,** 1480 (1985).
13. A. Wynshaw-Boris, J. M. Short, D. S. Loose and R. W. Hanson, *JBC* **261,** 9714 (1986).
14. K. Sasaki, T. P. Cripe, S. R. Koch, T. L. Andreone, D. D. Peterson, E. G. Beale and D. K. Grannr, *JBC* **259,** 15242 (1984).
15. E. G. Beale, N. B. Chrapkiewicz, H. A. Scoble, R. J. Metz, D. P. Quick, R. L. Noble, J. E. Donelson, K. Biemann and D. K. Granner, *JBC* **260** 10748 (1985).
16. D. S. Loose, D. K. Cameron, H. P. Short and R. W. Hanson, *Bchem* **24,** 4509 (1985).
17. D. S. Loose, A. Wynshaw-Boris, H. M. Meisner, Y. Hod and R. W. Hanson, *in* "Molecular Basis of Insulin Action" (M. Czech, ed.), p. 347. Plenum Press, New York, 1985.
18. Y. Hod, J. S. Cook, S. L. Weldon, J. M. Short, A. Wynshaw-Boris and R. W. Hanson, *NYAcad Sci.*, in press.
19. D. C. Dean, R. Gope, B. J. Knoll, M. E. Riser and B. W. O'Malley, *JBC* **259,** 9967 (1984).
20. R. C. Moen and R. D. Palmiter, *Dev. Biol.* **78,** 450 (1980).
21. S. A. Camper, Y. A. S. Yao and F. M. Rottman, *JBC* **260,** 12246 (1985).
22. A. Wynshaw-Boris, J. M. Short and R. W. Hanson, *Biotechniques* **4,** 104 (1986).
23. M. Wigler, S. Silverstein, L-S. Lee, A. Pellicer, Y-C. Cheng and R. Axel, *Cell* **11,** 223 (1977).
24. M. Perucho, D. Hanahan, L. Lipsich and M. Wigler, *Nature* **285,** 207 (1980).
25. F. Colbere-Garapin, F. Horodniceanu, P. Kourilsky and A-C. Garapin, *JMB* **150,** 1 (1981).
26. P. J. Southern and P. Berg, *J. Mol. Appl. Genet.* **1,** 327 (1982).
27. R. C. Mulligan and P. Berg, *Science* **209,** 1422 (1980).
28. R. C. Mulligan and P. Berg, *PNAS* **78,** 2072 (1981).
29. G. Ringold, B. Dieckmann and F. Lee, *J. Mol. Appl. Genet.* **1,** 165 (1981).
30. A. D. Miller, M-F. Law and I. M. Verma, *MCBiol* **5,** 431 (1985).
31. K. E. Mayo, R. Warren and R. D. Palmiter, *Cell* **29,** 99 (1982).
32. S. L. McKnight and R. Kingsbury, *Science* **217,** 316 (1982).
33. S. Geisse, C. Scheidereit, H. M. Westphal, N. E. Hynes, B. Groner and H. Beato, *EMBO J.* **1,** 1613 (1982).
34. M. Karin, A. Haslinger, H. Holtgreve, R. I. Richards, P. Krauter, H. M. Westphal and M. Beato, *Nature* **308,** 513 (1984).
35. F. Payvar, O. Wrange, J. Carlstedt-Duke, S. Okret, J-A. Gustafsson and K. R. Yamamoto, *PNAS* **78,** 6628 (1981).
36. M. Pfahl, *Cell* **31,** 475 (1981).
37. R. Renkawitz, G. Schutz, D. van der Ahe and M. Beato, *Cell* **37,** 503 (1984).
38. C. Scheidereit, S. Geisse, H. M. Westphal and M. Beato, *Nature* **304,** 749 (1983).
39. E. R. Mulvihill, J-P. LePennec and P. Chambon, *Cell* **24,** 621 (1982).
40. F. Payvar, D. DeFrance, G. L. Firestone, B. Edgar, O. Wrange, S. Oknet, J-A. Gustafsson and K. R. Yamamoto, *Cell* **35,** 381 (1983).
41. D. van der Ahe, S. Janich, C. Scheidereit, R. Renkawitz, G. Schutz and M. Beato, *Nature* **313,** 706 (1985).
42. N. Fasel, K. Pearson, E. Buetti and H. Digglemann, *EMBO J.* **1,** 3 (1982).

43. E. Buetti and H. Diggelmann, *Cell* **23,** 335 (1981).
44. N. E. Hynes, N. Kennedy, U. Rahmsdorf and B. Groner, *PNAS* **78,** 2038 (1981).
45. F. Lee, R. Mulligan, P. Berg and G. Ringold, *Nature* **294,** 228 (1981).
46. A. L. Huang, M. L. Ostrowski, D. Berard and G. L. Hager, *Cell* **27,** 245 (1981).
47. J. Majors and H. E. Varmus, *PNAS* **80,** 5866 (1983).
48. C. Scheidereit, S. Geisse, H. M. Westphal and M. Beato, *Nature* **304,** 749 (1983).
49. D. DeFranco, O. Wrange, J. Merriweather and K. R. Yamamoto, UCLA Symp. *Mol. Cell. Biol.* **20,** 305 (1985).
50. R. Renkawitz, H. Beug T. Graf, P. Matthias, M. Grez and G. Schutz, *Cell* **31,** 167 (1982).
51. D. van der Ahe, J-M. Renoir, T. Buchou, E-E. Baulieu and M. Beato, PNAS **83,** 2817 (1986).
52. D. D. Moore, A. R. Marks, D. I. Buckley, G. Kapler, F. Payvar and H. M. Goodman, *PNAS* **82,** 699 (1985).
53. E. P. Slater, O. Rabenau, M. Karin, J. D. Baxter and M. Beato, *MCBiol* **5,** 2984 (1985).
54. M. J. Birnbaum and J. D. Baxter, *JBC* **261,** 291 (1986).
55. D. DeFranco and K. R. Yamamoto, *MCBiol* **6,** 993 (1986).
56. R. Miksicek, A. Heber, W. Schmid, U. Danesch, G. Pooseckert, M. Beato and G. Schutz, *Cell* **46,** 283 (1986).
57. A. C. B. Cato, S. Geisse, M. Wenz, H. M. Westphal and M. Beato, *EMBO J.* **3,** 2771 (1984).
58. J. M. Short. A. Wynshaw-Boris, H. P. Short and R. W. Hanson, *JBC* **261,** 9721 (1986).
59. A. Wynshaw-Boris, T. G. Lugo, J. M. Short, R. E. K. Fournier and R. W. Hanson, *JBC* **259,** 12161 (1984).
60. A. M. Killary, T. G. Lugo and R. E. K. Fournier, *Biochem. Genet.* **22,** 201 (1984).
61. S. L. McKnight, E. R. Gavis, R. Kingsbury and R. Axel, *Cell* **25,** 385 (1981).
62. M. R. Montminy, R. H. Goodman, S. J. Horovitch and J. F. Habener, *PNAS* **81,** 3337 (1984).
63. Y. Hod, H. Yoo-Warren and R. W. Hanson, *JBC* **259,** 15609 (1984).
64. Y. Nagamine, D. Fearson, M. S. Altus and E. Reich, *NARes* **12,** 9525 (1984).
65. T. Tsukada, S. J. Horovitch, M. R. Montminy, G. Mandel and R. H. Goodman, *DNA* **4,** 293 (1985).
66. M. R. Montminy, K. R. Sevarino, J. A. Wagner, G. Mandel and R. H. Goodman, *PNAS* **83,** in press.
67. R. Treisman, *Cell* **46,** 567 (1986).
68. M. Comb, N. C. Birnberg, A. Seasholtz, E. Herbert and H. M. Goodman, *Nature* **323,** 353 (1986).
69. R. B. Darnell and I. Boime, *MCBiol* **5,** 3157 (1985).
70. R. A. Maurer, *DNA* **4,** 1 (1985).
71. J. P. Jost, M. Seldran and M. Geiser, *PNAS* **81,** 429 (1984).
72. R. B. Hammer, R. L. Idzerda, R. L. Brinster and G. S. McKnight, *MCBiol* **6,** 1010 (1986).
72a. L. Klein-Hitpab, M. Schorpp, U. Wagner and G. U. Ryffel, *Cell* **46,** 1053 (1986).
73. M. Page and M. Parker, *Cell* **32,** 495 (1983).
74. A. Bailly, M. Atger, P. Atger, M. A. Cerban, M. Alizon, M. T. V. Hai, F. Logeat and E. Milgrom, *JBC* **258,** 10384 (1983).
75. J. G. Compton, W. T. Schrader and B. W. O'Malley, *PNAS* **80,** 16 (1983).
76. S. C. Supowit, E. Potter, R. M. Evans and M. G. Rosenfeld, *PNAS* **81,** 2975 (1984).

77. E. Resendez, J. W. Attenello, A. Grafsky, C. S. Chang and A. S. Lee, *MCBiol* **5,** 1212 (1985).
78. A. Y. Lee, S. C. Chang and A. S. Lee, *MCBiol* **6,** 1235 (1986).
79. A. D. Miller, E. S. Ong, M. G. Rosenfeld, I. M. Verma and R. M. Evans, *Science* **225,** 993 (1984).
80. J. Casanova, R. P. Coop, L. Janocko and H. H. Samuels, *JBC* **260,** 11744 (1985).
81. S. Goodbourn, K. Zinn and T. Maniatis, *Cell* **41,** 509 (1985).
82. S. Goodbourn, H. Burstein and T. Maniatis, *Cell* **45,** 601 (1986).
83. K. Zinn and T. Maniatis, *Cell* **45,** 611 (1986).
84. P. B. C. Jones, D. R. Galeazzi, J. M. Fisher and J. P. Whitlock, *Science* **227,** 1499 (1985).
85. D. W. Melton, C. McEvan, A. V. McKie and A. M. Reid, *Cell* **44,** 319 1986).
86. R. Kahmann, F. Rudt, C. Koch and G. Mertens, *Cell* **41,** 771 (1985).
87. R. C. Johnson and M. Simon, *Cell* **41,** 781 (1985).
88. A. H. Brand, L. Breeden, J. Abraham, R. Sternglanz and K. Nasmyth, *Cell* **41,** 41 (1985).
89. Y. Gluzman and T. Shenk, "Enhancers and Eukaryotic Gene Expression." Cold Spring Harbor Laboratories, Cold Spring Harbor, N.Y.
90. S. D. Gilles, S. L. Morrison, V. T. Oi and S. Tonegawa, *Cell* **33,** 717 (1983).
91. J. Banerji, L. Olson and W. Schaffner, *Cell* **33,** 729 (1983).
92. C. Queen and D. Baltimore, *Cell* **33,** 741 (1983).
93. R. Grosschedl and D. Baltimore, *Cell* **41,** 885 (1985).
94. P. B. C. Jones, L. K. Durrin, D. R. Galeazzi and J. P. Whitlock, *PNAS* **83,** 2802 (1986).
95. P. D. Matthias, R. Renkawitz, M. Grez and G. Schutz, *EMBO J.* **1,** 1207 (1982).
96. T. Shenk, *Curr. Top. Microbiol. Immunol.* **93,** 25 (1981).
97. P. Dierks, A. van Ooyen, N. Mantei and C. Weissmann, *PNAS* **78,** 1411 (1981).
98. G. C. Grosveld, C. K. Shewmaker, P. Jat and R. Flavell, *Cell* **25,** 215 (1981).
99. S. McKnight and R. Tjian, *Cell* **46,** 795 (1986).

Transport of mRNA from Nucleus to Cytoplasm

Heinz C. Schröder,
Michael Bachmann,
Bärbel Diehl-Seifert,
and
Werner E. G. Müller

Institut für Physiologische Chemie
Abteilung Angewandte Molekularbiologie
Universität Mainz
6500 Mainz, Federal Republic of Germany

Transport of mRNP (messenger ribonucleoprotein) from nucleus to cytoplasm plays an important role in gene expression in eukaryotic cells. This view is supported by the fact that much more mRNA is synthesized in the nucleus than ever appears in the cytoplasm. There-

Progress in Nucleic Acid Research
and Molecular Biology, Vol. 34

fore, cells must be provided with control mechanisms that enable them to separate those RNA molecules to be exported to the cytoplasm from those to remain in the nucleus. In cancer cells, where mRNA molecules normally restricted to the nucleus also emerge in the cytoplasm, these control mechanisms may be impaired (*1–3*).

It is generally agreed that nucleocytoplasmic transport of mRNPs occurs through the nuclear envelope pore complexes (*4*). The effective diameter of the water-filled channel formed by such a complex is about 10 nm (*5*). Therefore, mRNPs (diameter about 20 nm) are too large to leave the nucleus by passive diffusion. Transport of mRNA seems to be associated with changes in mRNP three-dimensional structure, which can be observed in electron micrographs (*6*).

In this review we focus mainly on energy-(ATP)-dependent mRNP transport. Nucleocytoplasmic transport of ribosomal RNA can also be induced by ATP, but also occurs by varying $[Ca^{2+}] : [Mg^{2+}]$ (*7*). Release of ribosomal RNPs seems to be accompanied by an expansion of the nucleus (*8*). Nucleocytoplasmic transport of mRNA seems to be also distinct from the export of tRNA or the exchange of snRNPs and of proteins across the nuclear envelope. Nucleocytoplasmic transport of tRNA seems to involve a facilitated diffusion mechanism, showing saturability and sequence specificity (*9*); apparently, it does not depend on ATP (*9, 10*). In contrast to the transport of mRNPs through the nuclear pore, which appears strictly vectorial, snRNPs can shuttle between nuclear and cytoplasmic compartments (*11*). The nuclear uptake of at least some kinds of U-snRNAs[1] in oocytes seems to depend on their association with proteins stockpiled in the cytoplasm (*11, 12*). In contrast to the mRNA export, the import of most proteins into the nucleus seems to be energy-independent, although in some cases nucleotides promote this process (*13*; H. Fasold, personal communication). The accumulation of karyophilic proteins in the nucleus may be mediated by specific signal sequences recognizing the intranuclear binding sites of these proteins (*14–17*). However, some proteins (e.g., SV40 large T antigen; ref. *16*) seem to migrate into the nucleus via a transport mechanism (*18*).

Transport of mRNA from nucleus to cytoplasm can be subdivided into the following steps: (1) release of mRNA from the internal nuclear matrix structure, (2) translocation of mRNA through the nuclear envelope pore complex, and (3) binding of the transported mRNA to cytoplasmic cytoskeletal elements. Current evidence suggests that, during

[1] U-snRNA, uridylate-rich, small nuclear RNA; see Reddy and Busch in Vol. 30 of this series. [Eds.]

all these steps, mRNA never appears in a freely diffusible form. Transport of mRNA seems most likely to involve sequential attachment and detachment processes at specific sites of the nuclear matrix, the nuclear envelope (pore complex–lamina), and the cytoskeleton. At the stage of translocation, this binding site may be the poly(A)-recognizing transport-carrier within the nuclear envelope structure (*19, 20*). At the cytoplasmic site, microtubules (*21, 22*), actin filaments (*21*), and/or intermediate filaments (*23*) might be important for mRNA transport and function.

I. Sites of Transport

The intracellular structures involved in the transport of mRNA are the nuclear matrix, the nuclear envelope (or its pore complex–lamina substructure) and the cytoskeleton.

The nuclear matrix forms a dense intranuclear fibrillar network, which includes the nucleoli and which is connected at its periphery to the nuclear envelope. It is assumed that the nuclear matrix provides the organizational platform for most steps of gene expression occurring in the nucleus. Although aware of criticism of the matrix concept, we use this term throughout this review. The controversy about the *in vivo* relevance of this structure has been thoroughly discussed in some recent reviews (*24–26*).

The nuclear envelope, which forms the outer periphery of the nucleus, can be subdivided morphologically into four distinct components: the outer and the inner nuclear membranes, the lamina, and the pore complexes. By treatment of nuclear envelopes with Triton X-100, pore complex–laminae that differ from the envelopes in the absence of the nuclear membranes can be obtained.

The nuclear lamina consists of a fibrous network closely associated with the inner nuclear membrane and the nuclear pore complexes. In rat liver and many other tissues, it is composed of three major 60- to 80-kDa polypeptides, called lamins (A, B, and C) (*27, 28*); lamin B seems to be partially integrated in the inner nuclear membrane (*27, 29*).

The nuclear pore complexes (outer diameter about 80 nm; ref. *4*) are composed of eight granular components, of fibrils present in concentric and traversing arrangements, and of an inner annular granule (*30–32*). The inner annular granules are associated with intranuclear fibrils that appear to be associated with the nuclear envelope or the nuclear matrix. In some electron micrographs, an apparent continuity between the nuclear matrix and the cytoplasmic filaments (probably intermediate filaments) has been seen, formed by fibrils traversing the pore channel (*33*). In contrast to the lamina, the protein composition of the pore complexes is not known; one major component seems to be a 190-kDa glycoprotein (*34, 35*).

The cytoplasmic cytoskeleton consists of three major filamentous systems: microtubules, microfilaments, and intermediate filaments (for reviews see *36–40*). The proteinaceous networks formed by these filaments seem to be connected to additional fine filaments, not as well defined, termed microtrabeculae (*41*).

II. Some Methodological Aspects

Transport of mRNA is usually studied by measuring the efflux of mRNA (either prelabeled or identified by cDNA probes) from isolated nuclei into a suitable external transport medium containing ATP and an ATP-regenerating system, cations (Mg^{2+}, Mn^{2+}), and factors needed to maintain nuclear stability (Ca^{2+}, spermidine, salt) and to prevent RNA degradation (exogenous RNA, "RNasin")[2] (*10, 42*). Addition of dialyzed cytosol to the exogenous medium increased the efflux rate, but is not required because, in the absence of cytosolic proteins, efflux of mRNA (but not efflux of rRNA) also occurs (*10, 42, 43*). Efflux of mRNA, measured under these *in vitro* conditions, seems an adequate model for mRNA transport *in vivo*. This can be seen, for instance, by comparing the composition of the mRNA in the cytoplasm of cells from different tissues, or under special physiological conditions, with the mRNA composition in the respective efflux supernatants (*43–45*), or by the absence of nucleus-restricted sequences (*46, 47*) and hnRNP core proteins (*10*) from the efflux supernatants (hn is heterogeneous nuclear).

In the study of the efflux of RNA from isolated nuclei, only the first two steps of mRNA transport (release from the matrix and translocation through the pore complex) are measured. However, no evidence exists that the third step (binding to the cytoskeleton) is rate-limiting in overall mRNA transport. Using the recently developed technique for preparing resealed nuclear envelope vesicles, in which transportable RNAs or RNPs can be trapped during resealing, it is now possible to study the translocation of these macromolecules or particles through the nuclear pores independently of their detachment from the matrix structure (*48*).

[2] "RNasin": name given an RNase inhibitor from human placenta (*41a*). [Eds.]

Studies of the mRNA translocation system (protein composition and enzymology) are usually performed with isolated nuclear envelopes or nuclear ghosts (less pure nuclear envelopes that contain residual amounts of nuclear matrix; ref. *49*) or pore complex–laminae. Methods for preparing highly purified nuclear envelopes are now available (*50*). However, no suitable method for preparing nuclear pore complexes exists.

Because, for a long time, nucleocytoplasmic transport of mRNA has been taken for a simple exchange process (facilitated or energy-dependent) of soluble and freely diffusible mRNA molecules from a nuclear compartment to a cytoplasmic compartment, the translocation step through the nuclear envelope pore complexes has been the most studied step of mRNA transport. Only in the last few years has the importance of the release step and the step of cytoskeletal binding been recognized, and the investigation of these steps begun. Therefore, most of this review deals with the second step (translocation across the nuclear envelope) of mRNA transport, for which most data are available.

III. Importance of Posttranscriptional Processing for Transport

Only mature mRNAs are transported out of the nucleus (*46, 47, 51*). Therefore, posttranscriptional maturation of mRNA from hnRNA, which involves 5′ capping, 3′ polyadenylation, splicing, and internal methylation, seems to be essential for mRNA transport. Because of nuclear restriction of immature messengers, splicing seems to be a prerequisite for the transport of most mRNAs to the cytoplasm (*52*). However, nucleus-restricted sequences also appear, in addition to mature mRNA, in the cytoplasm of cancer cells (*1, 53*). It is not known whether 5′ capping is required for nucleocytoplasmic mRNA transport. However, mRNAs undermethylated at their 5′-cap structure are transported normally (*54, 55*). On the other hand, methylations of internal adenosine residues may influence nucleocytoplasmic mRNA transport; this has been shown for the transport of late SV40 mRNA (*56*). In the past, most workers focused their interest on the role of the 3′-terminal poly(A) in mRNA transport. Indeed, accumulating evidence suggests that polyadenylation is a prerequisite for nucleocytoplasmic translocation of most (but not all) mRNA species (*25, 57*). This assumption is supported by the finding that nuclear RNA transcripts from a late-region deletion mutant of SV40, which cannot be polyadenylated, are not transported out of the nucleus (*58*). Because of its importance for understanding the regulation of mRNA transport, some aspects of the regulation of poly(A) metabolism are noted in the following (for detailed recent reviews see refs. *57, 59, 60*).

Synthesis of the 3′ poly(A) seems to be an early step during hnRNA processing (*61, 62*) but it occurs also in the cytoplasm (*63, 64*); its initial size is about 250 AMP residues. Although most eukaryotic mRNAs seems to be polyadenylated, there are important exceptions. Most mRNAs coding for histones lack a 3′ poly(A) "tail" (*61*). In addition, a highly complex population of mRNAs devoid of poly(A) exists in the adult brain (*65*).

It is experimentally well proven that the 3′ poly(A) chain is involved in determining the stability of the mRNA (*66, 67*). Although this could influence the efficiency of mRNA transport, some other mechanisms are involved in facilitating nucleocytoplasmic mRNA translocation by poly(A); we discuss this in this review.

During aging of an individual mRNA molecule, a reduction of the initial size of the poly(A) sequence occurs; in consequence, a heterogeneous poly(A) chain-length distribution in the overall mRNA population in the cell results. The steady-state chain length of the poly(A) is adjusted by both poly(A) anabolic enzymes and poly(A) catabolic enzymes. The equilibrium between both groups of enzymes depends upon a series of physiological conditions, such as hormone status (*68*), nutrition (*69*), virus infection (*70*), or aging (*71*). In addition, it is controlled by posttranslational protein modification (*72–75*) and by specific poly(A)-associated proteins (*76, 77*).

The synthesis of the poly(A) segment at the 3′ terminus of hnRNA and mRNA is catalyzed by poly(A) polymerases [EC 2.7.7.19]. Poly(A) polymerases have been isolated from both nucleus (*68, 78–80*) and cytoplasm (*64*). Nuclear poly(A) polymerase has been reported to be bound to nuclear substructures [e.g., chromatin (*81*) or nuclear matrix (*82*)], or to be present in a soluble, nucleoplasmic state (*81*).

Poly(A) degradation is catalyzed by three enzymes, which we isolated and purified to homogeneity some years ago (for reviews see refs. *57, 59, 60*): endoribonuclease IV [EC 3.1.26.6], endoribonuclease V [EC 3.1.27.8], and poly(A)-specific 2′,3′-exoribonuclease [EC 3.1.13.4].

Endoribonuclease IV from chick oviduct, which has been purified to homogeneity (*83*), has a molecular weight of 45,000, is localized predominantly in the nucleus, and cleaves specifically $(A)_n$ to $(A)_{\overline{10}}$ fragments; the 3′ poly(A) tail of mRNA is degraded by the enzyme up to a residual $(A)_{\overline{5}}$, which is not hydrolyzed (*79, 83, 84*). It might be functionally significant that endoribonuclease IV has a high affinity for poly(A) polymerase; the two enzymes can form a stable complex

(*21*, *83*). Endoribonuclease V (*85*) and poly(A)-specific 2′,3′-exoribonuclease (*86*) have been purified to homogeneity from calf thymus gland. Endoribonuclease V has a mass of 52 kDa and hydrolyzes single-stranded poly(A) and poly(U) in an endo-exonucleolytic manner to 3′-AMP and 3′-UMP. The 2′,3′-exoribonuclease consists of two subunits of 58 and 41 kDa, is essentially localized in the nucleus, and specifically splits poly(A) [in the single-stranded and in the double-stranded form (poly(A)·poly(U)); with 3′,5′-internucleotide bonds; ref. *86*] or 2′,5′-linked oligo(A) (*87*) with formation of 5′-AMP.

Both poly(A) polymerase (*72*, *73*, *75*) and endoribonuclease IV (*74*, *75*) are regulated by posttranslational protein phosphorylation, resulting in higher specific activities (*73*, *74*). These reactions are mediated, at least in part, by nuclear "protein kinase NI" (*75*). With regard to the established association of the hnRNA and the mRNA with cytoskeletal structures, it might be important that the activities of both the poly(A) polymerase and endoribonuclease IV are strongly inhibited by tubulin and G-actin (*21*).

The following findings might be important for the altered nucleocytoplasmic transport under some physiological conditions. (1) In synchronized mouse L-cells, the poly(A) catabolic enzyme activities, but not the poly(A) polymerase, increase markedly during the transition from G_1 phase to S phase, indicating that, during DNA synthesis, a change of the balance between the poly(A) anabolic and poly(A) catabolic enzyme activities occurs in favor of poly(A) degradation (*79*). Interestingly, most histone mRNAs synthesized during S phase lack poly(A) (*88*, *89*). (2) The activity of poly(A)-specific 2′,3′-exoribonuclease in quail oviducts increases markedly during aging, while poly(A) polymerase activity remained essentially constant (*90*).

The conclusion drawn from these findings, that the poly(A) segment of mRNA is shorter in senescent animals compared to adult ones, is supported by analytical results. Determinations of the average sizes of the poly(A) sequences of mRNAs from oviducts of mature and old quails revealed that the average length of the poly(A) segment is shortened from approximately 130 (adult animals) to approximately 70 AMP residues (senescent animals) (*91*). Concomitantly, there is a drop of the average number of poly(A)-associated protein molecules bound to one poly(A) chain from approximately 4.7 (mature oviduct) to 1.9 molecules (old oviduct) (*91*).

The 3′ poly(A) segment of mRNA is associated with specific poly(A)-binding proteins with formation of a poly(A)–ribonucleoprotein [poly(A)–RNP] complex (*76*, *92–95*). This complex seems to contain at least one major polypeptide species of 73–78 kDa (p78) (*76*, *77*,

92, 93, 95, 96). There is also a 54-kDa polypeptide (*95*), and a 73- to 76-kDa polypeptide in hnRNP has been described (*97*). A 63-kDa polypeptide appears to be the major poly(A)-associated protein in the nucleus (*96*). From *in vivo* cross-linking studies (*98*), the association of p78 with the poly(A)$^+$ mRNA seems to be restricted exclusively to the cytoplasm.

Evidence has been presented that p78 is involved in nucleocytoplasmic mRNA transport (*99, 100*). Furthermore, p78 and p54 protect the poly(A) sequence in polysomal poly(A)$^+$ mRNP and in the isolated poly(A)-RNP complex against attack by endoribonuclease IV (*77*). Therefore, these proteins should maintain the ability of poly(A)$^+$ mRNA to interact with the mRNA translocation system. However, it seems possible that these proteins are also more directly involved in the poly(A)$^+$ mRNA transport (see Section V,D).

IV. Release from the Nuclear Matrix

The nuclear envelope seems not to act as a barrier preventing hnRNP from diffusing out of the nucleus: mechanical disruption of the nuclear envelopes from *Xenopus* oocytes does not result in changes of the nucleocytoplasmic RNA distribution, suggesting that the nuclear matrix is solely responsible for nuclear RNA restriction (*101*).

A. Association of mRNA Precursors with the Matrix

The nuclear matrix is involved in important nuclear events, such as DNA replication (*102*), transcription of RNA (*103, 104*), and RNA processing (*25, 105, 106*). Actively transcribed genes, like the ovalbumin gene of chicken oviduct cells, are associated preferentially with the nuclear matrix (*107*). In addition, enzymes involved in DNA or mRNA metabolism are bound to the matrix as DNA polymerase α (*108*), DNA topoisomerases (*109, 110*), and poly(A) polymerase (*82*).

It is now well established that essentially all hnRNA is bound to the nuclear matrix and remains associated with it throughout processing (*109–113*). The association of hnRNA with the matrix seems to occur as early as during transcription (*103*). As shown in Fig. 1, all of the ovalbumin mRNA precursors are associated nearly quantitatively with the nuclear matrix fraction from hen oviduct (*62, 114*). On the other hand, the portion of the nuclear, mature mRNA bound to the matrix was only about 30% (*114*). However, the fully processed mRNA displays no measurable difference in its binding affinity as compared to its precursors. The binding of both the hnRNA and the mRNA to the

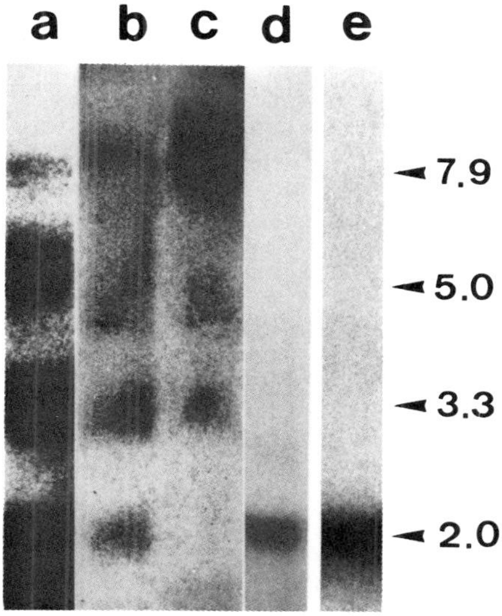

FIG. 1. Selective release of mature ovalbumin mRNA from hen oviduct nuclear matrices in the presence of ATP, from "Northern blot" analysis as follows. Total RNA isolated from whole nuclei (a), nuclear matrices (b and c), or the supernatants of RNA release experiments (d and e) was subjected to agarose gel electrophoresis and blot-transferred to nitrocellulose. Blots were analyzed by hybridization with ^{32}P-labeled, cloned plasmid pOV9.8 (containing the complete chicken ovalbumin gene; *292*) and autoradiography: (a) 2.4 μg of total RNA extracted from whole nuclei; (b) 1.5 μg of total RNA from untreated nuclear matrices; (c) 1.1 μg of total RNA from the matrix pellets obtained after incubation with 10 μM ATP; (d) 1.2 μg of total RNA from the supernatant obtained after incubation of nuclear matrices with 10 μM ATP; (e) 4.3 μg of total RNA from the efflux supernatant after incubation of whole, isolated nuclei in the presence of ATP. The kb (kilobase) values on the right refer to the length of the mature ovalbumin mRNA (2.0 kb) and of the three most abundant ovalbumin mRNA precursors (3.3, 5.0, and 7.9 kb). From (*114*).

nuclear matrix is very stable and resists treatment with high concentrations of salt, urea, EDTA, or nonionic detergents (*62, 114*).

The basis of the tight association of hnRNA and a part of the mRNA with the matrix is not known. Two polypeptides of 41.5 and 43 kDa, identified by RNA–protein cross-linking experiments (*113*), are probably involved in the binding of hnRNA to the nuclear matrix structure and appear to be related to the group C proteins in the RNP core particles (*115*). It has been proposed that snRNPs might mediate the binding of hnRNA or mRNA to the nuclear matrix (*116, 117*). However, under the high-salt buffer conditions used during matrix preparation, snRNPs were partially dissociated from L-cell nuclear matrices

(*116*). Moreover, Ciejek *et al.* (*62*) were unable to detect a preferential association of distinct classes of snRNAs with the oviduct nuclear matrix. It was also speculated that introns are involved in binding of mRNA to the nuclear matrix (*25*). The excision of the last intron should result in the release of the mature mRNA. However, detachment of mRNA also occurs in the presence of nonhydrolyzable ATP analogs (see below), while RNA processing absolutely requires hydrolysis of ATP or GTP (*118*).

B. Requirements for mRNA Release

Recently we found that detachment of mRNA from isolated matrices occurs in the presence of ATP (*114*). In this regard, mRNA release from the matrix resembles mRNA efflux from isolated nuclei (see Section V); however, it requires 1–10 μM ATP, whereas mRNA efflux requires higher concentrations. Most important, in the presence of ATP, only the mature mRNA was released from the matrix, whereas the immature mRNA precursors remained bound to this structure; in Fig. 1c and d, the selective release of the mature ovalbumin mRNA from the oviduct cell matrix in the presence of ATP is shown. This indicates that at the stage of mRNA release from the matrix (i.e., before mRNA translocation through the nuclear pore), a selection of the mature mRNAs occurs. Because only mature mRNAs are released from the matrix, only mature mRNA will also appear in the efflux supernatant from isolated nuclei *in vitro* (Fig. 1e) or in the cytoplasm *in vivo*.

Detachment of mature mRNA from the matrix occurs also in the presence of ATP analogs that contain nonhydrolyzable α,β and β,γ bonds (Table I). Therefore ATP hydrolysis seems not to be required for dissociation of mRNA from the matrix structure. In this property, RNA release from the matrix differs fundamentally from the efflux of RNA from isolated, whole nuclei (Table I). [β,γ-Methylene]ATP does not cause mRNA efflux (*119, 120*). It seems therefore unlikely that nuclear matrix-associated ATPase or protein kinase activities are essential for mRNA detachment from the matrix structure. A possible participation of nuclear matrix-associated poly(A) polymerase (*82*) in RNA release is excluded by the fact that the release also occurs in the presence of [α,β-methylene]ATP which does not serve as a substrate for poly(A) polymerase (*121*).

The release of RNA requires the presence of divalent cations and an elevated temperature (*114*). Maximal RNA release was found in the presence of 1 : 1 complexes of ATP with Mg^{2+} or Mn^{2+}. The reaction displays a broad pH optimum (pH 6–7.5). As shown in Table I, dATP

TABLE I
RELEASE OF RNA FROM ISOLATED NUCLEAR MATRICES AND EFFLUX OF RNA FROM ISOLATED NUCLEI IN THE PRESENCE OF DIFFERENT NUCLEOTIDES OR ANALOGS[a]

Nucleotide	RNA released: Nuclear matrices[b] (%)	RNA released: Nuclei[c] (%)
ATP	100	100
GTP	43.8	102.7
UTP	0.7	50.0
CTP	0.9	34.9
dATP	68.9	93.9[d]
dTTP	16.2	n.d.[e]
[β,γ-methylene]ATP	77.1	Zero[f]
[α,β-methylene]ATP	30.2	n.d.
[β,γ-imido]ATP	18.2	Zero[g]
[γ-thio]ATP	15.6	n.d.
Cordycepin 5′-triphosphate	12.9	n.d.

[a] RNA release in the presence of ATP was set at 100%. For substrate specificity of nuclear envelope NTPase, see Table IV.
[b] From Schröder *et al.* (*144*); L5178y mouse lymphoma cells.
[c] From Agutter *et al.* (*149*); rat liver.
[d] From Agutter *et al.* (*156*); SV40-3T3 cells
[e] n.d., not determined
[f] From Agutter *et al.* (*120*); rat liver.
[g] From Jacobs and Birnie (*46*); rat hepatoma tissue culture cells.

can substitute for ATP. GTP is less active and the pyrimidine nucleotides UTP and CTP display no significant effect. Quercetin, proflavine, oligomycin, and ouabain are strong inhibitors of the RNA release process (Table II). Some inhibition is also obtained in the presence of phalloidin, podophyllotoxin, and colchicine. Complexing of divalent cations by EDTA or EGTA strongly reduced the liberation of RNA (*114*).

A surprising finding was that ATP-induced RNA detachment from the matrix is strongly inhibited by low concentration of certain drugs (novobiocin, coumermycin A_1, m-AMSA, and others) known as inhibitors of the DNA topoisomerase reaction (*114*). The fact that nonintercalating inhibitors of type II DNA topoisomerase, such as etoposide or nalidixic acid (*122*), also affect RNA release from the matrix points particularly to an involvement of a topoisomerase-II-like activity in RNA detachment. This enzyme has recently been identified as a major component of the *Drosophila* nuclear matrix (*123*). Type I topoisomerase (EC 5.99.1.2), on the other hand, which is enriched in the

TABLE II
INFLUENCE OF DIFFERENT INHIBITORS ON RNA RELEASE FROM ISOLATED NUCLEAR MATRICES, RNA EFFLUX FROM ISOLATED NUCLEI, AND ON NUCLEAR ENVELOPE NTPase ACTIVITY[a]

Inhibitor	Concentration (μg/mL)	RNA released: Nuclear matrices[b] (%)	RNA released: Nuclei[c] (%)	Nuclear envelope NTPase activity[c] (%)
Control	—	100	100	100
Quercetin	20	22.2	7.9	zero
Proflavine	100	11.4	57.1	76.1
Oligomycin	20	38.1	35.7	52.1
Phalloidin	20	51.4	n.d.[d]	n.d.
Ouabain	1000	13.0	104.3	102.8
Podophyllotoxin	100	36.9	n.d.	n.d.
Colchicine	800	70.1	25.1[e]	78[f]

[a] RNA released and NTPase activity without inhibitors are set at 100%.
[b] From Schröder *et al.* (*114*); L5178y mouse lymphoma cells.
[c] From Agutter *et al.* (*156*); SV40-3T3 cells.
[d] n.d., not determined.
[e] From Agutter and Suckling (*191*); rat liver.
[f] From Agutter *et al.* (*181*); rat liver.

nucleolus, is mostly lost during isolation of the matrix (*124*). As shown in Fig. 2, we could detect both enzyme activities in hen oviduct nuclear matrices, although the latter at only a very low level (*114*).

The mechanism of inhibition of RNA liberation by DNA topoisomerase inhibitors is not yet understood. However, it is possible that topoisomerase-like activities also exist for double-stranded regions of RNA. Recently some hints were found for the existence of a RNA unwinding activity within the nuclear matrix structure (*125*). Similar activities are also present in the cap binding protein complex (*126*) and in eukaroytic initiation factors eIF-4A and eIF-4F (*127*). Perhaps changes in the three-dimensional structure of mRNP, which have been seen to occur during translocation through the pore complex (*6*), are also mediated by a topoisomerase-like activity.

Some experimental evidence indicates that actin is present in nuclei or as a component of the internal nuclear matrix structure (*117, 128–131*). Nuclear actin, which seems to be similar but not identical with cytoplasmic actin (*132*), occurs in an apparently monomeric or oligomeric form (*133*) and exchanges rapidly with cytoplasmic actin (*134*). Under artificial conditions, nuclear actin can be induced to form

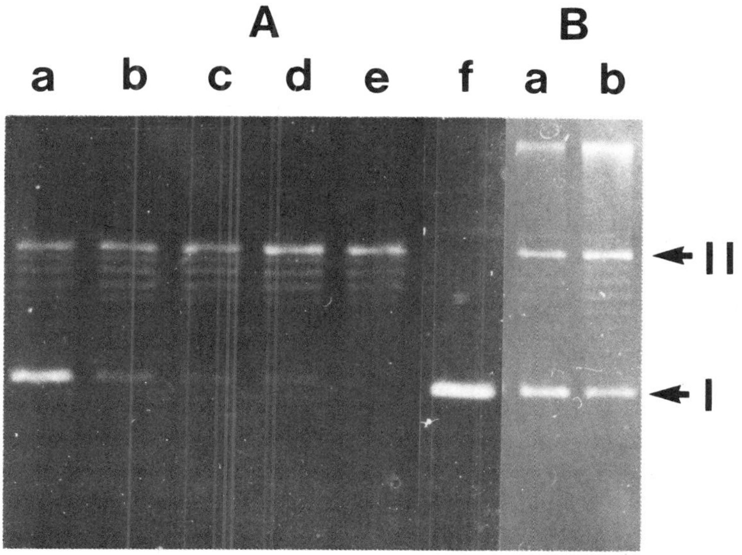

FIG. 2. Relaxation of negatively supercoiled circular pBR322 DNA by nuclear matrix-associated topoisomerase I and II from hen oviduct. Samples containing 0.6 μg DNA were electrophoresed in a 1.0% agarose gel. (A) Topoisomerase II assay; dependence on the ATP concentration. Reactions were for 20 minutes in the presence of 0.01, 0.05, 0.1, 0.5, and 1.0 mM ATP (lanes Aa to Ae). (B) Topoisomerase I assay. Incubations were for 10 minutes (lane Ba) and 60 minutes (lane Bb), respectively. Lane Af, supercoiled pBR322 DNA marker. (I) fully supercoiled circular DNA; (II) nicked circular DNA; intermediate to I and II, relaxed circular DNA. The upper band visible in lanes Ba and Bb represents endogenous DNA. From (*114*).

microfilament bundles within the nucleus, though it cannot be excluded that this bundle formation is caused by an influx of cytoplasmic actin (*135, 136*).

Recent evidence indicates that actin-containing filamentous structures are involved in binding and release of mRNA from nuclear matrices of L-cells (*137*). This conclusion has now been corroborated by inhibitor studies (*138*). In the presence of cytochalasin B, which depolymerizes F-actin, the immature ovalbumin mRNA, but not the fully processed ovalbumin mRNA, is released from the oviduct cell nuclear matrix (*138*). By this property, cytochalasin-induced RNA liberation contrasts with the release of RNA, caused by ATP or dATP, which is specific for mature messengers. On the other hand, detachment of mature mRNA by ATP is inhibited by cytochalasin B. Phalloidin, which stabilizes actin filaments, does not cause a RNA liberation by itself, but inhibits, like cytochalasin B, ATP-induced RNA release from L-cell matrices (*138*). A release of RNA was also achieved with a monoclonal antibody against actin but not with monoclonal antibodies against tubulin and intermediate filaments (unpublished results).

Although these results might indicate a participation of actin-containing filaments in the attachment of mRNA and its precursors to the nuclear matrix, some doubts exist that actin depolymerization is the normal way of mRNA release from the matrix structure. First, nuclear restriction of immature mRNA precursors is not maintained under conditions destabilizing actin-containing structures. Second, depolymerization of actin does not seem to be necessary for the ATP-induced release of matrix-bound mRNA under the conditions used by Schröder *et al.* (*114*) (see Fig. 1), since it also occurs in the presence of GTP, which does not bind to actin. Moreover, the presence of Mg^{2+} and of 100 mM KCl in the incubation mixture, which favor actin polymerization, does not prevent the release of RNA in the presence of ATP. Further, it could be demonstrated that under the low salt conditions (*137*), RNA release occurs also in the absence of ATP and Ca^{2+}, although at a lower rate. Nevertheless, these results suggest that intranuclear structures that are sensitive to agents affecting actin assembly and disassembly are important for nuclear restriction of immature mRNA. There are also some hints that nuclear actin is involved in the transcription mediated by RNA polymerase II (*139*) and in the regulation of the polyadenylation and deadenylation reactions catalyzed by the complex of endoribonuclease IV and poly(A) polymerase (*21*).

The 3′ poly(A) segment and the double-stranded regions of the mRNA and its precursors have been assigned a potential role in mRNA binding to the nuclear matrix structure (*110*). Interestingly detachment of rapidly labeled RNA from L-cell matrices occurs also in the presence of poly(A), but not in the presence of poly(U), poly(C), poly(G), or DNA (*114*). Double-stranded poly(A)·poly(U) was also ineffective. The mechanism by which poly(A) induces RNA liberation remains at present a matter of speculation. It is possible, for instance, that poly(A) interferes, by hybridization, with the binding of the hnRNA oligopyrimidine sequences to the "group C" proteins[3] of the hnRNP complex, which are assumed to be involved in the attachment of hnRNA to the matrix (*98*, *113*). Another possibility may be that RNA release by poly(A) is caused by stimulation of a nuclear matrix-associ-

[3] "Group C" proteins: those core proteins of 40-S hnRNP particles with M_r 42,000 and 44,000 (*139a*). [Eds.]

ated protein phosphatase activity, analogous to the situation in the nuclear envelope. In this structure, stimulation of an endogenous phosphatase by poly(A) results in dephosphorylation of a 110-kDa nuclear envelope polypeptide [the putative poly(A)$^+$ mRNA carrier] that is associated with a reduced binding affinity of the pore complex lamina fraction for poly(A) (see Section V,I). This phosphorylatable polypeptide may be related to or identical with a 100-kDa phosphoprotein identified in rat liver nuclear matrices (*140*).

Dissociation of mRNA from mouse L-cell matrices was also achieved in the presence of the copper chelator 1,10-phenanthroline (*114*). This effect could be specifically blocked by addition of Cu^{2+} or Ca^{2+}. Therefore, loosening of the mRNA·matrix interaction shows some similarities with the destabilization of the metaphase chromosome scaffold by metal chelators. This structure is also stabilized by Cu^{2+} or Ca^{2+} ions (*141*).

V. Translocation through the Nuclear Pore Complex

The mRNA is transported from the nucleus to the cytoplasm through the nuclear envelope pore complexes (*4, 33, 142*). In a first approximation, these structures can be regarded as simple, aqueous channels connecting the nuclear and the cytoplasmic compartments of the cell. This view may be sufficient for considering the import and export of small molecules such as nucleotides, but not of globular proteins with molecular weights over 40,000 (*143*). Microinjection studies on amphibian oocytes with colloidal gold (*144*), ferritin (*144*), tritiated dextran (*145*), and fluorescently labeled proteins indicate an effective pore diameter of about 9 nm. Recent measurements of the nuclear envelope permeability in single liver cells by the method of fluorescence microphotolysis ("photobleaching") (*146*) revealed an effective pore diameter of 10–11 nm (*5*). Therefore, the passage of an RNP particle with an estimated diameter of about 20 nm (*147*) through the nuclear pore is obviously not explainable by simple diffusion. This implies that a special translocation mechanism must exist allowing pore passage of these particles.

A. Specificity

It is experimentally well proven that only mature mRNA is released in the cytoplasm (*46, 47, 51*). There is evidence, however, that the translocation of mRNA through the pore complex does not present the step at which selection of mature mRNA occurs. First, almost all of the nuclear hnRNA is associated with the nuclear matrix fraction (see

Section IV). The matrix-free portion of the nuclear mRNA seems to consist exclusively of mature mRNA species (*62, 114*). In the presence of ATP (or dATP) only the mature mRNA is detached from the matrix (in contrast to mRNA translocation, this process does not require hydrolysis of the β,γ-phosphodiester bonds). Second, when isolated nuclei incubated with cytochalasin B, which releases the hnRNA from the matrix structure (see above), are used, a significant efflux of hnRNA occurs in the absence of an exogenous energy source. However, after addition of ATP to the external medium, both the mature and the immature mRNA precursors appear in the postnuclear supernatant (in contrast to the results obtained without cytochalasin pretreatment) (*138*). Third, and more striking: when total nuclear RNA that contains nucleus-restricted sequences is entrapped in resealed nuclear envelope vesicles (*48, 148*), it is exported from these vesicles in the presence of exogenous ATP just as well as is cytoplasmic mRNA (P. S. Agutter, personal communication). Therefore, we have good reasons to assume that the selectivity of the nucleocytoplasmic mRNA transport process is not provided by the nuclear envelope translocation system, but by the mechanisms involved in the release step of mRNA from the nuclear matrix.

B. Energy Dependence

From numerous studies (most striking are the recent vesicle studies), it seems now well-established that the translocation of poly(A)$^+$ mRNA through the nuclear pore complex is an energy-dependent process that requires hydrolysis of ATP or GTP (*48, 149–152*). However, there is some disagreement in the literature concerning the effect of the nonhydrolyzable ATP analog, [β,γ-methylene]ATP, on mRNA efflux. Under some conditions the analog causes RNA efflux reaching about half the efficacy of ATP (*150*). On the other hand, with a transport medium superior in preventing unspecific RNA leakage, it appears this nucleotide has no effect (*119, 120*). Moreover, there is no efflux from vesicles in the presence of nonhydrolyzable ATP (P. S. Agutter, personal communication). Based on our results (see Section IV), one explanation for the differing results might be that under the conditions used (*150*), the (most likely allosteric) effect of the nucleotide on the release of RNA from the matrix is what is measured, and not the effect on pore passage, which depends on hydrolysis of ATP.

In contrast to the transport of mRNA, the ATP dependence of the nucleocytoplasmic transport of ribosomal RNA is not well-established. Efflux of rRNP from isolated rat liver nuclei depends on the presence of exogenous ATP (*153*), while rRNP export from the macronuclei of the lower eukaryote *Tetrahymena* appears to occur even in the absence of ATP (*154*). This finding led to the assumption that rRNP transport in *Tetrahymena* requires nuclear expansion (*8*). The nuclear matrix from *Tetrahymena* has indeed been shown to expand and contract in a reversible manner in response to [Ca^{2+}] and [Mg^{2+}] (*154*). During expansion of the macronucleus, the attachment of RNA to the matrix is labilized (*155*). Interestingly, nuclear expansion has also been observed in rats in response to carcinogen feeding, combined with an enhanced mRNA efflux rate and a decreased nuclear restriction of mRNA (see Section VIII,B,5).

C. Evidence for Involvement of a Nucleoside Triphosphatase

Several lines of evidence indicate that the ATP-dependent pore passage of poly(A)$^+$ mRNA is mediated by a nucleotide-unspecific nucleoside-triphosphatase (NTPase) within the nuclear envelope (*151, 156–160*). This enzyme seems to be located at the inner face of the nuclear envelope, either in the inner nuclear membrane or in the lamina (*161–163*).

The evidence that the nuclear-envelope NTPase provides the energy for nucleocytoplasmic mRNA translocation is as follows:

1. In isolated nuclear envelopes, the activity of the NTPase is markedly increased when exogenous RNA [poly(A) or poly(A)$^+$ mRNA], but not poly(A)$^-$ mRNA, tRNA, rRNA, or other synthetic homopolymers, is added (see below).

2. The rate of mRNA efflux from isolated nuclei shows the same kinetics, the same substrate specificity, and the same sensitivity to inhibitors as the NTPase activity does (*120, 149, 151, 156*).

3. A correlation between alterations in mRNA transport and nuclear envelope NTPase activity, induced by carcinogen (*151, 164*), tryptophan (*165*) or insulin treatment (*166*), has been demonstrated. In addition, NTPase activity also displays age-dependent alterations paralleling the age-related changes of the mRNA transport rate (*158, 167*).

4. It is possible to prepare resealed vesicles from nuclear envelopes and to trap mRNA in these vesicles (*48, 148*). Efflux of mRNA from the vesicles also depends on ATP hydrolysis. Poly(A)$^+$ mRNA leaves the vesicles much more rapidly than poly(A)$^-$ mRNA (*48*).

5. Antibodies to pore complex–lamina components, which affect the NTPase, substantially change ATP-dependent efflux of mRNA from isolated nuclei (*168–170*).

Most important is the finding that the NTPase in intact nuclear envelopes is markedly stimulated by synthetic poly(A) or the 3′-

poly(A) tail of mRNA (*157–171*). From these results, it was concluded that synthesis of the 3′-terminal poly(A) sequence of mRNA is a prerequisite for nucleocytoplasmic translocation of this RNA species (*157*). As a consequence, poly(A) polymerase and poly(A) nucleases must play a crucial role in determining whether or not a defined mRNA species is transported out of the nucleus. On the other hand, some mRNAs [such as histone mRNAs, which lack poly(A)] are also transported efficiently to the cytoplasm (*172*).

The modulation of the mRNA translocation system by poly(A) is discussed in detail in the following sections. A brief survey about posttranscriptional poly(A) metabolism is given in Section III; some aspects of poly(A)$^-$ RNA transport are discussed in Section VII.

D. Dependence on Poly(A)

Agutter *et al.* (*173*) were the first to make the interesting observation that the activity of the NTPase in isolated, intact nuclear envelopes is enhanced by poly(A) or poly(G) (*173, 174*). Later, we determined that poly(A) is by far the most potent stimulator of NTPase activity from rat liver nuclear envelopes (*157*) (Table III). Stimulation by poly(A) requires a minimal chain length of 18 AMP residues (*157*). Poly(G), poly(dT), and poly(dA) stimulate the enzyme to a significantly lesser extent. Poly(U), poly(C), poly(I), and the alternating copolymer poly(A-U) were without any effect on the enzyme activity. Based on dose–response experiments, the concentration of poly(A) causing half-maximal stimulation of NTPase activity was 44 μM [with

TABLE III
Effect of Nucleic Acids on NTPase Activity Associated with the Nuclear Envelope[a]

Component added	Concentration (μM) of phosphate	Enzyme activity (%)
None (control)	—	100
Poly(A)	95	162
Poly(A-U)	93	97
Poly(A)·poly(U)	106	138
Poly(A)·poly(U)·poly(U)	59	123
Poly(A)$^+$ mRNA	44	182
Poly(A)$^-$ mRNA	44	103
Poly(A)$_{\overline{95}}$	34	159
Poly(A)·RNP complex	42	115

[a] From (*57, 157*).

respect to phosphate content; the size of poly(A) was 760 AMP units]. Double reciprocal plot analysis revealed a 2.8-fold stimulation of the enzyme by optimal poly(A) concentration. These polynucleotides increase the catalytic constant of the NTPase reaction, while the K_m for MgATP remains unchanged (*173*). The stimulation of the NTPase depends on the conformation of the polynucleotide. Poly(G) in the triple-helical form is more stimulatory than single-stranded poly(G) (*174*). As shown in Table III, the efficiency of single-stranded poly(A) to stimulate the NTPase is higher than that of poly(A) in the double-stranded form [poly(A)·poly(U)] or in the triple-stranded form [poly(A)·poly(U)·poly(U)] (*157*).

Naturally occurring RNA stimulates the NTPase, if it contains a poly(A) segment (*157*). Poly(A)$^+$ mRNA from rat liver [containing a poly(A)$_{\overline{95}}$ segment] strongly stimulates the NTPase within whole envelopes (Table III). The stimulatory effect of this macromolecule is abolished after selective digestion of the poly(A) segment with the poly(A) specific endoribonuclease IV [poly(A)$^-$ mRNA]. The conclusion drawn from this finding, that the poly(A) tract in the mRNA is responsible for its stimulatory potency, is supported by experiments with the poly(A)$_{\overline{95}}$ segment isolated from the same mRNA (Table III). The stimulatory effect of the poly(A)$_{\overline{95}}$ segment was of the same order of magnitude as that observed with synthetic poly(A) or poly(A)$^+$ mRNA. On the other hand, when the poly(A) segment [obtained from cytoplasmic poly(A)$^+$ RNP] was covered with poly(A)-associated proteins [poly(A) · RNA], only a very low stimulation was measured (Table III). This finding may indicate a possible role of poly(A)-associated proteins in controlling NTPase activity during nucleocytoplasmic mRNA transport. This assumption is supported by experimental evidence showing that the poly(A)-associated proteins within the nucleus differ from those found in the cytoplasm (*98*). Therefore, we conclude that only after the release of intranuclear poly(A)-associated proteins (*175*) can mRNP be transported to the cytoplasm. In the cytoplasm, the mRNA might then be charged with a novel set of poly(A)-associated proteins [p77 and p54 (*95*)]; they may prevent the cytoplasmic mRNA from being transported back to the nucleus. This might explain the fact that the transport of mRNA through the nuclear pore is strictly vectorial. However, some evidence suggests that the poly(A)$^+$ mRNA translocation system may be inherently vectorial. Recent studies with resealed nuclear envelope vesicles revealed that, in contrast to other RNA species, only a unidirectional exchange of poly(A) or poly(A)$^+$ mRNA occurs between the vesicles and the external medium (*176*).

E. The Nuclear Envelope Nucleoside Triphosphatase

Essential progress in elucidating mRNA translocation was made when it was possible to obtain an essentially homogeneous NTPase preparation. The NTPase can be solubilized from the nuclear envelopes in the presence of Triton X-100 (*177*) and purified (*178*).

The homogeneous enzyme from rat liver envelopes has an apparent mass of 40 kDa and displays a broad substrate specificity. ATP and GTP are hydrolyzed at similar rates and UTP and CTP at about half that rate (Table IV). The increased rate of hydrolysis of dATP by the NTPase in whole nuclear envelopes (Table IV) might be caused by an ATPase/dATPase (*179, 180*) in this preparation. This enzyme does not split GTP and is assumed not to be implicated in mRNA translocation. The purified NTPase as well as the NTPase in whole nuclear envelopes possesses an absolute requirement for divalent cations (Mg^{2+}, Mn^{2+}, or Ca^{2+}) (*120, 178*). Optimal activity is achieved at a 1 : 1 ratio of divalent cation to NTP. The pH optimum is around 8.0 (*151, 178, 181*). The apparent activation energy of the enzyme in the nuclear envelope was 13–14 kcal/mol (*182*); this value is nearly identical with that found for RNA efflux (*182, 183*).

The properties of the homogeneous NTPase seem identical to those of the NTPase previously identified within the total nuclear envelope protein from rat liver by photoaffinity labeling (*20, 184*). Combining this method with competition experiments using natural NTPase substrates and modulator studies with RNA (*184*), it was predicted (*184*) that the NTPase would be a 46-kDa polypeptide that is not phosphorylatable, as demonstrated by parallel labeling with [γ-^{32}P]ATP (*184*). These results were confirmed by the finding (*20*) of

TABLE IV
SUBSTRATE SPECIFICITY OF THE NTPase ACTIVITY MEASURED IN ISOLATED NUCLEAR ENVELOPES AND OF THE HOMOGENEOUS ENZYME (RAT LIVER)[a]

Nucleoside triphosphate	NTPase activity: Nuclear envelope (%)	NTPase activity: Homogeneous enzyme (%)
ATP	100	100
GTP	110	105
UTP	49	53
CTP	54	45
2′-dATP	106	34

[a] The rate of hydrolysis of ATP was set at 100%. From (*178*).

a 43-kDa band seemingly identical with the 46-kDa protein (*184*). A 47-kDa protein was also the major labeled polypeptide among the Triton-soluble proteins of rat liver nuclear envelopes (*148*). However, the purified NTPase is clearly distinguished from the ATPase/dATPase (a 174-kDa polypeptide) identified in the nuclear matrix–pore complex–lamina fraction from *Drosophila* and also from rat liver (*180*). This enzyme has a distinct substrate specificity that is incompatible with that found for mRNA efflux (*20*).

The nuclear envelope-associated NTPase exhibits non-Michaelis–Menten kinetics (*181*) (Fig. 3A). This has been attributed to the existence of two conformational states of the enzyme, which depend on its interaction with a phosphorylatable poly(A)-binding protein within the nuclear envelope (*20*). In contrast to the structure-bound NTPase, the homogeneous enzyme preparation yielded linear Lineweaver–Burk plots (*178*) (Fig. 3B). This finding is in agreement with the prediction (*20*) that the purified NTPase will display normal kinetic behavior due to the absence of the modulating protein (see Section V,3).

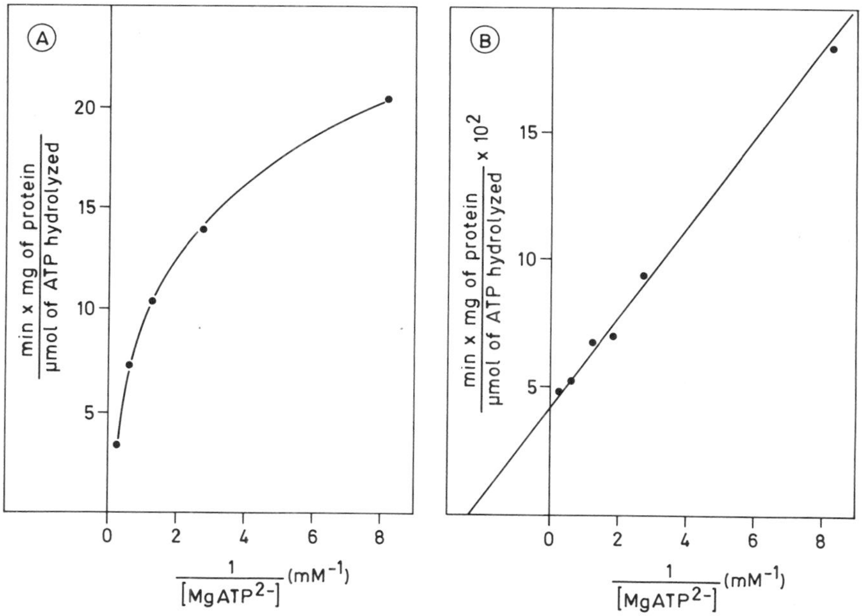

FIG. 3. Kinetics of hydrolysis of MgATP by the nuclear envelope NTPase; Lineweaver–Burk plot. (A) NTPase in whole nuclear envelopes; (B) homogeneous NTPase. From (*178*).

The K_m of the purified NTPase with MgATP as substrate is 0.42 mM. The maximal reaction velocity is 23.8 μmol of ATP hydrolyzed per minute per μg of protein. In comparison, NTPase in whole nuclear envelopes prepared according to the procedure of Kaufmann *et al.* (*50*) had an apparent K_m of 0.44 mM; its maximal reaction velocity was 0.18 μmol of ATP hydrolyzed per minute per μg of protein (determined in the presence of 150 mM KCl in the reaction mixture in order to obtain a linear Lineweaver–Burk plot; *160, 181*).

Although most of the characteristics of the purified NTPase resemble those of the nuclear envelope-associated enzyme—such as K_m, substrate specificity, pH optimum, dependence on divalent cation, and sensitivity to thiol reagents and specific inhibitors of RNA transport—there are some important differences (*178*). (1) In contrast to the enzyme associated with nuclear envelopes, which is stimulated by synthetic poly(A) and the poly(A) segment of the natural poly(A)$^+$ mRNA, homogeneous NTPase is not affected by this polynucleotide species. (2) In contrast to the NTPase associated with whole nuclear envelopes, the homogeneous enzyme is not inhibited by physiological concentrations of Mg^{2+}. (3) The homogeneous NTPase does not depend on phosphatidylinositol, in contrast to the results obtained with a partially purified NTPase preparation (*177*). This indicates that the modulation of the NTPase by these effectors in the intact envelope is mediated by one or more additional components of this structure, which are separated from the enzyme during the purification procedure. In the case of the stimulation by poly(A), this factor might be the putative mRNA carrier (see Section V,G,1).

F. Effectors

Several effectors of nucleocytoplasmic transport of mRNA are known. Most of these effectors exert their effect via a stimulation or an inhibition of the energy-delivering NTPase (directly or indirectly). However, some of these effectors may act on the nuclear envelope translocation system by a mechanism distinct from affecting NTPase.

1. Effectors of Nucleoside Triphosphatase

The NTPase in whole nuclear envelopes (*156, 181*) (Table II) and the homogeneous NTPase (*178*) are strongly inhibited by quercetin, oligomycin, and proflavine. Ouabain and actinomycin D show no inhibitory effect on either the purified or the envelope-associated NTPase (*156, 178*). The ATP analogs [β,γ-methylene]ATP and [γ-thio]ATP markedly inhibit the homogeneous NTPase (*178*). The [γ-thio]ATP is also a strong inhibitor of the structure bound NTPase

(*181*). On the other hand, when the effect of [β,γ-methylene]ATP on NTPase associated with nuclear envelopes was tested, no influence (*149*) or even a slight stimulation of the enzyme activity was obtained (*178*). This discrepancy is presently not understood. The nuclear envelope-associated NTPase was also strongly inhibited by the carcinostatic and antiviral agent 9-β-D-arabinofuranosyladenine 5′-triphosphate (ara-ATP), a known inhibitor of DNA polymerases α and β (*82*). This drug also markedly reduces the nuclear envelope protein kinase activity (*82*). The NTPase (homogeneous and envelope-associated enzyme) was further found to be sensitive to thiol-group reagents, such as *N*-ethylmaleimide, iodoacetamide, phenylarsine oxide, and *p*-hydroxymercuribenzoate (*156, 178, 181*). All of the drugs listed above (perhaps with the exception of [β,γ-methylene]ATP) also inhibit proportionally the ATP-dependent efflux from isolated nuclei (*120, 149, 156*) (Table II). In addition, some divalent cations affect mRNA transport and NTPase activity. Be^{2+} was found to inhibit strongly both efflux and the nuclear envelope-associated NTPase (*149, 181, 185*). Both the homogeneous NTPase (*178*) and the enzyme in whole nuclear envelopes as well as mRNA efflux were inhibited by Cu^{2+} and Co^{2+} (*120*).

2. Other Effectors

There are also examples of the inhibition of mRNA efflux by drugs that apparently do not affect the NTPase itself. Cordycepin (3′-deoxyadenosine) inhibits significantly the mRNA efflux from isolated nuclei (*186*). The conclusion drawn (*186*), that an inhibition of polyadenylation of mRNA might be responsible for the decreased efflux rate in the presence of the analog, was not supported by later studies (*159*); however, the mode of action of cordycepin on the mRNA translocation system is still not clear. The inhibition of mRNA transport by cordycepin seems not to occur by an inhibition of RNA synthesis (*187*), since actinomycin D also does not decrease the efflux rate (*159, 188*). α-Amanitin, on the other hand, inhibits mRNA transport (*189*). The inhibition of mRNA translocation by colchicine (*156, 190*) is not mediated through the NTPase (*178*) or nuclear envelope-bound tubulin, but is caused by a constriction of the pore complexes, which is independent of the NTPase reaction (*191*). Lumicolchicine, which does not depolymerize microtubules, also inhibits mRNA efflux (*156*), while the antimicrotubular drugs vinblastine and vincristine are inactive (*191*). Phenobarbital stimulates RNA transport, possibly by enhancing the efficiency of cytosolic transport stimulatory proteins (*192, 193*).

Despite some initial contradictory results (*155, 156, 159, 194*), it now seems clearly established that the nuclear membrane have no effect on the transport of mRNA. RNA efflux from isolated nuclei, from which the surrounding membranes have been removed by detergents, is not increased over that from intact nuclei (*195, 196*).

G. Other Components of the mRNA Translocation Apparatus

Nuclear envelopes contain, besides the NTPase (EC 3.6.1.15), protein kinase (EC 2.7.1.37) (*19, 197–200*) and phosphoprotein phosphatase (EC 3.1.3.16) activities (*19, 201, 202*). These enzymes are assumed to modulate NTPase activity, most likely via phosphorylation and dephosphorylation of the putative poly(A)$^+$ mRNA carrier, which in turn is modulated by poly(A) (*19, 200, 203*). Because of its important regulatory role in controlling NTPase activity, current interest has been focused on these enzymes.

1. The Poly(A)-Binding Transport Carrier

From the finding that the homogeneous NTPase is not stimulated by poly(A), it can be concluded that this enzyme is not identical with the putative poly(A)-binding carrier, but coupled to them in the intact nuclear envelope. This assumption is corroborated by the finding that the NTPase is solubilized from the envelope by Triton X-100 (*177, 178*), but the poly(A)-binding component is not. From Scatchard plot analyses of the results of poly(A)-binding experiments, it appears that only one class of poly(A)-binding sites in the nuclear envelope (pore complex lamina) exists (*200*). Poly(A) binding to this site is enhanced by nuclear-envelope protein-kinase-dependent phosphorylation; dephosphorylation of this site by nuclear-envelope phosphoprotein phosphatase results in a lower poly(A)-binding affinity (*19, 200*). In accordance with the assumed functional interaction of the poly(A)-recognizing transport "carrier" with the nuclear envelope NTPase, the dissociation constant for the binding of poly(A) to the phosphorylated poly(A)-binding site was almost identical with the concentration of poly(A) causing half-maximal stimulation of the NTPase in intact isolated envelopes (*173, 200*). The poly(A)-binding carrier protein seems to be a phosphorylatable nuclear envelope polypeptide of 106–110 kDa. This conclusion comes from the finding that the phosphorylation of this polypeptide by the endogenous kinase shows a marked sensitivity to poly(A), in contrast to other nuclear-envelope phosphoproteins (*19, 204*). However, it must be emphasized that this evidence is rather indirect and other proteins could represent the carrier as well. Recently, we were able to solubilize and purify two polypeptides of 55 and 64 kDa from rat liver nuclear envelopes that bind to poly(A) in an ATP-labile linkage (unpublished), suggesting that these proteins could be related to the carrier or to degradation products of it.

2. THE PROTEIN KINASE

The protein kinase activity in whole nuclear envelopes is strongly inhibited by poly(A) (*19*). Under some experimental conditions, only one (*198*) or only two major nuclear envelope polypeptides [64 kDa (p64) and 106 kDa (p106); ref. *19*] are phosphorylated. Under other conditions, a very complex phosphorylation pattern is obtained, but it is possible that some bands result from oligomerization or degradation of p64 or p106. We could demonstrate that the phosphorylation of p106, but not of p64, is strongly inhibited in the presence of low concentrations of poly(A) (half-maximal inhibition at about 3 μM; ref. *19*). Therefore, it was concluded (*20*) that this polypeptide might be the carrier.

We have identified both protein kinase C and a protein kinase NII-like activity in nuclear envelopes from rat liver (*163*). Both activities[4] could be separated by extraction with Triton X-100 (unpublished). Under these conditions, protein kinase C was solubilized, whereas the NII-like activity remained in the residual pore complex–lamina fraction; it could be extracted from this structure by treatment with 4 M urea. Interestingly, in the presence of phorbol esters which stimulate protein kinase C, the phosphorylation-related inhibition of the NTPase was increased, indicating that protein kinase C is involved in controlling NTPase activity. The following additional finding also strongly points to an involvement of protein kinase C in translocation of mRNA. Recently we identified a type II topoisomerase (EC 5.99.1.3) activity in rat liver nuclear envelopes. This enzyme is stimulated by the endogenous protein kinase C (Fig. 4B). Interestingly, relaxation of negatively supercoiled plasmid pBR322 DNA by this envelope-associated topoisomerase II activity was competitively inhibited by double-stranded poly(A)·poly(U) (Fig. 4A) but not by single-stranded poly(A). An intriguing possibility might be that this enzyme functions as an RNA unwinding activity responsible for the alteration in the mRNP three-dimensional structure, which seems to occur during pore passage (*6*).

[4] Protein kinases NI and NII (P. R. Desjardins *et al.*, *Can. J. Biochem.* **50**, 1249, 1972) are variants of "protein kinase" (EC 2.7.1.37) and/or "protein-tyrosine kinase" (EC 2.7.1.112). C is the Ca^{2+}- and phospholipid-dependent protein kinase (Y. Nishizuka, *Nature* **308**, 693, 1984). [Eds.]

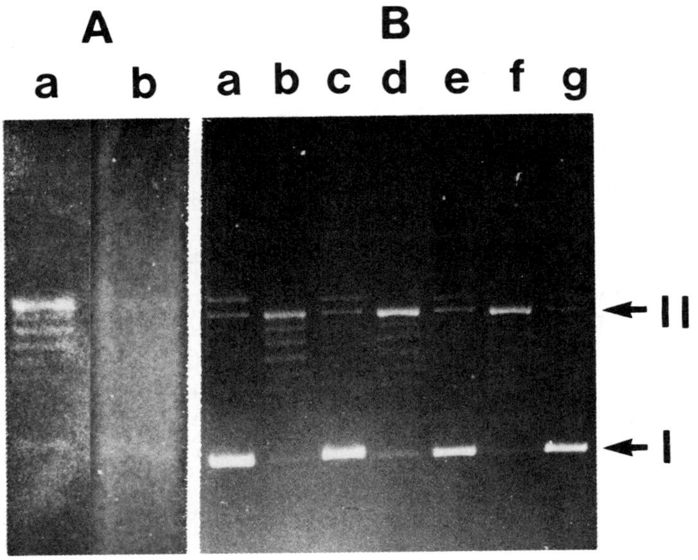

FIG. 4. Inhibition of topoisomerase II activity in whole nuclear envelopes by poly(A)·poly(U) and stimulation by phosphorylation mediated by protein kinase C. (A) Topoisomerase II assay in the absence (lane Aa) and in the presence of 40 μM poly(A)·poly(U) (lane Ab). (B) Time course (5, 10, and 20 minutes) of relaxation of negatively supercoiled pBR322 DNA by topoisomerase II in nuclear envelopes previously phosphorylated by endogenous protein kinase C (lanes Bb, Bd, and Bf) and in the controls (lanes Bc, Be, and Bg). Lane Ba, supercoiled pBR322 marker. For further details, see legend to Fig. 2.

H. Monoclonal Antibodies against Proteins Involved in mRNA Transport

We have elicited a series of monoclonal antibodies against defined nuclear envelope fractions and have screened them for their inhibitory or stimulatory effects on reactions involved in mRNA translocation. One monoclonal antibody (II-G7), which recognized two poly(A)-binding nuclear-envelope polypeptides of 65 and 83 kDa, attracted special attention (*163*). It markedly inhibited the efflux of ovalbumin mRNA from isolated oviduct cell nuclei (Fig. 5) and the efflux of rapidly labeled RNA from mouse L-cell nuclei. The homogeneous NTPase was not affected by the antibody but the V_{max} of the enzyme in the intact envelope was drastically increased, without altering its apparent K_m for MgATP and its capacity to be stimulated by poly(A). In phosphorylated envelopes, the antibody increases the affinity, but not

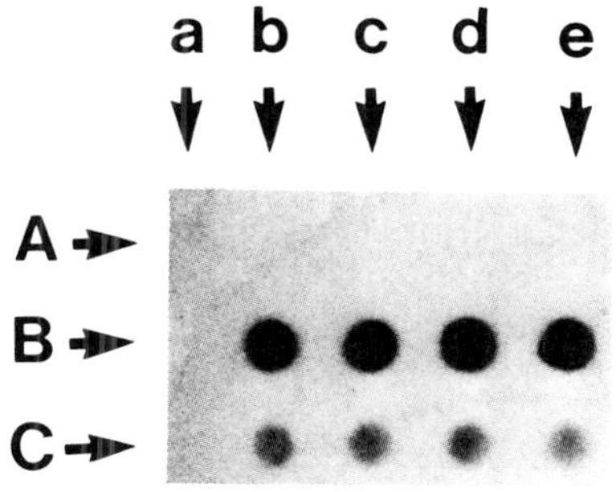

FIG. 5. Dot hybridization analysis of the efflux of ovalbumin mRNA from isolated oviduct cell nuclei in the absence and the presence of a monoclonal antibody (II-7) recognizing two nuclear envelope polypeptides. As an ovalbumin specific probe, the nick-translated plasmid pOV9.8 was used (see Fig. 1; ref. *292*). RNA efflux was measured in the absence (a) and in the presence of ATP (b, c) or GTP (d, e) after an incubation period of 0 minute (A) and 20 minutes (B, C). (A, B) Plus nonspecific IgG; (C) plus II-7.

the total numbers, of the poly(A)-binding sites. Interestingly, protein kinase C was also inhibited by the antibody; other nuclear-envelope-associated kinase activities remained unaffected. Immunofluorescence microscopy revealed that the antigen is associated with both the nuclear envelope and a filamentous network in the cytoplasm, which shows some superposition with the microfilament system (Fig. 6).

Both the ATP-dependent efflux of mRNA and the NTPase activity in nuclear envelopes are inhibited by a polyclonal antibody against lamin B (*170*). Therefore, one might speculate that, because of the close association of lamin B with the inner nuclear membrane, this protein might couple the energy-producing NTPase (in the lamina or in the inner nuclear membrane) with the poly(A)-binding protein (on the pore complex) and the protein kinase (possibly protein kinase C in the outer nuclear membrane).

I. Models for mRNA Translocation

The first kinetic scheme to explain nucleocytoplasmic mRNA transport was proposed by Agutter *et al.* (*120, 181, 200*). It was based on the assumption that the NTPase consists of a nuclear-envelope-associated protein kinase and phosphoprotein phosphatase. Following this model (Fig. 7a), phosphorylation and dephosphorylation of a "carrier structure" are the essential steps in the NTPase-mediated nucleocytoplasmic translocation of mRNA. In step 1, nuclear mRNA binds to a phosphorylated protein in the pore–lamina fraction of the

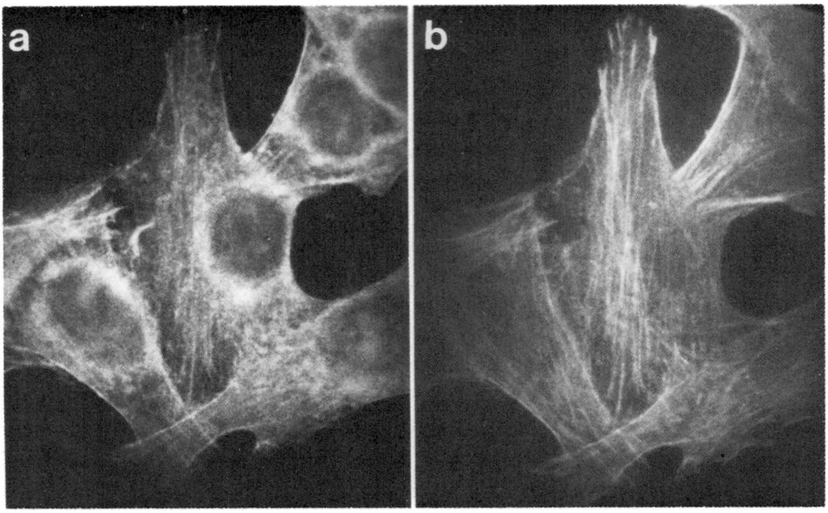

FIG. 6. Intracellular distribution of two nuclear envelope antigens (p65 and p83), recognized by a monoclonal antibody that inhibits mRNA transport *in vitro*, and visualized by this antibody in CV-1 cells in immunofluorescence microscopy. Visualization of the antibody with fluorescein isothiocyanate-labeled anti-mouse antibody reveals partial staining of cytoplasmic stress fibers as shown by double labeling with rhodamine isothiocyanate-labeled phalloidin (staining of microfilaments). (a) Fluorescein pattern; (b) rhodamine pattern of the same cells. ×1100.

nuclear envelope. During this binding process, P_i is liberated. The conclusion that the mRNA binds preferentially to a phosphorylated polypeptide in the nuclear envelope structure has been drawn from the results obtained by Scatchard plot analyses (see above; ref. *200*). The release of the bound mRNA into the cytoplasm is caused by binding of ATP (step 2); during this step no hydrolysis of ATP occurs (*200*). The conformational change in the carrier protein induced by ATP (*150*) and leading to the release of the mRNA is turned back in step 3. During this step, the carrier is rephosphorylated, mediated by a nuclear envelope protein kinase (*198*). Step 4 is a dephosphorylation of the carrier protein by a phosphoprotein phosphatase (*201*); this conversion is assumed to be the rate-limiting step in the process (*120, 150*), prone to stimulation by poly(A) to poly(A)$^+$ mRNA. Step 5 is the binding of ATP.

This original scheme was later refuted by a detailed enzymic study (*19*). Comparative determinations of the nuclear-envelope-associated enzyme activities revealed that the maximum catalytic rate of the

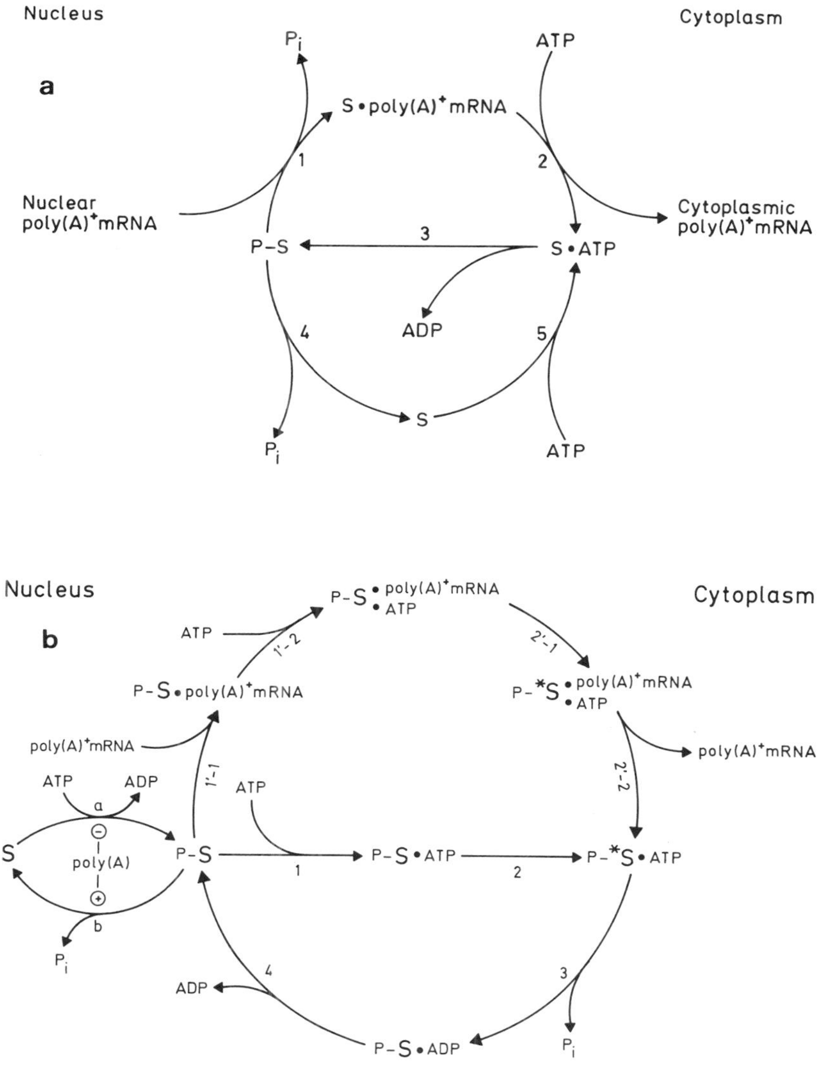

FIG. 7. Kinetic models of the poly(A)$^+$ mRNA translocation through the nuclear envelope pore complex. (a) Scheme proposed by Agutter (*120*). S, carrier structure; P-S, phosphorylated carrier structure. (b) Scheme proposed by Müller *et al.* (*19, 205*). S, complex of the poly(A)-recognizing carrier (possibly p106–110) and the NTPase which possesses two conformations, S and *S; −, inhibition of the protein kinase by poly(A); +, stimulation of the phosphoprotein phosphatase by poly(A). (Small p here indicates that all steps can occur also without prior phosphorylation of the "carrier.") (c) Revised scheme of Agutter (*20, 25*) and Schroder *et al.* (*171*). N, NTPase; S, poly(A) (or mRNA) binding site. Further details are described in the text.

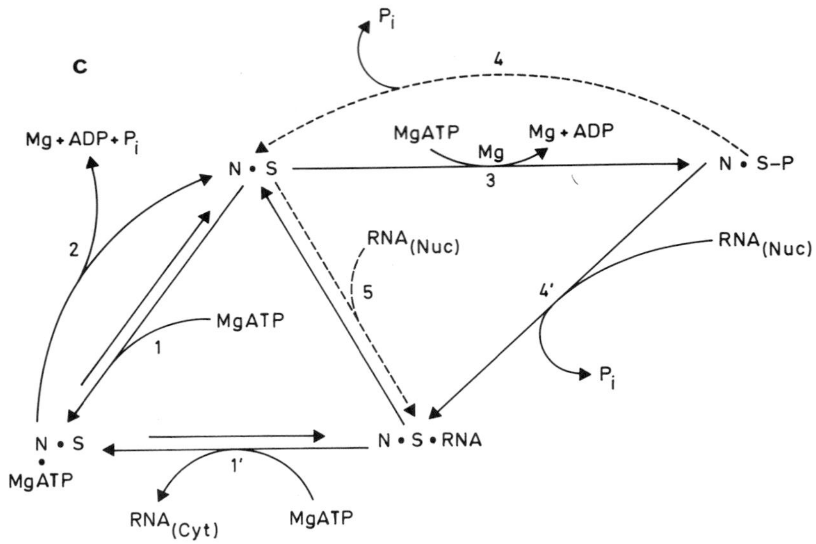

FIG. 7. (Continued)

NTPase is at least 500-fold above those of the kinase and of the phosphatase. These results clearly show that the NTPase activity of nuclear envelopes does not result from the combined action of nuclear-envelope-associated protein kinase with phosphoprotein phosphatase. However, the kinase and the phosphatase do modulate the NTPase activity, most likely via phosphorylation and dephosphorylation of the putative poly(A)$^+$ mRNA carrier (p106–110; refs. *19*, *20*). These phosphorylation and dephosphorylation processes are in turn modulated by poly(A) (*19*). An additional analytical result contradicting the model of Agutter was the finding that no marked dephosphorylation occurs after poly(A) binding to the nuclear envelope (*19*).

From these data we proposed (*19*, *205*) a new model for the NTPase-mediated translocation of poly(A)$^+$ mRNA (Fig. 7b). This model has some similarities to the general energy-transduction scheme introduced by Hill (*206*). In step 1, ATP binds to the phosphorylated carrier P-S. This binding results in a conformational change of P-S·ATP to P-*S·ATP (step 2). P-*S·ATP is a better enzyme for ATP hydrolysis than P-S·ATP. Splitting of phosphate from ATP (step 3) results in a conformational change from P-*S to P-S. ADP has a low affinity to P-S and dissociates from the carrier (step 4). The overall reaction velocity of the NTPase cycle is accelerated if the system is coupled with the poly(A)$^+$ mRNA molecule to be transported. According to the data (*167, 200*), poly(A)$^+$ mRNA binds preferentially to P-S rather than to S (step 1′-1). In step 1′-2, ATP binds to the P-S·poly(A)$^+$ mRNA complex which leads to a conformational change in the carrier from P-S to P-*S (step 2′-1). ATP in this complex causes a release of poly(A)$^+$ mRNA (step 2′-2), in accordance with the results of McDonald and Agutter (*200*). The carrier structure "S" in this scheme must not be a single polypeptide; it is quite possible that there are different polypeptides: one that binds ATP and another one that binds the poly(A)$^+$ mRNA. This view is supported by the finding that the nuclear envelope NTPase is solubilized in the presence of 2% Triton X-100 (*177, 178*), whereas the poly(A)-binding sites remain associated with the resulting pore complex–lamina structure. An important feature of scheme b is that, during NTPase-mediated mRNA translocation, no dephosphorylation of the carrier occurs. This means that, in contrast to scheme a, phosphorylation (reaction a) and dephosphorylation (reaction b) of the carrier are distinct reactions, which proceed separately from the events associated with the binding and release of poly(A)$^+$ mRNA. Phosphorylation and dephosphorylation of the carrier only modulate its affinity to poly(A)$^+$ mRNA and ATP, in accordance with the experimental results (*167*). Poly(A)$^+$ mRNA may also affect the ratio of S to P-S, either by inhibiting the kinase (*19, 207*) or by stimulating the phosphoprotein phosphatase (*200*).

In a further model (not shown) (*208*), the NTPase is modulated by a protein-kinase and phosphoprotein-phosphatase-dependent phosphorylation and dephosphorylation of the NTPase molecule itself. However, this assumption was disproved by photoaffinity labeling studies with 8-azido-ATP, showing that the NTPase represents a 43- to 47-kDa polypeptide that is not phosphorylatable by γ-^{32}P-labeled ATP (*184, 204*). This finding was also corroborated by use of the homogeneous NTPase (*178*).

The recent kinetic model (scheme c) of mRNA translocation has been proposed by (*20, 171*). This scheme also takes into consideration some experimental results incompatible with scheme a, but its major advantage over scheme b is that it can be applied successfully to explain the mode of action of the mRNA transport stimulatory proteins (*171*). This is possible because it is evident from this scheme (in accordance with the experimental results) that the NTPase itself is not phosphorylated and thus cannot directly be modulated by the kinase. In scheme c (Fig. 7), MgATP is the substrate of both the NTPase (reaction 1, 1′, and 2) and the protein kinase (reaction 3); in contrast to the NTPase, the protein kinase is additionally stimulated by excess Mg^{2+}. S can be phosphorylated by the kinase (reaction 3); when S is phosphorylated, N is inactive. Binding of poly(A)$^+$ mRNA to S stimulates its dephosphorylation (reaction 4′) by the phosphoprotein phosphatase (reaction 4 and 4′); the consequence is that N becomes active again. Because of a low but not negligible affinity of the unphosphorylated S for poly(A) (*207*), the normal route for mRNA binding and release can to some extent be "short-circuited" (reaction 5). A prediction from scheme c is that the following characteristics of the NTPase, measurable in whole nuclear envelopes, are lost when the enzyme is purified: (1) stimulation by poly(A) or poly(A)$^+$ mRNA, (2) inhibition by protein-kinase-dependent phosphorylation (via phosphorylation of S), and (3) non-Michaelis–Menten kinetics (in the presence of constant concentrations of free Mg, which stimulates protein kinase activity) (*12*). Indeed, none of these properties is found with the homogeneous NTPase, purified by the method of Schröder *et al.* (*178*).

VI. Binding to the Cytoskeleton

In the cytoplasm, the poly(A)$^+$ mRNA is again attached to cytoskeletal structures. This last step of mRNA transport is what had not so far been sufficiently studied.

Analysis of the mRNA association with the cytoskeleton by *in situ* hybridization indicates that the mRNA, after transport, apparently binds immediately to the cytoskeleton (*209*). Binding to the cytoskeleton seems to be a prerequisite for mRNA translation (*210, 211*). It has been suggested that the polyribosomal complex is fixed to the cytoskeleton via its mRNA component (*212, 213*). This conclusion was drawn because the ribosomes are released by ribonuclease treatment. However, more recent experiments indicate that intact mRNA is not required for ribosome association (*214*).

The p78 poly(A)-associated protein has been proposed to be involved in attaching polyribosomes to the cytoskeleton (*215*). On the other hand, histone mRNA or reovirus mRNA, which lacks a 3′-poly(A) sequence, or poliovirus RNA that does not contain a 5′-cap structure, are bound to the cytoskeleton to the same extent as are polyadenylated and capped mRNA (*210*).

It is assumed that the binding of the mRNA to the cytoskeletal framework provides the eukaryotic cell with a mechanism for controlling the polarity of mRNA distribution (*209*). This view is supported by the results from microinjection experiments with poly(A)$^+$ mRNA into *Xenopus* eggs, indicating a regional concentration of different mRNAs in the cytoplasm (*216*). A discrete distribution of specific mRNAs (actin, tubulin, and vimentin mRNA) has recently been demonstrated by *in situ* hybridization, suggesting that the proteins encoded by these mRNAs are synthesized at their site of function (*217, 218*). Moreover, the uneven regional distribution of poly(A) and RNA in the cytoplasm of *Xenopus* embryos is altered during early development (*219*).

At present, however, it is not clear which cytoskeletal system(s) is (are) involved in binding the mRNA. Poly(A) and poly(A)$^+$ mRNA have a high affinity for microtubule-associated "tau"[5] proteins (*220, 221*), provided their chain lengths are greater than 12 nucleotide units (*22*), but the *in vivo* relevance of these *in vitro* results is not known. On the other hand, an association of polyribosomes with the microfilaments in cultured lens cells has been reported (*222, 223*). Interestingly, cytochalasin, which causes detachment of hnRNA and mRNA from the nuclear matrix, also releases the cytoskeleton-bound mRNA (*212, 224*). In this context, the ability of actin and of microtubule protein to interact with poly(A)-metabolizing enzymes, poly(A) polymerase, and endoribonuclease IV might be important (*21*).

Possibly some small RNPs (snRNPs and scRNPs; sn = small nuclear; sc = small cytoplasmic) could be essential for RNA transport (*225*). It has been proposed that the Ro[6] scRNPs are also involved in the transport of mRNA in the cytoplasm (*226*), while the La[6] snRNPs are assumed to be involved in transcription, processing, or nucleocytoplasmic transport of RNA polymerase III (*226a*) transcripts (*227*). The Ro antigen of Ro scRNPs is associated with the cytoplasmic cytokeratin filaments of HEp-2 cells (*23*). The La antigen (contained in La snRNPs), on the other hand, is associated with nuclear structures that gave a speckled-type staining pattern with monoclonal anti-La antibodies (*23*). The association of the Ro antigen with cytokeratin filaments might indicate a possible participation of this cytoskeletal element in mRNA translation and distribution (*214, 228–230*). This view is supported by the finding (*231*) that the 50-kDa cap-binding protein is associated with intermediate filaments. The assumption that the intermediate filaments play an essential role during cytoplasmic trans-

[5] "tau" proteins: a group of microtubule-associated proteins with M_rs between 55,000 and 70,000 (P. Dustin, "Microtubules," 2nd ed., Springer, Berlin and New York, 1984).

[6] Ro and La are antigens of the Ro scRNPs and La snRNPs recognized by autoantibodies from patients with various rheumatic diseases (e.g., lupus erythematosus) (*23*). [Eds.]

port is also consistent with the finding that, in various mammalian cell types, these filaments terminate close to the area of the nuclear envelope and the nuclear envelope pore complexes (*232, 233*). There seems to be evidence for the existence of a continuous cytomatrix network throughout the cell, consisting of the intermediate filaments and the nuclear matrix, which are connected through the nuclear pore complexes (*230, 234*).

VII. Some Aspects of Poly(A)$^-$ mRNA Transport

In Section V, it is shown that poly(A)$^-$ mRNA fails to modulate nuclear-envelope-associated NTPase activity, suggesting that non-polyadenylated mRNAs are restricted to the nucleus due to their inability to stimulate the NTPase. Therefore, the question arises of how poly(A)$^-$ mRNAs, representing a major class of brain-specific mRNA, are exported out of the nucleus. An answer to this question might be important not only for understanding the transport of brain-specific mRNA, but also for understanding the transport of some mRNAs, which seem to be present in two forms, poly(A)$^+$ and poly(A)$^-$ [e.g., mRNAs for histones and β-actin (*235*)], from other tissues.

The mRNA from adult rat brain contains, in contrast to that from rat liver, a large proportion of mRNA molecules (about 50%) that are not polyadenylated [poly(A)$^-$ mRNA] (*65, 236, 237*). The brain-specific poly(A)$^-$ mRNAs, characterized by high sequence-complexity (*236–242*), appear postnatally in addition to the poly(A)$^+$ mRNAs (*65, 237*). Measurements of the nuclear-envelope-associated enzyme activities involved in mRNA transport, and their responsiveness to poly(A) during prenatal and postnatal rat brain and liver development, showed marked organ-dependent differences paralleling the appearance of the brain-specific poly(A)$^-$ mRNA (*243*). Thus, the brain NTPase activity decreases postnatally as soon as the poly(A)$^-$ mRNA emerges, while the liver NTPase significantly increases after birth (Fig. 8). The brain-specific poly(A)$^-$ mRNA sequences can be detected in the nucleus several days before they enter the cytoplasm (*65, 237*). Interestingly, this lag phase is also observed when the stimulation of the NTPase in brain envelopes by poly(A) is measured: the poly(A) stimulation of the brain enzyme is mostly lost exactly when the poly(A)$^-$ mRNA appears in the cytoplasm (*243*) (Fig. 8).

These results indicate that regulation of nucleocytoplasmic mRNA transport occurs somewhat differently in tissue with high poly(A)$^-$ mRNA content, e.g., brain, compared with tissue containing predomi-

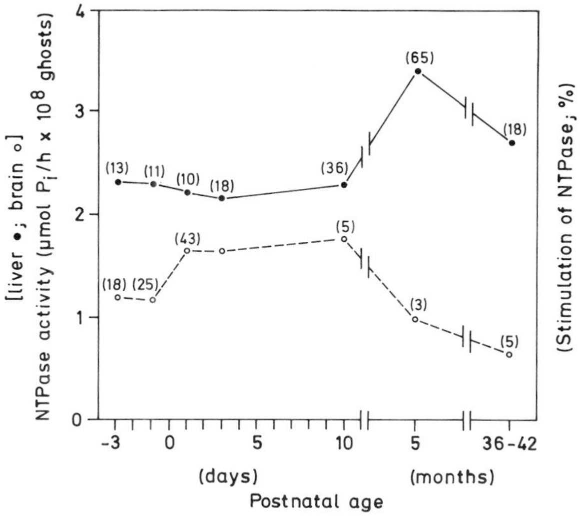

FIG. 8. Changes of the NTPase activity and its stimulation by poly(A) in whole nuclear envelopes from rat brain and liver during development. The extent of the stimulation of the NTPase in the presence of 90 μM poly(A) is given in percent (values in parentheses). From (*243*).

nantly poly(A)$^+$ mRNA, e.g., liver. However, the mechanism by which poly(A)$^-$ mRNAs are transported out of the nucleus is not known.

VIII. Regulation

Accumulating evidence suggests that much of the control of eukaryotic gene expression occurs at the levels of posttranscriptional hnRNA processing and mRNA transport. During these processes, there is a remarkable reduction of the RNA sequence complexity (*244*). Nucleic acid hybridization experiments revealed a 20-fold increase in complexity of nuclear over cytoplasmic RNA (*238*). It is now clearly demonstrated and has been the subject of numerous studies that the nucleocytoplasmic transport of mRNA is a key step in the control of eukaryotic gene expression. The multifactorial regulation of mRNA transport has been reviewed in detail (*44, 152, 245*).

Control of mRNA transport seems to occur both during the release stage and the translocation stage, both of which respond specifically to a series of physiological factors (see Section VIII,B). Some of this control can be exerted via protein factors that are recovered from soluble or polysomal preparations. Moreover, poly(A)-associated proteins and microtubule proteins possibly play a critical part in controlling nucleocytoplasmic poly(A)$^+$ mRNA transport (*157*).

A. Transport Stimulatory Proteins

It is well established that the cytoplasm contains proteins that stimulate or inhibit RNA efflux (both mRNA and rRNA) *in vitro* and presumably also RNA transport *in vivo* (*44, 245*). The proteins stimulating mRNA efflux from normal, but not neoplastically transformed, tissues seem to be tissue-specific (*243, 246*). The loss of nuclear RNA restriction occurring during carcinogenesis seems (*1, 53*) to be accompanied by a qualitative and/or quantitative change of the cytoplasmic, mRNA-transport-modulating proteins (*247, 248*).

At least two mRNA-transport-stimulatory proteins have been purified to apparent homogeneity from rat liver and other tissues by Webb's group (*247, 249, 250*). One of these proteins, with an M_r of 30,000–35,000, was purified from rat liver polysomes and shows a considerable tissue and species variability (*246, 249, 250*). The other protein, which stimulates mRNA efflux from isolated nuclei, is a 60-kDa polypeptide that has been purified from cancer cells (*247, 248*).

We have also purified two mRNA transport stimulatory proteins of 58 kDa (p58) and of 31 kDa (p31) from rat liver polysomes to apparent homogeneity (*171*). Both proteins show particular affinities for poly(A); p31 probably corresponds to the 30- to 35-kDa polypeptide already described (*250*), whereas p58 seems to be novel. In Fig. 9, the efficacy of the purified proteins to stimulate the efflux of mRNA from isolated nuclei is shown. The effect of p31 is enhanced by cAMP-dependent phosphorylation (in accordance with the results of Moffett and Webb; ref. *250*), but not the effect of p58 (Fig. 9). Both purified proteins stimulate the NTPase in whole isolated nuclear envelopes, particularly in the presence of poly(A) (Fig. 10). However, the homogeneous NTPase that does not respond to poly(A) is also not stimulated by p31 or p58, either in the absence or in combination with poly(A). This indicates that these proteins do not interact directly with the NTPase in the intact nuclear-envelope structure, but with some other component of the mRNA translocation system.

In subsequent studies, the targets for the action of the transport stimulatory proteins were identified (*171*). p31 increases the affinity of the poly(A)-binding site in the phosphorylated nuclear envelope. Poly(A) binding stimulates the phosphoprotein phosphatase activity and inhibits the protein-kinase activity in the envelope; in conse-

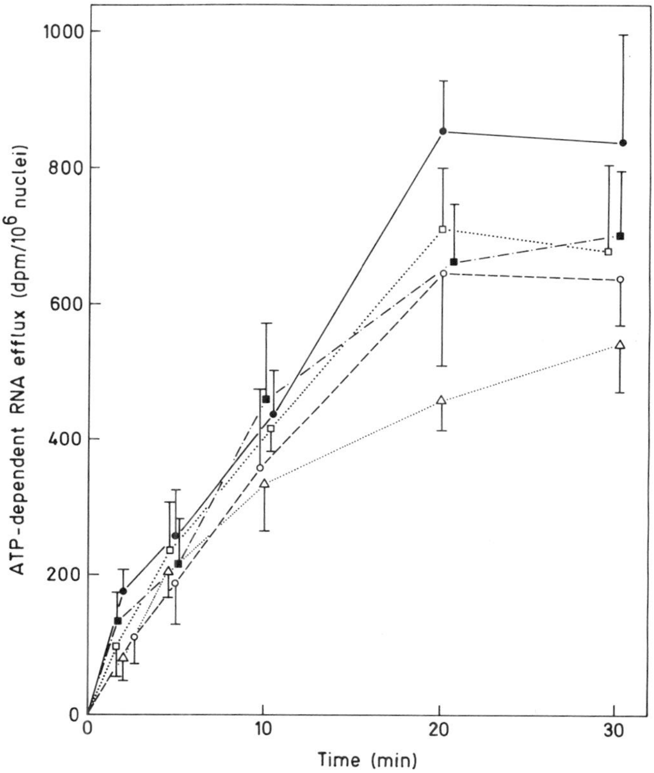

FIG. 9. Enhancement of ATP-dependent mRNA efflux from isolated rat liver nuclei by 1.5 μg/ml p31 (○, ●) or by 1.5 μg/ml p58 (□, ■) in the absence (○, □) and presence (●, ■) of 5 μM cAMP. △, Control. From *(171)*.

quence poly(A) stimulates the NTPase (*19, 20, 200, 251, 252*). The capacity of poly(A) to inhibit the kinase and to stimulate the phosphatase is enhanced in the presence of p31; without poly(A), p31 has no effect on these enzymes.

The other protein, p58, promotes the binding of poly(A) to unphosphorylated, but not to phosphorylated, nuclear envelopes; this contrasts with the effect of p31. It inhibits the protein kinase that down-regulates the NTPase even in the absence of poly(A); the NTPase activity is enhanced both in the presence and in the absence of poly(A). Since a proportion of the NTPase is inactivated by the protein kinase reaction (*20*), both p58 and p31 should increase the maximum catalytic rate of the NTPase in whole nuclear envelopes; this prediction is corroborated by the results shown in Fig. 11.

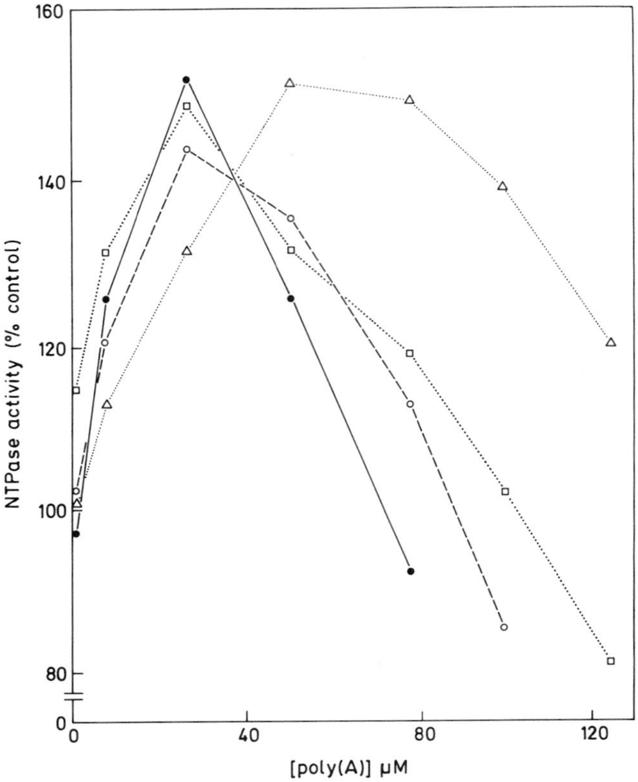

FIG. 10. Dependence of the NTPase activity in isolated rat liver nuclear envelopes on poly(A) in the presence or absence of purified mRNA efflux stimulatory proteins. △, No stimulatory protein added; □, 1 μg/ml p58 added; ○, 1 μg/ml p31 added; ●, 1 μg/ml p31, previously phosphorylated in the presence of cAMP. From (*171*).

The physiological significance of the purified mRNA transport stimulatory proteins is not definitely clear. P31 might promote the rate of translocation of low-abundance mRNAs through the nuclear pore complex (*171*). It is assumed that this polypeptide has the capacity to increase the probability of such an mRNA to interact with the poly(A)-binding site, since the affinity of this site for poly(A) is markedly enhanced by p31. Because p58 acts primarily by inhibiting the protein kinase, its regulatory role in the transport process could be related to the proposed participation of the phosphatidylinositol cycle in controlling the nuclear envelope mRNA translocation system (*177*). Another possibility might be that it "short-circuits" the kinase/phospha-

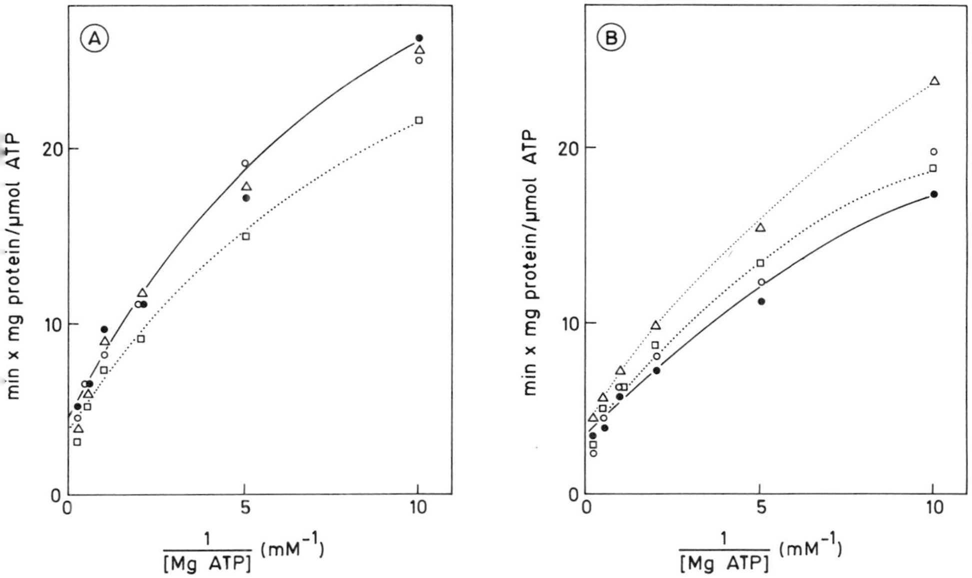

FIG. 11. Effect of the transport stimulatory proteins on the NTPase kinetics in isolated rat liver nuclear envelopes, in the absence (A) and the presence (B) of 30 μM poly(A). △, Control; □, 1 μg/ml p58; ○, 1 μg/ml p31; ●, 1 μg/ml p31 phosphorylated by a prior incubation in the presence of 5 μM cAMP. From (*171*).

tase loop (Fig. 7c) due to the enhancement of the affinity of poly(A) to the unphosphorylated poly(A)-binding site.

B. Dependence on Physiological Factors and Pathological Conditions

In the following, some important aspects of control of mRNA transport and its dependence on the physiological state of the cell are described; these include the influence of hormonal and nutritional factors, of lectins, of aging, of carcinogenesis, of virus infection, and of genetic diseases.

1. HORMONES

Nuclear envelopes from rat liver contain specific receptors for insulin (*253, 254*). Both the RNA efflux from isolated nuclei (*255*) and the nuclear envelope NTPase activity (*166*) respond to insulin. The hormone apparently acts by stimulating the phosphoprotein phosphatase in the nuclear envelope, and hence NTPase activity (*166, 208, 251, 252*). It is reasonable to assume that insulin also stimulates RNA transport *in vivo* (*256*). A possible influence on mRNA transport has also been claimed for corticosteroids (*257*) and estradiol (*258, 259*). There is a decrease of the NTPase activity in skin fibroblasts in response to dexamethasone, which is correlated with changes in protein synthesis *in vivo* (*260*).

Cyclic nucleotides (cAMP and cGMP) may be further effectors of mRNA transport. ATP-dependent efflux of mRNA from rat liver nuclei is stimulated by physiological concentrations of cAMP (*261*). The stimulation depends on the presence of exogenous cytosolic proteins and it seems that cAMP reacts with some cytosolic stimulators and/or inhibitors of mRNA efflux, whose efficiency is modulated by the nucleotide (*249*). This assumption was later corroborated when it was possible to study the effect of purified, homogeneous transport stimulatory proteins on mRNA efflux (*171*). However, experimental results from others (*151*) suggest that the nuclear envelope NTPase itself is also stimulated by cAMP, resulting in an enhanced rate of mRNA efflux.

2. Nutrition

The transport rate of mRNA and the nuclear envelope NTPase activity in the liver of rats also depend on dietary conditions (*165, 262–265*). Diets deficient in the essential tryptophan result in lower levels of the nuclear envelope NTPase activity (*165*) and of the nucleocytoplasmic RNA translocation *in vitro* and *in vivo* (*263, 264*). This effect is reversed by tryptophan refeeding. However, when tryptophan is added to isolated nuclear envelopes or to the homogeneous NTPase, no effect on the enzyme activity is observed. In contrast to normal liver, tryptophan does not affect transport and NTPase activity in a rat hepatoma (*266*).

3. Lectins

Binding sites for lectins [concanavalin A (Con A) and wheat germ agglutinin (WGA)] along the surfaces of the inner and outer nuclear membranes of rat liver cells (*267–269*) are localized on the side of the perinuclear space (*268, 269*). Immunoelectron microscopic studies showed that nuclear pore complexes from rat liver contain a high-molecular-weight glycoprotein that binds Con A (*35*). There is additional evidence that lectins are also present in the nucleus (*270*). RNA export from isolated nuclei and NTPase activity are inhibited by Con A and WGA (*170*), suggesting that a glycoprotein (possibly lamin B), is involved in nucleocytoplasmic RNA transport. However, this assumption is not corroborated by recent studies showing that neither these lectins nor some synthetic glycoproteins ("neoglycoproteins") display any influence on the ATP-dependent RNA efflux from isolated nuclei and the NTPase, protein kinase, or phosphoprotein phosphatase activity in whole nuclear envelopes (*243*).

4. Aging

It is well established that the amount of poly(A)$^+$ mRNA released from nuclei is markedly reduced during aging (*271, 272*). Experimental results indicate that the following factors contribute to this age-dependent change: (1) impaired polyadenylation of mRNA (*273*), (2) impaired hnRNA processing (*274*), and (3) impaired nucleocytoplasmic mRNA transport (*167*).

Inhibition of polyadenylation with cordycepin reduces the release of mRNA from hepatic nuclei (*186*); the differences observed are very similar to those found with juvenile and adult rat nuclei (*272*). The activity of the poly(A)–anabolic poly(A) polymerase in quail oviduct increases only little during aging, while the activities of the two poly(A) catabolic enzymes (endoribonuclease IV and 2′,3′-exoribonuclease) are significantly higher in senescent animals (*71, 90*). In consequence, and as confirmed by analytical studies (*91, 275*), the size of the poly(A) segment decreases with age. Furthermore, the pattern of the proteins specifically associated with the poly(A) tract changes during aging (*91*). Defined processing intermediates of ovalbumin mRNA accumulate during aging or are devoid of detectable poly(A) sequences (*273, 274*). However, increasing evidence suggests that at least some aging phenomena are correlated with an altered efficiency of the mRNA transporting system in the nuclear envelope (*205*).

The activity of the nuclear envelope NTPase as well as its capacity to be stimulated by poly(A) display drastic changes during development (*158, 167, 243*). As shown in Fig. 12, the activity of the NTPase in nuclear ghosts from quail oviduct is significantly altered during development, and responds to estrogen (diethylstilbestrol) and progesterone treatment of young animals; the liver enzyme from the same species does not change markedly between the different age groups and remains unaffected by hormone treatment (*158*). During estrogen-induced cell proliferation and differentiation, the activity of the oviduct enzyme strongly (7- to 8-fold) increases; concomitant administration of progesterone has no marked effect. During aging, the high level of NTPase reached in mature or hormone-treated immature animals is reduced by 50%. It was found that a close correlation exists between the level of NTPase and the extents of ovalbumin synthesis and avidin synthesis in quail oviduct (*158*). The capacity of the ovi-

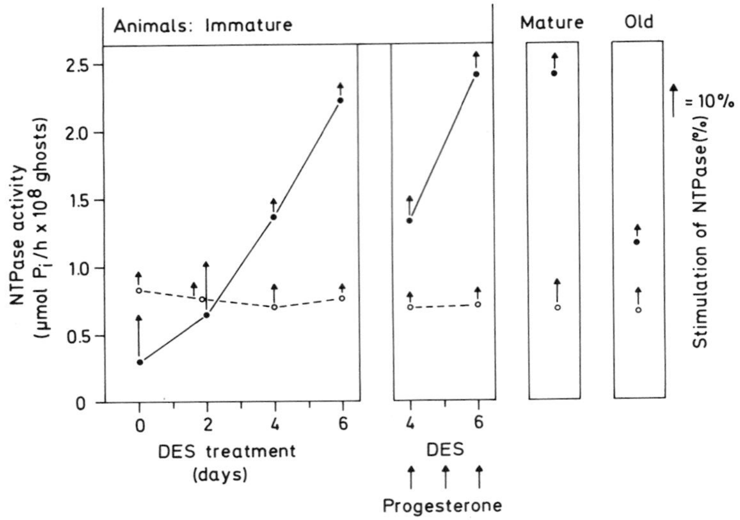

FIG. 12. Age dependence and hormone dependence of NTPase activity in whole nuclear envelopes from quail oviduct (●) and liver (○). Young animals were treated for 6 consecutive days with diethylstilbestrol (DES) alone or, in addition to DES, with progesterone at days 4, 5, and 6. The arrows indicate the efficacy of poly(A) to stimulate the enzyme (in %). From (*158*).

duct enzyme to be stimulated by poly(A) is generally low, but responds to aging and hormone treatment (Fig. 12). In rat liver, the NTPase is stimulated by poly(A) to about 170% in mature animals, while the enzyme from old animals is enhanced only to about 115% (*167*).

The extent of phosphorylation and dephosphorylation of nuclear envelope proteins is also markedly dependent on hormone treatment and development (*19*). In hormone-dependent quail oviduct the nuclear-envelope-associated protein-kinase and phosphoprotein-phosphatase activities were found to increase in immature animals in response to diethylstilbestrol and progesterone treatment (Fig. 13). During further development, the activities of these enzymes reached initially remain essentially unchanged (*203*). These results might indicate that age-dependent changes of nuclear envelope phosphorylation and dephosphorylation processes do not play an essential role during age-dependent reduction of nucleocytoplasmic mRNA efflux. However, Scatchard plot analyses revealed that the number of the poly(A)-binding sites is unphosphorylated, but not in phosphorylated, envelopes from old animals is markedly reduced, compared with

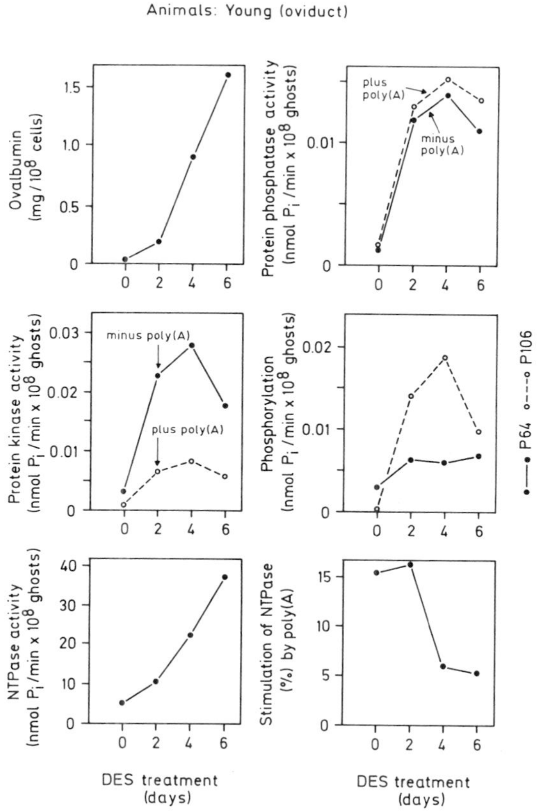

FIG. 13. Correlation between hormone-dependent changes of enzyme systems involved in nucleocytoplasmic mRNA translocation and the alteration of protein (ovalbumin) synthesis in oviducts from young quails stimulated with diethylstilbestrol. From (205).

those from mature animals (*167*). Interestingly, the extent of phosphorylation of the two major phosphoproteins, p64 and p106, of nuclear envelopes changes differently during development. In oviducts from untreated, immature animals, no phosphorylation of p106 is observed. During aging, the phosphorylation of this protein strongly increases and exceeds that of p64 (*203*). The same alterations in the extent of phosphorylation of p106 and p64 are also observed during hormone treatment of immature quails (*19*). These results offer good reasons to assume that the age-dependent decrease of overall protein synthesis (*276*) is partially caused by an age-dependent impairment of the mRNA translocation apparatus in the nuclear envelopes. With age, the functioning of this system might additionally be reduced due to a shorter size of the poly(A) tract of the mRNA to be transported (*91*), and due to a decreased efficiency of cytoplasmic transport stimulatory proteins (*243, 272*).

5. Carcinogens

It is well established that drastic alterations in transport and nuclear restriction of RNA occur during neoplastic transformation. Carcinogen-induced changes have been observed at the levels of the nuclear matrix (*277*), of the nuclear envelope mRNA translocation system (*151, 278*) and of the cytosolic transport modulating proteins (*43, 279, 280*). A series of studies (*1, 281, 282*) demonstrate that hepatocarcinogens induce an increase in both the size and the complexity of the cytoplasmic mRNA population in rat liver. This loss of nuclear RNA restriction, which has also been observed by others (*283–285*), is accompanied by a rapid swelling of the nuclei. At the level of translocation, carcinogenesis seems to be associated with marked changes in the energy dependence of this process, although some contradictory results have been obtained: an increase of NTPase activity (*151, 278*), and at least a partial loss of ATP dependence of mRNA transport (*2, 286*). At the cytoplasmic level, cytosols from liver and from hepatoma promote the efflux of different mRNA sequences from nuclei of normal liver (*43*).

6. Virus Infection

In some virus-infected cells, e.g., adenovirus-infected HeLa cells (*287*) or erythroblastosis-virus-infected avian erythroblasts (*288*), transport of host mRNA is markedly reduced; however, it cannot be excluded that these viruses may have interfered with the host mRNA at the processing stage.

7. Genetic Diseases

In hereditary congenital goiter in goats suffering from thyroglobulin deficiency, 33-S thyroglobulin mRNA sequences almost disappear from the cytoplasm, although these mRNA sequences are present at normal levels in the nucleus (*289*), indicating that some step in the transport of this specific mRNA is impaired.

IX. Concluding Remarks

We have good reason to assume that maturation, transport, and function of poly(A)$^+$ mRNA is associated with the structures of the nuclear matrix, the nuclear envelope, and the cytoskeleton. Ten years ago, Lichtenstein and Shapot (*290*) presented a model for eukaryotic gene regulation that postulated that each pore complex is specialized for the translocation of only one specific mRNA or a small set of specific mRNAs coding for different proteins. This intriguing idea, implicating a "gene-gating" mechanism intimately linked with the three-dimensional organization of the cell, has recently been well-formulated by Blobel (*291*). This hypothesis implies that the existence of a cytoskeletal matrix throughout the whole cell provides this cell with the ability for a sectorial distribution of distinct mRNAs and hence the realization of the genetic information at its respective destination. It is assumed that a direct connection exists between the sites of transcription and processing and the site of translation within the cell, provided by cytomatrix fibrils forming special routes along which the mRNA is transported. The movement of the mRNA through the nuclear pore along such fibrils, continuous with both nuclear and cytoplasmic skeletal structures, might occur in a treadmilling- or cable-car-like fashion (*33*). Attachment to the fibrillar structures and pore passage seem to be mediated by special signals in the transported molecule. Such signals should be newly formed or disappear during the lifetime of this molecule, e.g., during hnRNA processing. Based on the facts that the 3′-poly(A) segment of mRNA facilitates pore passage (resealed vesicle experiments; *48*) but does not cause release of mRNA from the matrix (for instance, all ovalbumin mRNA precursors are polyadenylated; *62*, *273*), it is reasonable to assume that, besides the signal for mRNA translocation ["poly(A)"], at least one further signal exists that prevents release of immature messengers into the cytoplasm. However, this signal has not yet been identified.

The phenomenon of nuclear restriction of immature messengers implies a selection of the mature mRNA molecules during transport from the nucleus to the cytoplasm. Therefore nucleocytoplasmic transport of mRNA must have an important regulatory function in gene expression. It seems to be well established that nucleocytoplasmic mRNA transport is regulated not only during passage through the pore complex but also at the level of RNA release from the nuclear matrix structure. In Fig. 14 we have summarized the well-established and the putative components involved in the release and the translocation mechanism. Both steps of nucleocytoplasmic transport can be clearly distinguished by their different energy and nucleotide requirements (Fig. 14). While mRNA traverses the nuclear envelope pore complexes by a mechanism that depends on ATP/GTP hydrolysis, detachment of mRNA from the nuclear matrix seems to be medi-

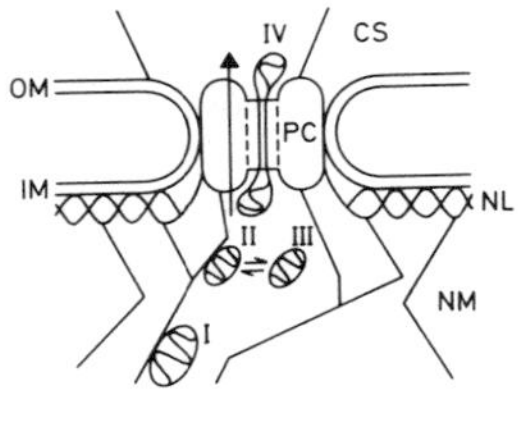

Stage	Requirements	Components involved
Translocation (Pore passage)	ATP (→ADP+P) (or GTP) Mg^{2+}	NTPase (p40), Poly(A)-recognizing carrier protein (p110), PK, PPH, Lamin B (?), Transport-stimulating proteins (p31 and p58)
Release (Matrix-bound→ matrix-free)	ATP (no cleavage) (or dATP) Mg^{2+}	Topoisomerase (?) Actin (?)

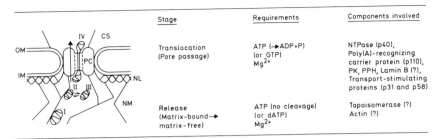

FIG. 14. Requirements and components involved in the release and translocation stages of nucleocytoplasmic poly(A)$^+$ mRNA transport. PC, pore complex; OM, outer nuclear membrane; IM, inner nuclear membrane; CS, cytoskeleton; NL, nuclear lamina; NM, nuclear matrix; PK, protein kinase; PPH, phosphoprotein phosphatase; I, immature poly(A)$^+$ hnRNP; II, matrix-bound mature poly(A)$^+$ mRNP; III, free mature poly(A)$^+$ mRNP; IV, poly(A)$^+$ mRNP undergoing pore passage.

ated most likely by a conformational change induced by ATP or dATP. The release of mRNA from the matrix seems to be the stage of nucleocytoplasmic transport at which the selection of mature mRNAs occurs.

Although intense studies of nucleocytoplasmic mRNA transport during the last decade resulted in a partial elucidation of the essential requirements and control factors involved in this process, and in an identification (and purification) of some proteins of the mRNA translocation apparatus, a full understanding of the molecular mechanism of this process requires the reconstitution of the whole system from the isolated and purified components. With regard to the mRNA translocation apparatus, these include: (1) the NTPase, (2) the poly(A)-recognizing "carrier" protein, (3) the protein kinase, (4) the phosphoprotein phosphatase, and possibly (5) lamin B, and (6) an RNA unwinding enzyme. Nevertheless, our present stage of knowledge about the mRNA translocation process through the nuclear envelope exceeds by far that about the release of mRNA from the matrix and that about the binding of mRNA to the cytoskeleton. Future work must be increasingly devoted to these aspects.

ACKNOWLEDGMENTS

We thank the members of our laboratory for their collaboration over the past 10 years, especially our former or present students R. Becker, A. Bernd, U. Friese, R. Messer, K. Pfeifer, M. Rottmann, D. Trölltsch, R. Wenger, and T. Zaubitzer. We are also grateful to P. S. Agutter (Napier College, Edinburgh, U.K.) and H. Fasold (Universität, Frankfurt, Federal Republic of Germany) for invaluable discussions. H.C.S. and W.E.G.M. acknowledge support by Grants Schr 277/2-1 and Mu 348/7-6 from the Deutsche Forschungsgemeinschaft.

References

1. R. W. Shearer, *Bchem* **13,** 1764 (1974).
2. D. E. Schumm, M. Hanausek-Walasek, A. Yannarell and T. E. Webb, *Eur. J. Cancer* **13,** 139 (1977).
3. E. A. Smuckler and G. A. Clawson, *in* "Nuclear Envelope Structure and RNA Maturation" (E. A. Smuckler and G. A. Clawson, eds.), UCLA Symposium Vol. 26, p. 645. Alan R. Liss, New York, 1985.
4. G. G. Maul, *Int. Rev. Cytol.* Suppl. **6,** 75 (1977).
5. R. Peters, *JBC* **258,** 11427 (1983).
6. U. Skoglund, K. Andersson, B. Bjorkroth, M. M. Lamb and B. Daneholt, *Cell* **34,** 847 (1983).
7. F. Wunderlich and G. Herlan, *J. Cell Biol.* **73,** 271 (1977).
8. G. Herlan, G. Giese and F. Wunderlich, *Bchem* **19,** 3960 (1980).
9. M. Zasloff, *PNAS* **80,** 6436 (1983).
10. P. S. Agutter, *in* "The Nuclear Envelope and the Nuclear Matrix" (G. G. Maul, ed.), p. 91. Alan R. Liss, New York, 1982.
11. E. M. DeRobertis, S. Lienhard and R. F. Parisot, *Nature* **295,** 572 (1982).
12. I. W. Mattaj and E. W. DeRobertis, *Cell* **40,** 111 (1985).
13. L. Gerace, *Nature* **318,** 508 (1985).
14. M. C. Dabauville and W. W. Franke, *PNAS* **79,** 5302 (1982).
15. C. Dingwall, S. V. Sharnick and R. A. Laskey, *Cell* **30,** 449 (1982).
16. D. Kalderon, W. D. Richardson, A. F. Markham and A. E. Smith, *Nature* **311,** 33 (1984).
17. J. Davey, N. J. Dimmock and A. Colman, *Cell* **40,** 667 (1985).
18. C. M. Feldherr, *in* "Nuclear Envelope Structure and RNA Maturation" (E. A. Smuckler and G. A. Clawson, eds.), UCLA Symposium Vol. 26, p. 645. Alan R. Liss, New York, 1985.
19. M. Bachmann, A. Bernd, H. C. Schröder, R. K. Zahn and W. E. G. Müller, *BBA* **773,** 308 (1984).
20. P. S. Agutter, *in* "Nuclear Envelope Structure and RNA Maturation" (E. A. Smuckler and G. A. Clawson, eds.), UCLA Symposium Vol. 26, p. 539. Alan R. Liss, New York, 1985.
21. H. C. Schröder, R. K. Zahn and W. E. G. Müller, *JBC* **257,** 2305 (1982).
22. H. C. Schröder, A. Bernd, R. K. Zahn and W. E. G. Müller, *Mol. Biol. Rep.* **8,** 233 (1982).
23. M. Bachmann, W. J. Mayet, H. C. Schröder, K. Pfeifer, K. H. Meyer zum Büschenfelde and W. E. G. Müller, *PNAS* **83,** 7770 (1986).
24. P. S. Agutter and J. C. W. Richardson, *J. Cell Sci.* **44,** 395 (1980).
25. P. S. Agutter, *Prog. Mol. Subcell. Biol.*, in press (1987).
26. S. A. Comerford, A. G. McLennan and P. S. Agutter, *in* "Nuclear Structures; Their Isolation and Characterisation" (G. D. Birnie and A. S. MacGillivray), in press. Butterworth, London, 1987.
27. L. Gerace and G. Blobel, *Cell* **19,** 277 (1980).
28. K. R. Shelton, *in* "The Nuclear Matrix" (R. Berezney, ed.), in press. Plenum, New York, 1987.
29. S. Lebel and Y. Raymond, *JBC* **259,** 2693 (1984).
30. W. W. Franke, *Phil. Trans. R. Soc. London, Ser. B* **268,** 67 (1974).
31. F. Wunderlich, *in* "The Cell Nucleus" (H. Busch, ed.), Vol. IX, p. 249. Academic Press, New York, 1981.

32. P. N. Unwin and R. A. Milligan, *J. Cell Biol.* **93,** 63 (1982).
33. G. G. Maul, *in* "The Nuclear Envelope and the Nuclear Matrix" (G. G. Maul, ed.), p. 1. Alan R. Liss, New York, 1982.
34. A. J. Filson, A. Lewis, G. Blobel and P. A. Fisher, *JBC* **260,** 3164 (1985).
35. L. Gerace, Y. Ottaviano and C. Kondor-Koch, *J. Cell Biol.* **95,** 826 (1982).
36. P. Dustin, "Microtubules," 2nd ed. Springer, Berlin and New York, 1984.
37. K. Weber and M. Osborn, *Natl. Cancer Inst. Monogr.* **60,** 31 (1982).
38. M. Clarke and J. A. Spudich, *ARB* **46,** 797 (1977).
39. E. Lazarides, *ARB* **51,** 219 (1982).
40. P. M. Steinert and D. A. D. Parry, *Annu. Rev. Cell Biol.* **1,** 42 (1985).
41. J. J. Wolosewick and K. R. Porter, *J. Cell Biol.* **82,** 114 (1976).
41a. P. Blackburn, *JBC* **254,** 12484 (1979).
42. P. S. Agutter, *BJ* **214,** 915 (1983).
43. D. E. Schumm, H. P. Morris and T. E. Webb, *Cancer Res.* **33,** 1821 (1973).
44. P. S. Agutter, *Subcell. Biochem.* **10,** 281 (1984).
45. E. A. Smuckler and M. Koplitz, *Cancer Res.* **34,** 827 (1974).
46. H. Jacobs and G. D. Birnie, *EJB* **121,** 597 (1982).
47. C. Otegui and R. J. Patterson, *NARes* **9,** 4767 (1981).
48. N. Riedel, H. Fasold, M. Bachmann and D. Prochnow, *BJ*, in press (1987).
49. S. H. Kaufmann, D. S. Coffey and J. H. Shaper, *Exp. Cell Res.* **132,** 105 (1981).
50. S. H. Kaufmann, W. Gibson and J. H. Shaper, *JBC* **258,** 2710 (1983).
51. I. Kindas-Mügge and G. Sauermann, *EJB* **148,** 49 (1985).
52. P. Gruss, C.-J. Lai, R. Dhar and G. Khoury, *PNAS* **76,** 4317 (1979).
53. R. W. Shearer, *Vestn. Akad. Med. Nauk SSSR* **3,** 64 (1977).
54. M. Kaehler, J. Coward and F. Rottmann, *NARes* **6,** 1161 (1979).
55. K. Dimock and C. M. Stolzfus, *JBC* **254,** 5591 (1979).
56. D. Finkel and Y. Groner, *Viroloy* **131,** 409 (1983).
57. W. E. G. Müller, A. Bernd and H. C. Schröder, *MCBchem* **53/54,** 197 (1983).
58. L. P. Villarreal and R. T. Whyte, *MCBiol* **3,** 1381 (1983).
59. H. C. Schröder, M. Bachmann, R. Messer and W. E. G. Müller, *Progr. Mol. Subcell. Biol.* **9,** 53 (1985).
60. M. Bachmann, H. C. Schröder, R. Messer and W. E. G. Müller, *FEBS Lett.* **171,** 25 (1984).
61. U. Z. Littauer and H. Soreq, this series **27,** 53 (1982).
62. E. M. Ciejek, J. L. Nordstrom, M.-J. Tsai and B. W. O'Malley, *Bchem* **21,** 4945 (1982).
63. R. P. Perry, D. E. Kelley and J. LaTorre, *JMB* **82,** 315 (1974).
64. C. M. Tsiapalis, J. W. Dorson and F. J. Bollum, *JBC* **250,** 4486 (1975).
65. N. Chaudhari and W. E. Hahn, *Science* **220,** 924 (1983).
66. D. Sheiness, L. Puckett and J. E. Darnell, *PNAS* **72,** 1077 (1975).
67. U. Nudel, H. Soreq, U. Z. Littauer, G. Marbaix, G. Huez, M. Leclercq, E. Hubert and H. Chantrenne, *EJB* **64,** 115 (1976).
68. W. E. G. Müller, A. Totsuka, I. Kroll, I. Nusser and R. K. Zahn, *BBA* **383,** 147 (1975).
69. R. L. Matts and F. L. Siegel, *JBC* **254,** 11228 (1979).
70. W. E. G. Müller, D. Falke, R. K. Zahn and J. Arendes, *Virology* **87,** 89 (1978).
71. W. E. G. Müller, R. K. Zahn and J. Arendes, *Mech. Ageing Dev.* **14,** 39 (1980).
72. K. M. Rose and S. T. Jacob, *JBC* **254,** 10256 (1979).
73. K. M. Rose and S. T. Jacob, *Bchem* **19,** 1472 (1980).

74. C. M. Tsiapalis, T. Trangas and A. Gounaris, *FEBS Lett. 140,* 213 (1982).
75. H. C. Schröder, P. Schenk, H. Baydoun, K. G. Wagner and W. E. G. Müller, *Arch. Gerontol. Geriatr.* **2,** 349 (1983).
76. G. Blobel, *PNAS* **70,** 924 (1973).
77. W. E. G. Müller, J. Arendes, R. K. Zahn and H. C. Schröder, *EJB* **86,** 283 (1978).
78. M. A. Winters and M. Edmonds, *JBC* **248,** 4756 (1973).
79. W. E. G. Müller, H. C. Schröder, J. Arendes, R. Steffen, R. K. Zahn and K. Dose, *EJB* **76,** 531 (1977).
80. K. M. Rose and S. T. Jacob, *EJB* **67,** 11 (1976).
81. K. M. Rose, F. J. Roe and S. T. Jacob, *BBA* **478,** 180 (1977).
82. H. C. Schröder, D. E. Nitzgen, A. Bernd, B. Kurelec, R. K. Zahn, M. Gramzow and W. E. G. Müller, *Cancer Res.* **44,** 3812 (1984).
83. W. E. G. Müller, *EJB* **70,** 241 (1976).
84. W. E. G. Müller, G. Seibert, R. Steffen and R. K. Zahn, *EJB* **70,** 249 (1976).
85. H. C. Schröder, K. Dose, R. K. Zahn and W. E. G. Müller, *JBC* **255,** 5108 (1980).
86. H. C. Schröder, R. K. Zahn, K. Dose and W. E. G. Müller, *JBC* **255,** 4535 (1980).
87. W. E. G. Müller, H. C. Schröder, R. K. Zahn and K. Dose, *ZpChem.* **361,** 469 (1980).
88. M. Adesnik and J. E. Darnell, *JMB* **67,** 397 (1972).
89. J. L. Stein, C. L. Thrall, W. D. Park, R. J. Mans and G. S. Stein, *Science* **189,** 557 (1975).
90. W. E. G. Müller, R. K. Zahn, H. C. Schröder and J. Arendes, *Gerontology* **25,** 61 (1979).
91. A. Bernd, E. Batke, R. K. Zahn and W. E. G. Müller, *Mech. Ageing Dev.* **19,** 361 (1982).
92. S. W. Kwan and G. Brawerman, *PNAS* **69,** 3247 (1972).
93. W. R. Jeffery, *JBC* **252,** 3525 (1977).
94. A. Vincent, S. Goldenberg and K. Scherrer, *EJB* **114,** 179 (1981).
95. A. Bernd, R. K. Zahn, A. Maidhof and W. E. G. Müller, *ZpChem.* **363,** 221 (1982).
96. T. Tomcsányi, L. Komáromy and A. Tigyi, *EJB* **114,** 421 (1980).
97. V. M. Kish and T. Pederson, *JMB* **95,** 227 (1975).
98. C. A. G. van Eekelen, T. Rieman and W. J. van Venrooij, *FEBS Lett.* **130,** 223 (1981).
99. H. Schwartz and J. E. Darnell, *JMB* **104,** 833 (1976).
100. R. K. Roy, S. Sarkar, C. Guha and H. N. Munro, *in* "The Cell Nucleus" (H. Busch, ed.), Vol. IX, p. 289. Academic Press, New York, 1981.
101. C. M. Feldherr, *Cell Tissue Res.* **205,** 157 (1980).
102. D. M. Pardoll, B. Vogelstein and D. S. Coffey, *Cell* **19,** 527 (1980).
103. D. A. Jackson, S. J. McCready and P. R. Cook, *Nature* **292,** 552 (1981).
104. D. A. Jackson and P. R. Cook, *EMBO J.* **4,** 919 (1985).
105. E. Mariman, A.-M. Hagebols and W. J. van Venrooij, *NARes* **10,** 6131 (1982).
106. E. C. M. Mariman, C. A. G. von Eekelen, R. J. Reinders, A. J. M. Berns and W. J. van Venrooij, *JMB* **154,** 103 (1982).
107. S. I. Robinson, B. D. Nelkin and B. Vogelstein, *Cell* **28,** 99 (1982).
108. H. C. Smith and R. Berezney, *BBRC* **97,** 1541 (1980).
109. T. E. Miller, C. Y. Huang and A. O. Pogo, *J. Cell. Biol.* **76,** 675 (1978).
110. R. Herman, L. Weymouth and S. Penman, *J. Cell Biol.* **78,** 663 (1978).
111. I. Faiferman and A. O. Pogo, *Bchem* **14,** 3808 (1975).
112. B. H. Long, C.-Y. Huang and A. O. Pogo, *Cell* **18,** 1079 (1979).

113. C. A. G. van Eekelen and W. J. van Venrooij, *J. Cell Biol.* **88,** 554 (1981).
114. H. C. Schröder, D. Trölltsch, U. Friese, M. Bachmann and W. E. G. Müller, *JBC*, in press (1987).
115. A. L. Beyer, M. E. Christiensen, W. Walker and W. M. LeStourgeon, *Cell* **11,** 127 (1977).
116. H. Nakayasu, H. Mori and K. Ueda, *Cell Struct. Funct.* **7,** 253 (1982).
117. H. Nakayasu and K. Ueda, *Cell Struct. Funct.* **9,** 317 (1984).
118. P. R. DiMaria, G. Kaltwasser and C. J. Goldenberg, *JBC* **260,** 1096 (1985).
119. L. C. Yu, J. Racevskis and T. E. Webb, *Cancer Res.* **32,** 2314 (1972).
120. P. S. Agutter, *BJ* **188,** 91 (1980).
121. C. L. Moore and P. A. Sharp, *Cell* **41,** 845 (1985).
122. G. L. Chen, L. Yang, T. C. Rowe, B. D. Halligan, K. M. Tewey and L. F. Liu, *JBC* **259,** 13560 (1984).
123. M. Berrios, N. Osheroff and P. A. Fisher, *PNAS* **82,** 4142 (1985).
124. M. T. Muller, W. P. Pfund, V. B. Mehta and D. K. Trask, *EMBO J.* **4,** 1237 (1985).
125. B. Ruskin and M. R. Green, *Cell* **43,** 131 (1985).
126. A. J. Shatkin, *Cell* **40,** 223 (1985).
127. B. K. Ray, T. G. Lawson, J. C. Kramer, M. H. Cladaras, J. A. Grifo, R. D. Abramson, W. C. Merrick and R. E. Thach, *JBC* **260,** 7651 (1985).
128. K. Maundrell, E. S. Maxwell, E. Puvion and K. Scherrer, *Exp. Cell Res.* **136,** 435 (1981).
129. A. S. Douvas, C. A. Harrington and J. Bonner, *PNAS* **72,** 3902 (1975).
130. H. Nakayasu and K. Ueda, *Exp. Cell Res.* **143,** 55 (1983).
131. H. Nakayasu and K. Ueda, *Cell Struct. Funct.* **10,** 305 (1985).
132. J. W. Bremer, H. Busch and L. C. Yeoman, *Bchem* **20,** 2013 (1981).
133. P. Gounon and E. Karsenti, *J. Cell Biol.* **88,** 410 (1981).
134. L. Goldstein, R. Rubin and C. Ko, *Cell* **12,** 601 (1977).
135. Y. Fukui, *J. Cell Biol.* **76,** 146 (1978).
136. J. Wehland, K. Weber and M. Osborn, *Biol. Cell.* **39,** 109 (1980).
137. H. Nakayasu and K. Ueda, *Exp. Cell Res.* **160,** 319 (1985).
138. H. C. Schröder, D. Trölltsch, M. Bachmann, R. Wenger and W. E. G. Müller, *EJB*, in press.
139. S. S. Smith, K. H. Kelly and B. M. Jockusch, *BBRC* **86,** 161 (1979).
139a. A. L. Beyer, M. E. Christensen, B. W. Walker and W. M. LeStourgeon, *Cell* **11,** 127 (1977). See also LeStourgeon *et al.*, *CSHSQB* **42,** 885 (1978). [Eds.]
140. A. Schweiger and G. Kostka, *BBA* **782,** 262 (1984).
141. C. D. Lewis and U. K. Laemmli, *Cell* **29,** 171 (1982).
142. J. B. Gurdon and D. A. Melton, *ARGen* **15,** 189 (1981).
143. K. J. Mills and L. G. E. Bell, *Exp. Cell Res.* **136,** 469 (1981).
144. C. M. Feldherr, *in* "Advances in Cell and Molecular Biology" (E. J. Dupraw, ed.), p. 273. Academic Press, New York, 1972.
145. P. L. Paine, L. C. Moore and S. B. Horowitz, *Nature* **254,** 109 (1985).
146. R. Peters, *Naturwissenschaften* **70,** 294 (1983).
147. J. Dubochet, C. Morel, B. LeBleu and M. Merzberg, *EJB* **36,** 465 (1973).
148. C. Kondor-Koch, N. Riedel, R. Valentin, H. Fasold and H. Fischer, *EJB* **127,** 285 (1982).
149. P. S. Agutter, H. J. McArdle and B. McCaldin, *Nature* **263,** 165 (1976).
150. K. Ishikawa, S. Sato-Odani and K. Ogata, *BBA* **521,** 650 (1978).
151. G. A. Clawson, J. James, C. H. Woo, D. S. Friend, D. Moody and E. A. Smuckler, *Bchem* **19,** 2756 (1980).

152. T. E. Webb, D. E. Schumm and T. Palayoor, *in* "The Cell Nucleus" (H. Busch, ed.), Vol. 9, p. 199. Academic Press, New York, 1981.
153. D. E. Schumm, M. A. Niemann, T. Palayoor and T. E. Webb, *JBC* **254**, 12126 (1979).
154. F. Wunderlich and G. Herlan, *J. Cell Biol.* **73**, 271 (1977).
155. G. Herlan, G. Giese and F. Wunderlich, *Exp. Cell Res.* **118**, 305 (1979).
156. P. S. Agutter, B. McCaldin and M. J. McArdle, *BJ* **182**, 811 (1979).
157. A. Bernd, H. C. Schröder, R. K. Zahn and W. E. G. Müller, *EJB* **129**, 43 (1982).
158. A. Bernd, H. C. Schröder, G. Leyhausen, R. K. Zahn and W. E. G. Müller, *Gerontology* **29**, 394 (1983).
159. P. S. Agutter and B. McCaldin, *BJ* **180**, 371 (1979).
160. P. S. Agutter and I. Thomson, *in* "Membrane Structure and Function" (E. E. Bittar, ed.), Vol. 6, p. 43. Wiley, New York, 1984.
161. A. Vorbrodt and G. G. Maul, *J. Histochem. Cytochem.* **28**, 27 (1980).
162. G. G. Maul and F. A. Baglia, *Exp. Cell Res.* **145**, 285 (1983).
163. G. A. Clawson, D. S. Friend and E. A. Smuckler, *Exp. Cell Res.* **155**, 310 (1984).
164. G. A. Clawson, M. Koplitz, D. E. Moody and E. A. Smuckler, *Cancer Res.* **40**, 75 (1980).
165. C. N. Murty, E. Verney and H. Sidransky, *PSEBM* **163**, 155 (1980).
166. F. Purrello, R. Vigneri, G. A. Clawson and I. D. Goldfine, *Science* **216**, 1005 (1982).
167. A. Bernd, H. C. Schröder, R. K. Zahn and W. E. G. Müller, *Mech. Ageing Dev.* **20**, 331 (1982).
168. F. A. Baglia and G. G. Maul, *in* "The Nuclear Envelope and the Nuclear Matrix" (G. G. Maul, ed.), p. 129. Alan R. Liss, New York, 1982.
169. H. C. Schröder, B. Diehl-Seifert, M. Rottmann, D. Trölltsch, B. A. Bryson, P. S. Agutter and W. E. G. Müller, submitted.
170. F. A. Baglia and G. G. Maul, *PNAS* **80**, 2285 (1983).
171. H. C. Schröder, M. Rottmann, M. Bachmann, W. E. G. Müller, A. R. McDonald and P. S. Agutter, *EJB* **159**, 51 (1986).
172. O. Georgiev, J. Mous and M. L. Birnstiel, *NARes* **12**, 8539 (1984).
173. P. S. Agutter, J. R. Harris and I. Stevenson, *BJ* **162**, 671 (1977).
174. P. S. Agutter and I. Ramsay, *Biochem. Soc. Trans.* **7**, 720 (1979).
175. T. Pederson, *Am. Scient.* **69**, 76 (1981).
176. N. Riedel, M. Bachmann and H. Fasold, 15th FEBS Meeting, Abstract book no. S-03 TU-144 (1983).
177. C. D. Smith and W. W. Wells, *JBC* **259**, 11890 (1984).
178. H. C. Schröder, M. Rottmann, M. Bachmann and W. E. G. Müller, *JBC* **261**, 663 (1986).
179. M. Berrios, A. J. Filson, G. Blobel and P. A. Fisher, *JBC* **258**, 13384 (1983).
180. M. Berrios, G. Blobel and P. A. Fisher, *JBC* **258**, 4548 (1983).
181. P. S. Agutter, J. R. Cockrill, J. E. Lavine, B. McCaldin and R. B. Sim, *BJ* **181**, 647 (1979).
182. G. A. Clawson and E. A. Smuckler, *PNAS* **75**, 5400 (1978).
183. E. A. Smuckler and G. A. Clawson, *in* "The Nuclear Envelope and the Nuclear Matrix" (G. G. Maul, ed.), p. 271. Alan R. Liss, New York, 1982.
184. G. A. Clawson, C. H. Woo, J. Button and E. A. Smuckler, *Bchem* **23**, 3501 (1984).
185. K. A. Perevoshchikova, Kh. Prokop, B. Hering, Ya. M. Koen and I. B. Zbarskii, *Bull. Exp. Biol. Med.* **87**, 565 (1979).
186. D. E. Schumm and T. E. Webb, *BJ* **139**, 191 (1974).

187. G. Maale, G. Stein and R. Mans, *Nature* **255**, 80 (1975).
188. E. Egyhazi, *Nature* **250**, 221 (1974).
189. M. C. Chisick, B. A. Brennessel and D. K. Biswas, *BBRC* **91**, 1109 (1979).
190. D. E. Schumm and T. E. Webb, *BBRC* **58**, 354 (1974).
191. P. S. Agutter and K. E. Suckling, *BBA* **698**, 223 (1982).
192. N. Hazan and R. McCauley, *BJ* **156**, 665 (1976).
193. A. Kumar, M. R. Satyanarayana Rao and G. Padmanaban, *BJ* **186**, 81 (1980).
194. F. Wunderlich, W. Batz, V. Speth and D. F. H. Wallach, *J. Cell Biol.* **61**, 633 (1974).
195. S. E. Stuart, G. A. Clawson, F. M. Rottmann and R. Y. Patterson, *J. Cell Biol* **72**, 57 (1977).
196. P. S. Agutter and K. E. Suckling, *BBA* **696**, 308 (1982).
197. R. C. Steer, S. A. Goueli, M. J. Wilson and K. Ahmed, *BBRC* **92**, 919 (1980).
198. K. S. Lam and C. B. Kasper, *Bchem* **18**, 307 (1979).
199. R. C. Steer, M. J. Wilson and K. Ahmed, *Exp. Cell Res.* **119**, 403 (1979).
200. J. R. McDonald and P. S. Agutter, *FEBS Lett.* **116**, 145 (1980).
201. R. C. Steer, M. J. Wilson and K. Ahmed, *BBRC* **89**, 1082 (1979).
202. K. Ahmed and R. C. Steer, *in* "The Nuclear Envelope and the Nuclear Matrix" (G. G. Maul, ed.), p. 31. Alan R. Liss, New York, 1982.
203. H. C. Schröder, M. Bachmann, A. Bernd, R. K. Zahn and W. E. G. Müller, *Mech. Ageing Dev.* **27**, 87 (1984).
204. P. S. Agutter, *in* "Nuclear Envelope Structure and RNA Maturation" (E. A. Smuckler and G. A. Clawson, eds.), p. 561. Alan R. Liss, New York, 1985.
205. W. E. G. Müller, P. S. Agutter, A. Bernd, M. Bachmann and H. C. Schröder, *in* "Thresholds in Aging" (M. Bergener, M. Ermini and H. B. Stähelin, eds.), The 1984 Sandoz Lectures in Gerontology, p. 21. Academic Press, London, 1985.
206. T. L. Hill, *PNAS* **69**, 267 (1969).
207. H. C. Schröder, M. Rottmann, M. Bachmann and W. E. G. Müller, submitted.
208. F. Purello, D. B. Burnham and I. D. Goldfine, *PNAS* **80**, 1189 (1983).
209. W. R. Jeffery, *J. Cell Biol.* **95**, 1 (1982).
210. A.-M. Bonneau, A. Darveau and N. Sonenberg, *J. Cell Biol.* **100**, 1209 (1985).
211. R. T. Moon, R. F. Nicosia, C. Olsen, M. B. Hille and W. R. Jeffery, *Dev. Biol.* **95**, 447 (1983).
212. R. Lenk, L. Ransom, Y. Kaufmann and S. Penman, *Cell* **10**, 67 (1977).
213. W. J. van Venrooij, P. T. G. Sillekens, C. A. G. van Eekeln and R. J. Reinders, *Exp. Cell Res.* **135**, 79 (1981).
214. J. G. Howe and J. W. B. Hershey, *Cell* **37**, 85 (1984).
215. C. Milcarek and S. Penman, *JMB* **89**, 327 (1974).
216. D. G. Capco and W. R. Jeffery, *Nature* **294**, 255 (1981).
217. R. H. Singer and D. C. Ward, *PNAS* **79**, 7331 (1982).
218. J. B. Lawrence and R. H. Singer, *Cell* **45**, 407 (1986).
219. C. R. Phillips, *J. Exp. Zool.* **223**, 265 (1982).
220. H. C. Schröder, A. Bernd, R. K. Zahn and W. E. G. Müller, *Mech. Ageing Dev.* **24**, 101 (1984).
221. J. Bryan, B. W. Nagle and K. H. Doenges, *PNAS* **72**, 3570 (1975).
222. F. C. S. Ramaekers, A. M. E. Selten-Versteegen, E. L. Benedetti, I. Dunia and H. Bloemendal, *PNAS* **77**, 725 (1980).
223. F. C. S. Ramaekers, E. L. Benedetti, I. Dunia, P. Vorstenbosch and H. Bloemendal, *BBA* **740**, 441 (1983).
224. D. A. Ornelles, E. G. Fey and S. Penman, *MCBiol* **6**, 1650 (1986).
225. R. Reddy and H. Busch, this series **30**, 127 (1983).

226. S. L. Wolin and J. A. Steitz, *Cell* **32,** 735 (1983).
226a. R. G. Roeder, *in* "RNA Polymerase" (R. Losick and M. Chamberlin, eds.), p. 293. Cold Spring Harbor Laboratory, Cold Spring Harbor, N.Y., 1976. Also, P. Chambon, *ARB* **44,** 613 (1975).
227. J. Rinke and J. A. Steitz, *Cell* **29,** 149 (1982).
228. M. Cervera, G. Dreyfuss and S. Penman, *Cell* **23,** 113 (1981).
229. E. G. Fey, K. M. Wan and S. Penman, *J. Cell Biol.* **98,** 1973 (1984).
230. P. Nilsen, S. Goelz and H. Trachsel, *Cell Biol. Int. Rep.* **7,** 245 (1983).
231. A. Zumbe, C. Staehli and H. Trachsel, *PNAS* **79,** 2927 (1982).
232. J. C. R. Jones, A. E. Goldman, H.-Y. Yang and R. D. Goldman, *J. Cell Biol.* **100,** 93 (1985).
233. V. L. Lehto, I. Virtaanen and P. Kurki, *Nature* **272,** 175 (1978).
234. D. G. Capco, K. M. Wan and S. Penman, *Cell* **29,** 847 (1982).
235. P. K. Katinakis, A. Slater and R. H. Burdon, *FEBS Lett.* **116,** 1 (1980).
236. D. M. Chikaraishi, *Bchem* **18,** 3249 (1979).
237. W. E. Hahn, N. Chaudhari, L. Beck, K. Wilber and D. Peffley, *CSHSQB* **48,** 465 (1983).
238. D. M. Chikaraishi, S. S. Deeb and N. Sueoka, *Cell* **13,** 111 (1978).
239. S. S. Deeb, *Cell. Mol. Biol.* **29,** 113 (1983).
240. S. L. Beckmann, D. M. Chikaraishi, S. S. Deeb and N. Sueoka, *Bchem* **20,** 2684 (1981).
241. J. van Ness, I. H. Maxwell and W. E. Hahn, *Cell* **18,** 1341 (1979).
242. B. B. Kaplan, B. S. Schachter, H. H. Osterbury, T. S. Villis and C. E. Finch, *Bchem* **17,** 5516 (1978).
243. H. C. Schröder, R. Becker, M. Bachmann, M. Gramzow, A.-P. Seve, M. Monsigny and W. E. G. Müller, *BBA* **868,** 108 (1986).
244. B. J. Wold, W. H. Klein, B. R. Hough-Evans, R. J. Britten and E. H. Davidson, *Cell* **14,** 941 (1978).
245. A. V. Lichtenstein, M. M. Zaboykin, V. L. Mojseev and V. S. Shapot, *Subcell. Biochem.* **8,** 185 (1982).
246. S. Smart-Nixon, D. E. Schumm and T. E. Webb, *Comp. Biochem. Physiol.* **75,** 655 (1983).
247. B. T. French, M. Hanausek-Walaszek, Z. Walaszek, D. E. Schumm and T. E. Webb, *Cancer Lett.* **23,** 45 (1984).
248. Z. Walaszek, M. Hanausek-Walaszek, D. E. Schumm and T. E. Webb, *Cancer Lett.* **20,** 277 (1983).
249. R. B. Moffett and T. E. Webb, *Bchem* **20,** 3253 (1981).
250. R. B. Moffett and T. E. Webb, *BBA* **740,** 231 (1983).
251. I. D. Goldfine, G. A. Clawson, E. A. Smuckler, F. Purrello and R. Vigneri, *MCBchem* **48,** 3 (1982).
252. I. D. Goldfine, F. Purrello, G. A. Clawson and R. Vigneri, *J. Cell Biochem.* **20,** 29 (1982).
253. I. D. Goldfine, R. Vigneri, D. Cohen, N. B. Pliam and C. R. Kahn, *Nature* **269,** 698 (1977).
254. R. Vigneri, I. D. Goldfine, K. Y. Wong, G. J. Smith and V. Pezzino, *JBC* **253,** 2098 (1978).
255. D. E. Schumm and T. E. Webb, *ABB* **210,** 275 (1981).
256. L. S. Jefferson, T. A. Boyd, K. E. Flaim and D. E. Peavy, *Biochem. Soc. Trans.* **8,** 282 (1980).
257. T. Lund-Larsen and T. Berg, *ZpChem* **354,** 1334 (1973).

258. R. B. Church and B. J. McCarthy, *BBA* **199,** 103 (1970).
259. G. H. Vazquez-Nin, O. M. Echeverria and J. Pedron, *Biol. Cell.* **35,** 221 (1979).
260. A. Bernd, P. Altmeyer, W. E. G. Müller, H. C. Schröder and H. Holzmann, *J. Invest. Dermatol.* **83,** 20 (1984).
261. D. E. Schumm and T. E. Webb, *JBC* **253,** 8513 (1978).
262. C. N. Murty, E. Verney and H. Sidransky, *in* "The Nuclear Envelope and the Nuclear Matrix" (G. G. Maul, ed.), p. 111. Alan R. Liss, New York, 1982.
263. C. N. Murty, E. Verney and H. Sidransky, *BBA* **474,** 117 (1977).
264. C. N. Murty, E. Verney and H. Sidransky, *Biochem. Med.* **22,** 98 (1979).
265. J. Luetzeler, E. Verney and H. Sidransky, *Exp. Mol. Pathol.* **31,** 261 (1979).
266. H. Sidransky, E. Verney and C. N. Murty, *Exp. Mol. Pathol.* **35,** 124 (1981).
267. A. Monneron and D. Segretain, *FEBS Lett.* **42,** 209 (1974).
268. I. Virtanen and J. Wartiovaara, *J. Cell Sci.* **22,** 335 (1976).
269. C. M. Feldherr, P. A. Richmond and K. D. Noonan, *Exp. Cell Res.* **107,** 439 (1977).
270. A. P. Seve, J. Hubert, D. Bouvier, M. Bouteille, C. Maintier and M. Monsigny, *Exp. Cell Res.* **157,** 533 (1985).
271. R. E. Moore, T. L. Goldsworthy and H. C. Pitot, *Cancer Res.* **40,** 1449 (1980).
272. A. Yannarell, D. E. Schumm and T. E. Webb, *Mech. Ageing Dev.* **6,** 259 (1977).
273. R. Messer, H. C. Schröder, H.-J. Breter and W. E. G. Müller, *Mol. Biol. Rep.* **11,** 81 (1986).
274. H. C. Schröder, R. Messer, H.-J. Breter and W. E. G. Müller, *Mech. Ageing Dev.* **30,** 319 (1985).
275. J. Arendes, R. K. Zahn and W. E. G. Müller, *Mech. Ageing Dev.* **14,** 49 (1980).
276. A. Richardson, *in* "Handbook of Biochemistry of Aging" (J. R. Florini, ed.), p. 79. CRC Press, Boca Raton, Florida, 1981.
277. G. A. Clawson and E. A. Smuckler, *BBRC* **108,** 1331 (1982).
278. G. A. Clawson, C. H. Woo and E. A. Smuckler, *BBRC* **95,** 1200 (1980).
279. D. E. Schumm, D. J. McNamara and T. E. Webb, *Nature NB* **245,** 201 (1973).
280. M. Lemaire, W. Bayens and L. Baugnet-Mahieu, *Biomedicine* **34,** 47 (1981).
281. R. W. Shearer, *BBRC* **57,** 604 (1974).
282. R. W. Shearer, *Chem.-Biol. Interact.* **27,** 91 (1979).
283. E. A. Smuckler and M. Koplitz, *BBRC* **55,** 499 (1973).
284. A. V. Lichtenstein, R. P. Alechina and V. S. Shapot, *Eur. J. Cancer* **14,** 939 (1978).
285. D. E. Schumm and T. E. Webb, *in* "Nuclear Envelope Structure and RNA Maturation" (E. A. Smuckler and G. A. Clawson, eds.), UCLA Symposium Vol. 26, p. 483. Alan R. Liss, New York, 1985.
286. D. E. Schumm and T. E. Webb, *BBRC* **67,** 706 (1975).
287. G. A. Beltz and S. J. Flint, *JMB* **131,** 353 (1979).
288. A. Therwath and K. Scherrer, *PNAS* **75,** 3776 (1978).
289. W. F. van Voorthuizen, C. Dinsart, R. A. Flavell, J. J. M. DeVijlder and G. Vassart, *PNAS* **75,** 74 (1978).
290. A. V. Lichtenstein and V. S. Shapot, *BJ* **159,** 783 (1976).
291. G. Blobel, *PNAS* **82,** 8527 (1985).
292. R. Breathnach, J. L. Mandel, P. Gerlinger, A. Krust, M. LeMeur, P. Humphries, M. Cochet, F. Gannon and P. Chambon, *in* "Genetic Engineering" (N. W. Boyer and S. Nicosia, eds.), p. 77. Elsevier/North-Holland, Amsterdam, 1978.

Foreign Gene Expression in Plant Cells

PAUL F. LURQUIN

Program in Genetics and Cell Biology
Washington State University
Pullman, Washington 99164

Scientific breakthroughs can sometimes be dated with great accuracy. Inasmuch as plant-cell genetic transformation belongs to the category "discoveries," one might wonder whether the demonstration of this phenomenon can be precisely localized in time. Reflecting back on the 10 years or so since the publication of the first truly critical articles reviewing the topic (*1*, *2*), the answer to this question must be a negative one. The road to plant transformation has been tortuous, dotted with dead ends; in fact, it is characteristic of scientific progress in general.

Quite certainly, this discovery has had a very long gestation period marked by false hopes, lukewarm acceptance of perfectly reasonable data, heated debates, and finally the demonstration that after all, plant cells do not behave so differently from mammalian cells when supplied with purified DNA under the right conditions.

In sum, plant cell genomes display a definite propensity to tolerate, recombine, and express genes of foreign origin.

The objective of this review is to show how the above conclusion came about, and what experimental systems can be used to alter plant cells genetically through uptake of exogenous DNA. The use of these techniques in the study of gene regulation is also discussed, as well as some of the practical applications of plant genetic engineering.

Progress in Nucleic Acid Research
and Molecular Biology, Vol. 34

As already said, it would be foolish to attempt to pinpoint exactly the studies that demonstrated conclusively that genes present outside plant cells can cross several biological barriers and become established inside plant cell nuclei. Nevertheless, it is quite clear that without the elucidation of the dicotyledonous plant disease known as crown gall, the genetic manipulation of plant cells through DNA insertion would not be what it is today.

Indeed the phenomenon of genetic transformation of plant cells has generated an enormous amount of controversy, starting a little over a decade ago, and has only very recently gained wide acceptance. That this happened certainly derives from the discovery that gene transfer can occur naturally between bacteria belonging to the genus *Agrobacterium* and many dicotyledonous plants (thereby causing crown gall) and a few monocots. It is therefore not surprising that plant cell transformation by foreign genes was first demonstrated through the use of a natural vector and a natural gene transfer system, namely the Ti plasmid present in its host, *Agrobacterium*.[1]

I. The Crown-Gall Saga

It is not the purpose of this article to review all that has been done and said about the molecular biology of crown gall, this field having been recently and authoritatively reviewed (3). Suffice it to say that this neoplasmic disease of plants is caused by the transfer to and integration within the plant cell genome of a portion—the T-DNA (T-transferred)—of a large plasmid (pTi) harbored by *Agrobacterium tumefaciens*. The T-DNA, whose length is approximately 20 kilobasepairs (kbp) contains genes coding for the production of phytohormones. When these genes are integrated and expressed in a host plant cell, the hormonal balance of the latter is disrupted and non-self-limiting proliferation of undifferentiated or incompletely differentiated tissue occurs. In addition, the T-DNA carries genes coding for the production in plant cells of specific compounds, opines, not found in normal plant tissues. Thus, *Agrobacterium tumefaciens* can be considered a natural plant genetic engineering system since it can transfer functional genes to plant cells. Therefore, the question was asked whether the T-DNA could be manipulated in such a way that, for example, foreign genes inserted within it could also be transferred to plant cells. The bacterial transposon Tn7 inserted into the T-DNA of

[1] See Section IX (Glossary) for definitions of terms and abbreviations used in this article.

A. tumefaciens T37 can be detected in tumors incited by this bacterial strain in tobacco plants (*4*). Thus, it was shown that the Ti plasmid may be used as a vector to introduce and stably maintain foreign genes in higher plant tissues. However, expression of these genes had yet to be demonstrated.

This was accomplished once the functional organization of the T-DNA was understood and eventually the knowledge gained made it possible to build a disabled (nononcogenic) Ti plasmid still able to transfer and integrate foreign genes into plant cells but no longer causing undifferentiated cell proliferation. In parallel, plant expression vectors containing 5′- and 3′-control sequences recognizable by plant RNA polymerase II (responsible for mRNA synthesis) were being constructed. The necessity of such plant expression vectors became apparent when it was demonstrated that genes isolated from prokaryotes or other eukaryotes such as yeast (*4a*) and animal cells (*4b*) were either not at all or incorrectly transcribed in plant cells. This phenomenon is presumably due to the absence in plants of specific transcription signals or elements required for heterologous gene expression. These expression vectors, into which desirable coding sequences had been cloned, could then be recombined *in vivo* with complete or deleted Ti plasmids via homologous recombination within the T-DNA. *Agrobacterium*-mediated transfer of this engineered T-DNA region still occurred normally.

Cloned plant control sequences usually separated by a short polynucleotide containing one or several restriction sites useful for cloning foreign coding sequences were also used in direct gene transfer experiments. This approach did not rely either on *A. tumefaciens* as a biological transfer system or on the ability of the T-DNA to become integrated within a plant genome. Rather, these plant expression vectors had little or no homology with the T-DNA. Interestingly, these plasmids became randomly integrated in the host plant genome, just as well as A. tumefaciens T-DNA. Thus, it turned out that nonliving delivery systems were also able to achieve genetic transformation of plant cells, and in some cases quite efficiently. These two approaches are discussed in detail in this article.

Figure 1 clearly shows the enormous interest in the field during the decade 1977–1986.

II. General Nature of Plant Expression Vectors

As indicated above, the integrative property of the *A. tumefaciens* T-DNA makes it a candidate of choice in order to maintain genetic

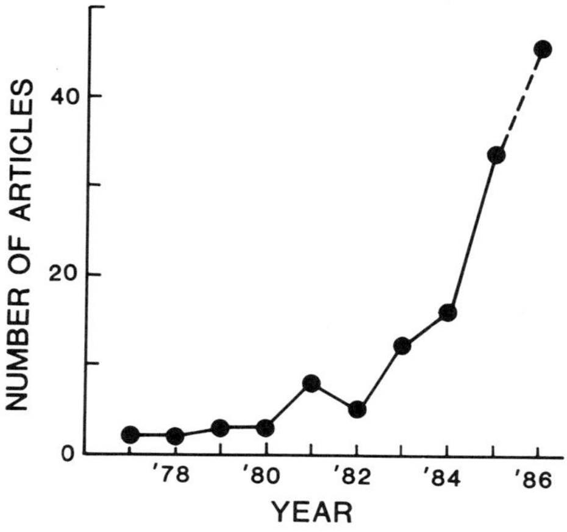

FIG. 1. Number of published research articles in which topics such as DNA transfer to protoplasts, plant selectable markers, plant cell transformation, etc. are investigated. Papers dealing strictly with the molecular biology of crown gall are not included in these figures. The 1986 data point is a projection based on the number of articles published in the first 6 months of that year. It is assumed that during the second half of 1986 researchers will be as prolific as in the course of its first half.

markers stably within transformed plant cells. However, there may be cases where it is desirable to include the foreign genetic information in an episomal vector, i.e., one whose fate is not necessarily to become covalently integrated in the plant genome. Such a vector exists in the double-stranded circular DNA genome of the cauliflower mosaic virus, and potentially in the single-stranded DNA genome of geminiviruses. The rationales used to develop these two types of vectors from these systems are described below.

A. Sequencing of the T-DNA; Assignment of Control Regions

The complete nucleotide sequence of the T-DNA from the octopine[2] tumor-inducing *A. tumefaciens* (pTi15955) is known (5), as is the primary structure of the TL region of the *A. tumefaciens* octopine plasmid pTiAch5 (6) and the sequence of the nopaline[3] synthase (EC 1.5.1.19) gene (*nos*) of pTiC58. Sequence determination coupled with S1 nuclease (EC 3.1.30.1) analysis have allowed the identification of mRNA initiation and poly(A) addition signals of T-DNA genes with

[2] Octopine is N^2-(1-carboxyethyl)arginine. [Eds.]

[3] Nopaline is N^2-(1,3-dicarboxypropyl)arginine. [Eds.]

great accuracy, thereby allowing the cloning of both promoter and terminator regions useful in the design of plant expression vectors. However, the most popular plant vectors based on T-DNA control regions contain the *nos* promoter and terminator. Conceivably 5′- and 3′-flanking regions corresponding to other T-DNA genes could also be used in the construction of plant expression vectors. Nevertheless, the *nos* promoter is among the two or three most active in the T-DNA (*8*); in addition, it lies immediately next to the right T-DNA border, which plays a crucial role in pTi transfer and T-DNA insertion (*3*).

The independently integrating TR-DNA of octopine Ti plasmids is known to code for five different transcripts (*9*). Interestingly, two of those transcripts are under the control of a dual promoter fragment about 500-bp long that directs transcription in opposite orientations. Chimeric genes based on this promoter fragment are expressed in plants in either direction (*10*). The advantages of such dual promoters in the construction of plant selection–expression vectors are clear: they allow coupled expression of a selectable and a nonselectable marker in plant transformants and thus allow isolation of transgenic plants containing genes for which there is presently no phenotypic selection system.

A conceptually similar approach based on sequence and S1 nuclease protection data has been used (*11*) to build other chimeric genes present in the TR-DNA portion of *A. tumefaciens* pTiA6.

B. Sequencing of the Cauliflower Mosaic Virus (CaMV) Genome; Determination of Open Reading Frames

The molecular biology of CaMV has been concisely reviewed (*12*). Sequencing has shown that this double-stranded DNA genome of approximately 8 kbp possesses eight potential reading frames. There is good evidence that this molecule is replicated by reverse transcription of a 35-S RNA template. The origin of transcription of this template has been mapped as well as that of gene VI, which codes for a 19-S RNA species known to be responsible for the synthesis of cytoplasmic inclusion bodies. The 35-S and the 19-S RNAs share a poly(A) addition signal. Insertion of an *E. coli lac* operator fragment into a nonessential (gene II) region of the CaMV genome does not impair viral multiplication (*13*); the piece of foreign DNA was stably maintained after several transfers, but was eventually lost. In a subsequent study, the coding sequence of gene II was replaced by a bacterial gene coding for a methotrexate-resistant dihydrofolate reductase (EC 1.5.1.3) without interfering with viral replication (*14*). Furthermore, the chimeric gene was expressed in infected turnip plants at the protein level and conferred methotrexate resistance. This chimeric gene was stably maintained for up to three cycles of infection.

Some of the drawbacks to using whole or deficient CaMV DNA as a plant vector are that the virus has a narrow host range, has limited packaging ability, and is not known to become integrated within the host genome (which in itself can be an advantage, depending on one's objectives).

The development of direct gene transfer techniques (see Section V) has prompted the construction of plant expression vectors based on the use of CaMV promoters and terminators but no longer relying on the ability of the virus to systemically infect plants. Basically, these vectors comprise the promoter of either gene VI or the one controlling the synthesis of the 35-S RNA. That these promoters are active in determining the expression of foreign genes in plant cells was initially demonstrated by inserting the chimeric genes into the *A. tumefaciens* Ti plasmid and relying on the bacterium-mediated transfer (*15, 16*).

Subsequently, more versatile plant expression vectors, such as pDOB612 and pCIB710 (both containing cloning sited positioned between CaMV promoters and the polyadenylation signal), were constructed. pDOB612 (*17*) is a pBR322 derivative equipped with a sequence recognized by *Sma*I and *Bam*HI, positioned just downstream from the gene VI promoter, while pCIB710 (a CIBA-GEIGY vector) is a pUC19 derivative with a single *Bam*HI site located next to the 35-S promoter. Both vectors contain the shared gene VI/ 35-S RNA terminator sequence.

C. Structure of Geminivirus Genomes

To date, the potential of geminivirus DNAs as plant vectors has not yet been demonstrated. In fact, the molecular biology of these viruses is just beginning to be understood. Two examples of such viruses are briefly discussed here.

The two single-stranded circular DNA molecules composing the genome of tomato golden mosaic virus (TGMV) have been sequenced (*18*), as has the bipartite genome of cassava latent virus (CLV) (*19*). The TGMV and CLV genomes both show a total of six conserved open reading frames distributed over the (+) and (−) strands of the two components. Putative 5′-control regions and poly(A) addition signals have been identified and could possibly be used in the construction of plant expression vectors.

However, it is not at all clear whether geminiviruses would have significant advantages over CaMV. For example, it is not known whether foreign inserts will or will not annihilate virus multiplication.

Here also, there is no evidence that geminivirus DNA (in its double-stranded form) can recombine with the host genome despite a nuclear replication phase. Interestingly, however, it has been demonstrated that the A component of TGMV and component 1 of CLV are capable of independent replication in transgenic pTi-transformed petunia plants (*20*) or transfected tobacco protoplasts (*21*), indicating that these single components could be used alone in gene transfer experiments.

Even more interesting is the finding that an independently replicating A component does not produce any symptoms of infection in transgenic plants (*20*).

If it turns out that foreign gene insertion into this component can be achieved without perturbing its replication, it should be possible to use it as an independently replicating and nonpathogenic vector.

III. Dominant Selectable Markers Used in Transformation Studies

Powerful methods of selection are required to detect a small number of transformed cells from a large number of untransformed ones. Dominant selectable marker genes that have no equivalent function in host cells and confer an easily recognizable phenotype upon transformants are best suited for such studies.

Many such markers, including phytohormone independence genes encoded by *Agrobacterium* T-DNA as well as a series of bacterial coding sequences determining antibiotic resistance via enzymatic modification of the latter, are presently available. The use of a gene determining the synthesis of a methotrexate-insensitive dihydrofolate reductase has also been considered. As far as the first example is concerned, it is well known that normal plant cells are unable to grow in tissue culture in the absence of phytohormones, whereas acquisition of T-DNA through agrotransformation makes them hormone-autotrophic. Thus, selection on hormone-free medium was used (*22*) to demonstrate for the first time that cell-wall-regenerating tobacco protoplasts could be transformed *in vitro* by cocultivation with *Agrobacterium* cells.

This method of selection has also been used successfully by others in transformation experiments (see Section IV,C). Plant cells can sometimes spontaneously acquire a hormone-independent phenotype (a phenomenon called habituation), which constitutes a drawback in the use of this particular selection technique. However, since the T-DNA also carries one or several genes coding for easily assayable opines, it is not difficult to differentiate true from false transformants. The biggest problem here is that a T-DNA-induced hormone-independent phenotype renders plant cells extremely recalcitrant to regeneration and therefore precludes the production of transformed normal plants. It follows that this selection technique is not used in cases where plant regeneration is sought.

Plant cells are sensitive to certain antibiotics, such as chloramphenicol (*23*), those belonging to the aminoglycoside family (kanamycin, G418, neomycin) (*24, 25*), methotrexate (*26*), and hygromycin B, an aminocyclitol antibiotic (*27*). Bacteria contain genes that confer resistance to these compounds. For example, transposon Tn9 carries a gene coding for chloramphenicol acetyltransferase (*cat*) while transposons Tn601 and Tn5 carry genes coding for aminoglycoside phosphotransferases APH(3′)I and APH(3′)II, respectively (*neo*). Furthermore, transposon Tn7 contains a gene coding for a methotrexate-resistant dihydrofolate reductase (DHFR MtxR). Finally, a strain of *E. coli* contains a gene conferring resistance to hygromycin B by directing the synthesis of a hygromycin phosphotransferase (APHIV).

All the above genes have been sequenced and their coding regions used to construct plant selection vectors after insertion between appropriate promoters and terminators.

Expression of these chimeric genes was initially demonstrated (*27–32*) in crown gall cells: intermediate vectors containing the bacterial coding sequences flanked by T-DNA 5′- and 3′-control regions were conjugated into *A. tumefaciens* that usually contained an intact Ti plasmid. *A. tumefaciens* exconjugants carrying a recombinant Ti plasmid following single or double crossover with the intermediate vector were selected and used to induce crown gall on suitable hosts. Tumor cells expressed the chimeric genes as evidenced either by enzyme assay, growth on the relevant antibiotic, specific mRNA production, or a combination of these techniques. In addition, the presence of these genes in the transformed tissues was ascertained by Southern blot hybridization. It should be noted that the strategy used to demonstrate transformation of tobacco cells to hygromycin B resistance was slightly different from that used by earlier investigators. In this case, the coding sequence of the APHIV gene cloned between appropriate control regions was transferred via a binary vector system (see Section IV,B).

Table I summarizes strategies and systems used initially to demonstrate selectable gene expression in plant cells.

At the present time, selection on kanamycin rather than on any other antibiotic or on hormone-free medium is favored by most investigators.

TABLE I
CHIMERIC GENES FOR ANTIBIOTIC RESISTANCE EXPRESSED IN PLANT TISSUES[a]

Marker	Control regions	Vector	Phenotypic expression	Host
cat	Pnos + Tnos	pTiC58	Enzyme activity	Tobacco (*28*)
neo	Pnos + Tnos	pTiC58	Enzyme activity Kanamycin resistance	Tobacco (*29*)
DHFR MtxR	Pnos + Tnos	pTiC58	Methotrexate resistance	Tobacco (*29*)
neo	Pnos + Tnos	pTiB6S3	mRNA Kanamycin resistance	Petunia (*30*)
neo	Pnos + Tnos	pTiT37	mRNA G418 resistance	Tobacco (*31*)
APHIV	Pocs + Tnos	Binary	Hygromycin resistance	Tobacco (*27*)
APHIV	Pnos + Tnos	Binary	Hygromycin resistance	Tobacco (*32*)

[a] Pnos, 5′-control region of the nopaline synthase gene containing the promoter sequence. Pocs, 5′-control region of the octopine synthase gene containing the promoter sequence. Tnos, 3′-control region of the nopaline synthase gene containing the polyadenylation and termination signals. Binary, indicates the use of a binary vector system where the chimeric selectable marker is present between T-DNA borders, in a wide host range plasmid that does not contain the *vir* region. DHFR, dihydrofolate. MtxR, methotrexate resistance.

IV. *Agrobacterium*-Based Transformation Systems

This section emphasizes the great flexibility of *Agrobacterium* as a gene delivery system and describes strategies designed to produce nononcogenic Ti plasmids.

A. Disarming the T-DNA: Deletion of Oncogenes and Regeneration of Transformed Cells

As seen in the previous section, chimeric gene expression was initially demonstrated in crown gall tissues incapable of regeneration into plants. From both a practical and a fundamental point of view, it is highly desirable to study foreign gene expression in fully differentiated and fertile plants. Such a goal cannot be attained if the transformation vector is T-DNA equipped with a full set of oncogenes.

Basing their strategy on the observation that the T-DNA contains separate loci which prevent either root (*tmr*) or shoot (*tms*) formation (*33*), Barton *et al.* (*4a*) introduced the yeast alcohol dehydrogenase I (EC 1.1.1.1.) gene into the *tmr* locus of *A. tumefaciens* pTiT37. The technique used first involved the cloning of the foreign gene in the middle of a T-DNA fragment encompassing the *tmr* locus, the whole construct being carried by a wide-host-range plasmid. After introducing this recombinant plasmid into *A. tumefaciens* pTiT37 (a nopaline strain), these authors selected exconjugants in which a double crossing-over had occurred between regions of homology located on the resident Ti plasmid and the recombinant molecule carrying the foreign gene. The result of such a double crossing-over was to specifically replace a complete *tmr* locus by one containing a chimeric gene in its center.

Such an engineered *A. tumefaciens* strain could induce tumorous growth on inverted tobacco stems. Cloned tumorous cells had lost their requirement for auxin but required a cytokinin for growth. Very slow growth was observed on hormone-free medium. Shoots regenerated from transformed tissue formed roots and produced normal, fertile plants. This experiment demonstrated that the inactivation of a single oncogene (gene 4) in the T-DNA *tmr* locus led to the induction of attenuated crown gall tumors from which it was possible to regenerate plants. Moreover, these regenerated plants contained full copies of the engineered T-DNA still harboring the yeast alcohol dehydrogenase gene. The latter could be transmitted to the progeny of self-pollinated transformed plants. Why the products of oncogenes 1 and 2 still present in the *tms* locus of this engineered T-DNA did not interfere with plant regeneration is not understood.

In order to produce a Ti plasmid from which all T-DNA oncogenes have been removed Zambryski *et al.* (35) constructed a T-DNA deletion mutant of pTiC58. For this, they took advantage of a clone, pAcgB, containing only the right and left border fragments of the nopaline T-DNA cloned in pBR322. This plasmid contained a full nopaline synthase gene located immediately adjacent to the right border. After mobilization of this plasmid into *A. tumefaciens* carrying a pTiC58 derivative containing a *neo* gene near the center of its T-DNA and after selection on appropriate antibiotics, a recombinant between pAcgB and the Ti plasmid was isolated. This new recombinant, pGV3850, was shown to have originated from a double crossover event between the right and left T-DNA borders of pAcgB and the right and left borders of the host Ti plasmid. The result of such a double crossing-over was to introduce a large T-DNA deletion into pTi and, in fact, led to the replacement of all oncogenes by pBR322, the cloning vector of pAcgB.

A. tumefaciens (pGV3850) was, as expected, avirulent on a variety of plant hosts. However, when the site of infection on young tobacco plants was excised and cultivated on hormone-containing medium, callus cells expressing the nopaline synthase gene were recovered. In addition, these callus cells could be regenerated into plants still expressing the *nos* gene and containing integrated copies of the deleted pGV3850 T-DNA. These experiments clearly demonstrated the possibility of transferring foreign genes (in this case, pBR322) into plant cells via a completely nononcogenic T-DNA, thereby allowing the full differentiation of transformed cells into transgenic plants.

In addition, the T-DNA structure of pGV3850 made it possible to recombine any foreign gene cloned in pBR322 with the pBR322 portion of the deleted T-DNA by a single crossover event (Fig. 2). This was demonstrated by insertion, by homologous recombination into pGV3850, of several chimeric genes under the control of the *nos* promoter (*36*). Plants regenerated after infection with *A. tumefaciens* (pGV3850::*nos–neo*) or *A. tumefaciens* (pGV3850::*nos–cat*) contained and expressed the foreign genes.

To date, pGV3850 is one of the most versatile pTi-based plant transformation vectors. For example, it has also been used to transfer and express simultaneously in plant cells the *neo* and *cat* genes under

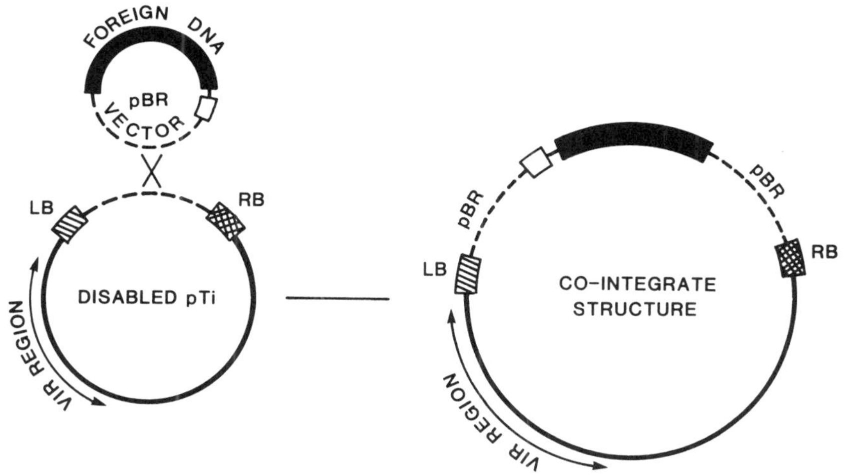

FIG. 2. Schematic representation of a disabled Ti plasmid accompanied by its intermediate cloning vector. Plasmids are not drawn to scale. The pBR vector contains an origin of replication as well as a *bla* gene. The open box represents an additional marker selectable in *A. tumefaciens*. Since pBR vectors cannot replicate in *Agrobacterium*, exconjugants carrying this marker are the result of a crossover between regions of homology in the pBR-based intermediate vector and the pBR sequence present between the left (LB) and right (RB) T-DNA borders. The closed box represents a foreign gene equipped with a promoter active in plant cells. Other intermediate vectors may contain a *nos* promoter and a *nos* terminator separated by a DNA segment containing one or several unique restriction sites. These vectors are useful in the transfer of promoterless coding sequences. The disabled pTi contains an intact *vir* region and a replication origin. Oncogenes have been deleted. Interpreted from (*35*).

the control of the dual pTiAch5 T-DNA promoter described in Section II,A (37).

Another pTi-derived vector system involves the acceptor Ti plasmid pGV2260, which is an octopine pTiB6S3 derivative from which the T-DNA has been completely deleted as a result of a double crossing-over with a pBR322-based vector containing complete TL and TR regions plus flanking non-T-DNA sequences. Thus, pGV2260 carries a copy of pBR322 substituting the T-DNA, including its border fragments. The intermediate cloning vector pGV831 is a pBR325 derivative containing an additional streptomycin-resistance gene for selection in *A. tumefaciens,* the left and right border sequences of the TL-DNA flanking a *nos–neo* chimeric gene, and a single *Bam*HI site upstream from the *nos–neo* gene, near the left T-DNA border. This unique site can be used for the cloning of foreign genes (equipped with an appropriate promoter) and transfer to plant cells. Cointegrates between pGV2260 and pGV831 containing a foreign gene can be isolated after mobilization of the latter into *A. tumefaciens* (pGV2260) and are the result of a single recombination event between the homologous portion of the two plasmids, i.e., pBR322.

The resulting engineered *A. tumefaciens* cells can be used to infect plants, and selection of transformants can be achieved on kanamycin, due to the presence of the *nos–neo* gene located between the T-DNA borders. Since the *Bam*HI cloning site containing the foreign gene is also present between the T-DNA borders, kanamycin-resistant plants will of course contain the foreign gene itself. This system evidently allows for the isolation of nonselectable markers in transformed plant cells. In addition, since the pBR322 sequence used for homologous recombination is located outside the T-DNA borders, it will not be transferred into plant cells, contrary to the case of pGV3850. This considerably reduces the length of the engineered T-DNA and potentially increases the amount of useful foreign DNA that can be integrated into a plant genome. Since pGV2260 lacks oncogenes, plants can be regenerated from transformed cells.

A less-direct approach was first used to produce transgenic *Nicotiana plumbaginifolia* plants (39). Here also, selection of transformed plant cells was performed on kanamycin medium, after transfer of a *nos–neo* chimeric gene. These authors used the intermediate vector pMON128 notably containing the *nos–neo* gene as well as the right border of a nopaline T-DNA and a region of homology with the left-hand region of the TL-DNA of the octopine pTiA6 plasmid.

After mobilization of pMON128 into *A. tumefaciens* (pTiB6S3) (also an octopine strain) a single crossover event can occur that produces a cointegrate between the two plasmids. Since the intermediate vector contains a nopaline T-DNA right-hand border, the structure of the T-DNA in the cointegrate will be sequentially B6S3 T-DNA left border/*nos–neo–nos* plus T37 T-DNA right border/*tms*/*tmr*/B6S3 T-DNA right border. Thus, this T-DNA contains two right borders and a single left border. Therefore, two integrative events in plant cells are possible: the "long" transfer which will involve the left and right borders of the pTi B6S3 T-DNA while the "short" transfer will be limited to the left border of the pTiB6S3 T-DNA and the right border of the pTiT37 T-DNA.

Evidently, the "long" transfer will result in the formation of kanamycin-resistant crown-gall cells (since the *tms* and *tmr* loci will also be transferred), whereas the "short" transfer will result in the production of kanamycin-resistant nontumorous cells. This was indeed shown to occur, as out of the four kanamycin-resistant calli obtained after infection with engineered *A. tumefaciens*, one turned out to be morphogenetic and regenerated a kanamycin-resistant fertile plant. Later on, a disarmed pTiB6S3, devoid of all oncogenes as well as the right TL-DNA border and the TR-DNA, was constructed (*40*). Thus, the only residual portion of the T-DNA remaining in this plasmid (called pTiB6S3-SE) is the left TL-DNA border. Intermediate vectors such as pMON120 or pMON200, which contain an intact nopaline synthase gene included in the nopaline right T-DNA border, a chimeric *nos–neo* gene (pMON200), and a polylinker region (pMON200), were used to clone desired foreign genes. Since the intermediate vectors contained a short region of homology with the pTiB6S3 T-DNA left border, a single crossover event between pTiB6S3-SE and the intermediate vector generated an artificial T-DNA consisting of pTiB6S3 left border followed by the intermediate vector itself and ending with the right border of the nopaline T-DNA present in this vector.

In summary, nontumorigenic Ti plasmids that can transfer and integrate artificial T-DNAs into a variety of dicot cells have been constructed and are now available. Sophisticated experimental designs make it possible to use pTi as a plant transformation vector despite its large size (about 200 kbp) and the impossibility of directly engineering it *in vitro*. The versatility of Ti plasmids totally lacking oncogenes like pGV3850, pGV2260, or pTiB6S3-SE is such that only simple cloning steps of foreign genes into adequate intermediate vectors like pGV2385, pGV831, or pMON200 are necessary to produce a highly efficient plant transformation system with full regenerative capacity.

B. Development of Binary Transformation Vectors

As indicated above, the large size of the Ti plasmid precludes the direct genetic manipulation of its T-DNA region. We have also seen that the production of pTi's where the oncogenes or the whole T-DNA are replaced by pBR322 or one of its derivatives allows the formation of recombinant Ti plasmids with small vectors containing the foreign gene.

However, this is not the only possible approach to generating convenient *Agrobacterium*-based plant transformation vectors. Apart from the T-DNA, another region of the Ti plasmid, the *vir* region, is necessary for tumor induction in plants (*3*). The *vir* regulon presumably codes for, among other things, functions involved in the formation of linear single-stranded T-DNA intermediates found in *Agrobacterium* cells stimulated by acetosyringone[4] or α-hydroxyacetosyringone (*41*). These molecules are thought to be intermediates in DNA transfer from *A. tumefaciens* cells to plant cells (*42*).

If the *vir* genes are capable of acting in trans on the T-DNA, one could develop a binary plant transformation system in which the *vir* region (but not the T-DNA) would be present on one plasmid, whereas the T-DNA (but not the rest of the Ti plasmid) would be present in a second replicon. In this way, it would be much easier to engineer the T-DNA region, for example in *E. coli*, and reintroduce it into an *A. tumefaciens* strain cured from its intact Ti plasmid and harboring instead a plasmid containing the *vir* region and the origin of replication of pTi. Indeed, the *vir*-carrying plasmid pAL4404 can achieve the transfer of the pTi Ach5 T-DNA region cloned in a wide host range vector (pAL1050) (*43*). *Agrobacterium* cells containing both pAL1050 and pAL4404 are fully virulent on a variety of plant hosts. Neither plasmid alone is virulent.

This particular strategy has been further developed; shuttle vectors containing a *cos* site for cosmid cloning in *E. coli* as well as the origin of replication of a wide host range plasmid have been constructed (*44, 45*). In addition to the above, these vectors also contain a ColE1 origin of replication and a chimeric *nos–neo* gene for the selection of plant transformants. The *cos* site, the ColE1 origin, and the *nos–neo* gene are present in a DNA sequence, located between T-DNA borders, that contains single restriction endonuclease sites for foreign DNA insertion. No oncogenes are present in these vectors.

Using such a system, An *et al.* (*44*) showed that the *tmr* locus of pTiA6 cloned into one of their shuttle vectors led to the expected

[4] Acetosyringone is 4-acetyl-2,6-dimethoxyphenol. [Eds.]

production of shooty kanamycin-resistant tumors in *Nicotiana glauca* and was thus able to achieve cloned DNA transfer. In addition, the helper Ti plasmid (supplier of *vir* functions) need not be one from which the T-DNA had been deleted. Indeed, 80–90% of the transformants obtained by infecting tobacco cells with *Agrobacterium* harboring a virulent pTi plus their shuttle vector did not become hormone-independent. Therefore, the cotransformation to both hormone independence (mediated by wild-type T-DNA transfer) and kanamycin resistance (mediated by the nontumorigenic shuttle vector) was not sufficient to impair efficient plant regeneration from transgenic cells. Figure 3 is an idealized representation of a binary vector system.

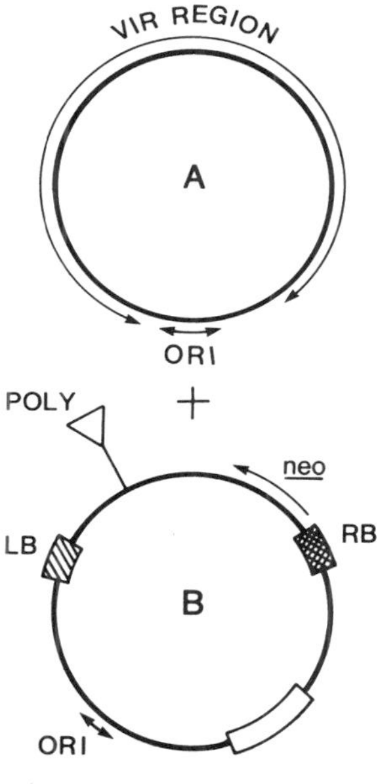

FIG. 3. Schematic representation of a binary plant vector. Plasmids are not drawn to scale. The A plasmid is a severely deleted pTi such as pAL4404 (*43*) still containing its replication origin and an intact *vir* region necessary to effect T-DNA transfer. The B plasmid is the foreign DNA cloning vector containing a wide host range origin (for replication in *Agrobacterium*), a marker selectable in *Agrobacterium* (open box) and the two T-DNA borders (legend as in Fig. 2). In this example, the *nos* gene coding sequence has been deleted and replaced by *neo*, now under the control of the *nos* promoter. This marker is used for the selection of plant transformants. "Poly" represents a polylinker region used for the cloning of foreign DNA. If promoterless genes are to be cloned, the polylinker region can be flanked by a promoter and terminator such as those from the *nos* gene. Plasmid A can be replaced by normal, virulent pTi. Interpreted from (*43*) and (*44*).

As it now stands, the binary system has no definitive advantages over Ti plasmids from which oncogenic functions have been deleted. Both approaches require cloning in *E. coli* of the foreign DNA in either a shuttle vector as above or an intermediate vector such as pMON200. Both types of vectors must then be conjugated or transformed into an appropriate *A. tumefaciens* strain selected for either the presence of the shuttle vector or that of a pTi::intermediate vector cointegrate.

The decision to use one versus the other system to transform plants may well be largely a question of taste at present.

C. *Agrobacterium tumefaciens* as a Gene Delivery System *in Vivo* and *in Vitro*

Not only has it been possible to do elegant genetic manipulations with the T-DNA of *Agrobacterium*, but *A. tumefaciens* cells turned out to contain an exquisitely flexible gene delivery system, as explained below.

1. Gene Delivery in Whole Plants

In nature, *A. tumefaciens* incites crown gall by colonizing a wound site usually situated, as the name indicates, at the crown of the plant. Evidently, such a process can easily be reproduced in the laboratory, and *A. tumefaciens* will indeed produce tumors on a variety of wounded tissues, such as stems, leaves, and roots. Therefore, it is not surprising that this technique was used in the first demonstration of T-DNA-mediated foreign gene transfer into plant cells. For example, the initial demonstrations that functional chimeric genes could be found in crown-gall tissue were achieved with wounded plants infected with engineered virulent *A. tumefaciens* (*27–31*).

The same experimental procedure was used with avirulent *A. tumefaciens* containing truncated T-DNA to demonstrate plant regeneration and maintenance of the disabled T-DNA (*35*). In this case, since no tumors were formed, the primary wound sites on decapitated tobacco and petunia plantlets were excised and cultivated on hormone-containing media.

2. The Leaf Disk Transformation Method

Large-scale experiments involving wounded or decapitated plantlets can be cumbersome. Basing their approach on the fact that explants of the leaves of many plant varieties show regeneration capabilities, Horsch *et al.* (*46*) designed a leaf disk/*Agrobacterium* cocultivation technique that allows rapid regeneration of transgenic plants. Disks are punched from surface-sterilized leaves and incubated with an avirulent *A. tumefaciens* strain harboring a disabled Ti plasmid whose T-DNA has been replaced by a selectable marker (such as *nos–neo*) flanked by T-DNA borders. After washing, the leaf disks are directly transferred to a shoot-regeneration medium. Selection on the antibiotic is performed 2 or 3 days after transfer. With this technique, well-developed transformed shoots were seen on *Petunia hybrida* after 3 weeks. Rooting of these shoots on an appropriate medium was observed 2 weeks later. Similar results were obtained with *Nicotiana tabacum* (*46*) and a number of commercial tomato varieties (*47*).

Because of the limited number of manipulations involved and the small space needed to accommodate leaf disks, this technique has very good potential for the regeneration of large numbers of plants, and, in fact, large-scale experiments will be necessary to "shotgun" nonselectable and ill-identified genes from one plant species to another.

3. Cocultivation with Cell-Wall-Regenerating Plant Protoplasts

It is often important—and sometimes crucial—in experiments of the type described in this article to be able to isolate clones of cells originating from single transformation events. The two transformation techniques described above do not lend themselves easily to such cloning since many cells clustered in the same area of a plant tissue are simultaneously subjected to infection by *Agrobacterium*.

A more satisfactory approach would involve infection of isolated plant cells or protoplasts, preferably the latter, given their decreased tendency to clump. That *Agrobacterium* can indeed transfer T-DNA in such an *in vitro* system was first demonstrated by Marton *et al.* (*22*), who incubated cell-wall-regenerating tobacco mesophyll protoplasts and virulent or avirulent *A. tumefaciens* together. Hormone-independent clones could be selected only when pathogenic bacterial strains were used. The presence of octopine synthase (EC 1.5.1.11) in many of the hormone-autotrophic calli ascertained their transformed nature.

This cocultivation technique was later extended by others to *Petunia hybrida* (*49*) and *Hyoscyamus muticus* (*50*) mesophyll protoplasts and has now gained wide acceptance as an extremely efficient plant cell transformation technique.

Improvements in the plating and selection methods (*51*) allowed transformation frequencies as high as 10% or more of the surviving protoplast population to be achieved. When dominant selectable markers in plants became available, the cocultivation procedure was extensively used to isolate and characterize dozens of independently obtained transformants after T-DNA transfer in single or binary vector systems (*29, 30, 32, 36–38, 44*).

Aside from being a very efficient transformation system, the cocultivation technique has been used in an attempt to understand the early events in the transfer and integration of the T-DNA into plant DNA (*52*).

4. Fusion of Protoplasts with *Agrobacterium* Spheroplasts

Whenever plant protoplasts are involved, scientists have an immoderate tendency to fuse them with themselves, with other cells, and sometimes even with strictly nonbiological materials. Whether these operations should be called fusion or endocytosis is still an issue that is not relevant to the present topic.

A variation of the cocultivation technique involving freshly isolated protoplasts and *Agrobacterium* spheroplasts has been devised. To improve early *in vitro* transformation frequencies of protoplasts, it was reasoned that the efficiency of Ti plasmid transfer could possibly be increased by protoplast–spheroplast fusion (*53*). This method does not rely on the ability of the bacterial cells to adhere naturally to a newly sythesized plant cell wall. *Vinca rosea* protoplasts isolated from suspension cultures did indeed acquire a hormone-independent phenotype coupled with octopine production in some cases. Even though the transformation frequences achieved (*53*) were higher than previously reported (*22*), they were several orders of magnitude lower than the current, optimized values attained through cocultivation of protoplasts and bacteria.

Transformation by fusion with *Agrobacterium* spheroplasts was confirmed by the transformation of tobacco mesophyll protoplasts to kanamycin resistance after T-DNA-mediated *nos–neo* transfer (*54*).

From a practical point of view, it does not seem that the fusion technique should supersede the cocultivation method, given its much lower efficiency. However, one of the most severe limitations of the *Agrobacterium* system is that none of the commercially important monocotyledonous plant species can be infected by this bacterium. Nevertheless, flow cytometry shows that *Agrobacterium* spheroplasts can become tightly bound to protoplasts isolated from monocots (*55*). Hence, rice protoplasts isolated from suspension cells—and having the ability to undergo cell division—did acquire a hormone-independent phenotype after fusion with virulent octopine or nopaline *A.*

tumefaciens spheroplasts (*56*). Among these hormone-autotrophic lines, about 20% produced opines. It is of paramount importance to repeat these experiments with a dominant selectable marker to ascertain that this valuable monocotyledonous crop species can indeed be transformed by *Agrobacterium*.

5. Cocultivation with Callus Cells

Plant protoplast transformation gains its full significance only if the latter can undergo sustained division. It is common knowledge that this point still presents nagging problems in a very large number of plants. On the other hand, totipotent callus or suspension cultures have been derived from many plant species. Thus, it was of interest to determine whether or not such cells could also be transformed by cocultivation with *Agrobacterium*. Crown gall transformation at low frequency of *N. tabacum* callus cells was first demonstrated by experiments in which calli were squashed, coincubated with *A. tumefaciens*, and plated on hormone-free medium (*57*). The transformant nature of some of the hormone-autotrophic clones selected was ascertained by the presence of octopine. [The low transformation efficiency (10^{-5}) was possibly due to the limited access of plant cells to bacteria and to the heterogeneous physiological state of the cells in the tissue.] These results were confirmed by the demonstration of crown-gall transformation (at a frequency of 10^{-4}) of haploid *N. plumbaginifolia* cells growing in fine suspension cultures (*58*). Finally, rapidly growing *N. tabacum* suspension cells were efficiently transformed (50% transformation efficiency) in mid-logarithmic phase by means of a supervirulent *A. tumefaciens* strain harboring a *nos–neo* binary vector and a helper Ti plasmid (*59*).

It would be interesting to see whether callus cells derived from a genus other than *Nicotiana* can also be transformed by cocultivation.

D. Gene Delivery Mediated by *Agrobacterium rhizogenes*

A. rhizogenes is a close relative of *A. tumefaciens*. Contrary to the latter, however, *A. rhizogenes* does not necessarily induce undifferentiated tumorous growth on wounded dicots. Rather, it incites root proliferation at wound sites, hence its name. Root induction occurs through the transfer of a T-DNA region carried by a large Ri plasmid (*60, 61*). Moreover, root cells harboring the T-DNA can readily be regenerated into fertile plants capable of sexually transmitting the T-DNA (*62*). Therefore, *A. rhizogenes* transformation does not present the problems that had to be surmounted in the disabling of *A. tumefa-*

ciens pTi. Thus, the *A. rhizogenes* T-DNA (which bears little resemblance to the *A. tumefaciens* T-DNA) is a good candidate to achieve easy gene transfer into plants.

Surprisingly little work has been done along these lines and it is only very recently that *A. rhizogenes*-based transformation systems have been explored. *Solanum nigrum* mesophyll protoplasts can be transformed by pRI T-DNA after cocultivation with *A. rhizogenes* (*63*). Moreover, the transformed tissues could be regenerated into plants at very high frequency (70%). There is no reason to doubt that this cocultivation technique can be extended to other plants susceptible to *A. rhizogenes*.

The Ri plasmid as a vector in gene transfer experiments was used to study the regulation of a chimeric soybean leghemoglobin gene in transgenic lotus (*corniculatus*) nodules (*64*). For this, the authors first constructed intermediate vectors containing (1) a region of homology with *A. rhizogenes* 15834 TL-DNA, (2) a chimeric *cat* gene containing the coding region of the chloramphenicol acetyltransferase gene flanked on one end by a 2-kb promoter region from the soybean *1bc3* gene and on the other end by a soybean *1bc3* gene 3′-downstream sequence, and (3) antibiotic markers for the selection of transconjugant bacteria. These intermediate vectors were thus conjugated into *A. rhizogenes* 15834 and transconjugants containing a cointegrate between pRi15834 and the intermediate vectors were isolated. These modified bacteria were then used to incite root formation on wounded lotus plants. Such roots were subsequently regenerated into whole plants that were used for inoculation with *Rhizobium loti*, a specific lotus symbiont. Results showed that the chimeric *1bc3-cat* gene was specifically expressed in the nodules induced by *Rhizobium* on the roots of the transgenic plants, at least at the transcription level. This type of study opens the way for the analysis of genes specifically expressed either in roots or in root nodules.

V. Direct Gene Transfer

This section reviews the studies that culminated in the demonstration that plant protoplasts can be efficiently transformed without the help of a natural gene carrier (such as *Agrobacterium*), and that integration of dominant selectable markers into plant genomes can occur without the help of a natural vector (the T-DNA border fragments).

The fact that higher plant protoplasts are permeable to recombinant DNA vectors was shown a decade ago (*65, 66*). Subsequently, *Petunia* mesophyll protoplasts were transformed into hormone-autotrophic cell lines, at low frequency, after incubation with a pTiAch5 DNA/poly-L-ornithine complex (*67*, *68*). The detection of octopine in some of these clones indicated that they were bona fide transformants. In the following years, a variety of DNA uptake techniques clearly demonstrated that DNA can be used to achieve *in vitro* genetic transformation of protoplasts isolated from both dicotyledonous and monocotyledonous plants.

A. Liposome-Mediated Transformation

The usefulness of encapsulating nucleic acids into large liposomes for subsequent transfer into protoplasts was shown by Lurquin (*69*). Plasmid DNA trapped in liposomes was well-protected from nuclease attack, and there were good indications that part of this DNA was intracellular after fusogen treatment of a protoplast/liposome mixture. It was later amply demonstrated that large unilamellar liposomes composed of phosphatidylserine and cholesterol could very efficiently transfect plant protoplasts with plant viral RNA [reviewed in (*70*)].

However, it is only very recently that this technique has been applied to effect plant protoplast DNA transformation. Tobacco mesophyll protoplasts were transformed at a frequency of 4×10^{-5} by plasmid pLGV23*neo* encapsulated in negatively charged unilamellar liposomes (*71*). This plasmid contained the chimeric *nos–neo* gene, including the right border of the nopaline T-DNA. The *E. coli* vector became integrated within the genome of the transformed cells and was stably maintained and expressed in regenerated plants. Interestingly, the Southern blot data suggested that at least three copies of pLGV23*neo* were integrated in tandem at a single site in the plant genome. Whether the right T-DNA border present in this vector played a role in the integration process is still an open question. A general methodology for achieving liposome-mediated transfection of protoplasts has been described (*72*).

B. Fusogen-Induced DNA Transformation

The *A. tumefaciens* Ti plasmid is all that is necessary to confer a crown gall phenotype upon plant cells (*73*). To demonstrate this, a tobacco protoplast *in vitro* transformation system based on naked pTi DNA was devised. Protoplasts were treated with a mixture of pTiAch5 DNA, calf thymus DNA as carrier, polyethylene glycol 6000 (PEG), and $CaCl_2$, and then selected for growth on hormone-free medium. Transformation to hormone autotrophy occurred at frequencies of 10^{-6} to 10^{-5}, and several transformed lines contained integrated T-

DNA as well as octopine synthase activity. Thus, *A. tumefaciens* itself is not a prerequisite for tumor induction in plant cells. However, it was also observed that some of the transformants could regenerate shoots (not a normal property of octopine tumors), some did not produce octopine, and some contained integrated pTi fragments not associated with the T-DNA.

Later, the same group extensively analyzed the structure of the T-DNA in several *in vitro* transformants (*74*). Their conclusions are fascinating and can be summarized as follows. Contrary to the very precise integration of the octopine T-DNA in *A. tumefaciens*-induced crown gall cells, which occurs via a 23-bp direct repeat present within the T-DNA borders, the T-DNA of *in vitro* transformants was often shorter, scrambled, or longer than the normal T-DNA. In addition, the variation in phenotypic traits was well correlated with the presence or absence of specific transcripts with known functions. Furthermore, pTi fragments located outside the T-DNA, such as the *vir* region, were also integrated within the genome of some of the transformants. The fact that the 23-bp repeat located within the T-DNA borders played no role in the integration process was further supported by the observation that a cloned T-DNA fragment lacking such a repeat still could transform protoplasts via stable integration. Finally, this report also showed cotransformation of tobacco protoplasts by T-DNA fragments located on separate cloning vectors.

One of the major points of this research is certainly that no special sequences are required for the integration of foreign DNA within the host genome, making the use of T-DNA as a plant vector obsolete in cases where sustained protoplast division is possible.

Confirming the above, carrier calf thymus DNA sequences can be cotransformed with T-DNA markers, obviously in the absence of a particular selection pressure and without any homology with the T-DNA (*75*).

In fact, direct gene transfer to plant protoplasts followed by *A. tumefaciens*-independent integration had actually been demonstrated (*76*). Instead of relying on the upstream and downstream control regions of the T-DNA *nos* gene, these workers constructed a chimeric gene consisting of the *neo* coding sequence flanked by the 5′- and 3′-expression signals of the cauliflower mosaic virus gene VI. This chimeric gene was cloned in the *E. coli* vector pUC8 and this plasmid, now called pABD-I, was transferred into tobacco protoplasts essentially according to the PEG/$CaCl_2$ method described in *73*. Kanamycin-resistant lines recovered at a frequency of about 10^{-5} where shown to express the *neo* gene at the enzyme level as did the whole plants regenerated from these lines.

Integration of the chimeric gene was demonstrated by Southern blot hybridization at a level of 3–5 copies per haploid genome. Restriction endonuclease analysis suggested that integration within the plant genome was random, regarding both the length of the integrated vector and the site of integration.

These results were essentially confirmed by use of a chimeric gene consisting of the *neo* coding sequence flanked by two *nos* promoter regions and an *ocs* terminator (*77*). Tandem integration of the vector was suggested as in *71*. In addition, these authors improved the transformation frequency to a maximum of about 10^{-4} by combining the actions of a fusogen (polyvinyl alcohol), a high pH treatment, and DNA trapping by microcrystalline calcium phosphate.

All these studies left no doubt that tobacco protoplasts can be transformed by direct gene transfer.

However, at this point, DNA-mediated transformation frequencies were still about three orders of magnitude or more below those observed in *Agrobacterium* cocultivation experiments. This problem has been solved in several ways. Synchronized tobacco protoplasts were transformed by a PEG-based treatment at an efficiency of about 3%, i.e., more than 10,000 times more efficiently than the basal level (*78*). Here also, selection was based on the expression of a chimeric *neo* gene as in *77*. The protoplasts were treated with 2,6-dichlorobenzonitrile to prevent cell wall formation, and with aphidicolin to block nuclear DNA synthesis in early S phase. These high transformation frequences were observed only up to 8 hours after release of the aphidicolin block, indicating that only cells in S or M phase were highly competent. Such values compare very favorably with transformation efficiencies obtained by cocultivation with *Agrobacterium*.

Another fact that generated interest in direct gene transfer is that graminaceous monocots, which comprise several important agronomical species, have never been shown to be transformable by *Agrobacterium*. Chimeric *neo* genes have been transferred via PEG-stimulated uptake and stably expressed in protoplasts isolated from cultured cells of *Triticum monococcum* and *Lolium multiflorum*, two *Graminae* (*79, 80*), showing that the *nos* and CaMV VI promoters are active in monocots, and also that the direct gene-transfer technique can some day be applied to the improvement of monocotyledonous crop species.

Also, protoplasts isolated from *Brassica campestris* (a crucifer) were transformed by direct transfer and integration of a modified

CaMV genome (*80a*). This was the first demonstration that dicot cells not related to the family *Solanaceae* can be transformed by a foreign marker gene. In these experiments, the coding sequence of gene VI was replaced by the *neo* coding region to produce a chimeric gene. This CaMV hybrid (pCaMV6km) could no longer infect young plants systemically but was integrated and its *neo* gene expressed after PEG-mediated transfer to *Brassica* protoplasts. This occurred only when the modified genome was complemented in trans by a cloned wild-type CaMV genome. Integration of a single copy of pCaMV6km per diploid genome was detected, in the absence of scrambling or vector rearrangement, which is a rarity in direct gene transfer. This fact was attributed to the low deoxyribonuclease activity found in *Brassica* protoplasts (*66*).

C. DNA Transfer by "Electroporation"

Electroporation was coined (*81*) to designate hole or pore formation in cell membranes after subjecting living cells to a brief electric impulse. According to this model, the electric field induces dipole formation in the hydrophilic portion of membrane lipids. The interactions between opposite charges in adjacent dipoles then led to the formation of a hole in the cell membrane. This phenomenon is reversible. A thermodynamic and kinetic analysis of the opening and annealing processes, in the presence and in the absence of an electric field, respectively, shows that pore formation during the electric discharge is extremely rapid (in the microsecond range) while pore closing after the impulse is much slower (in the second or minute range).

These authors demonstrated that such temporary perturbations of mouse L cells dramatically enhanced DNA transport across the membrane, as shown by a large increase in transformation frequency.

This technique was modified for use with higher plant protoplasts (*82*), and used to show that carrot, corn, and tobacco protoplasts express chimeric genes when DNA uptake is stimulated by an electric pulse of 350 V/cm. This pulse was generated by the spontaneous discharge of a capacitor with an RC constant of 14 msec (calculated according to their data). The chimeric genes used in this study were based on the Tn9 *cat* gene flanked either by the *nos* promoter and terminator or by the CaMV 35-S promoter and the *nos* terminator. The *nos* promoter was much weaker than the CaMV 35-S promoter in directing *cat* gene expression in corn protoplasts. Our observations also show that after PEG-mediated transfer, a chimeric *nos–cat* gene is expressed in wheat and corn mesophyll protoplasts much more weakly than in *N. plumbaginifolia* mesophyll protoplasts (C. Paszty and P. F. Lurquin, unpublished).

There is stable transformation of a callus-forming line of corn protoplasts after electroporation at 500 V/cm with RC constants of 2 or 4 msec (*83, 84*), shown with a chimeric CaMV 35-S promoter-*neo–nos* terminator gene cloned in the *E. coli* vector pUCPiAN7. Transformants, selected on kanamycin, appeared at a frequency of up to 1% of the dividing cells. Vector integration of up to five copies per haploid genome was detected by Southern blot hybridization, as well as the presence of rearranged copies of the chimeric gene. These results demonstrated the superiority of the electroporation technique over all other means of directly transferring DNA into plant protoplasts, at least in the absence of synchronization (*78*).

Carrot protoplasts electroporated in the presence of pTiC58 at up to 3.8 kV/cm with an RC constant of 6 μsec can be transformed to hormone independence and nopaline production without losing the ability to regenerate teratomas by somatic embryogenesis, with transformation frequencies of 1–2% (*85*).

Finally, in the most comprehensive study so far, Shillito *et al.* (*86*) designed an "electroporation" method producing transformation frequencies ranging between 1 and 3%. Instead of using the low voltage/high RC constant system developed by Fromm *et al.* (*82*) they used three pulses at 1.25 kV/cm with an RC constant of 10 μsec. These parameters were calculated from the theoretical equations derived in *81*. Optimal transformation efficiency was observed at a pABD I (the selection vector carrying a chimeric *neo* gene) concentration of 1–2 μg/ml in the presence of 50 μg/ml carrier calf thymus DNA and 7.8% PEG 6000. They also observed that protoplasts given a 5-minute heat-shock at 45°C yielded 5 to 20 times more transformants than untreated ones. Also, linearized pABD I DNA was more efficient than its supercoiled counterpart.

Clearly, the electroporation technique is almost as efficient as the *Agrobacterium* cocultivation procedure. It is even possible to isolate transformants without selection, simply by randomly picking and testing enough electroporated clones propagated on nonselective culture medium (*86*). This observation makes it reasonable to envision experiments in which genomic plant DNA from a heterologous source is used to transfer as yet unidentified genes known, for example, to code for disease or herbicide resistance. In such a scenario, cloning in bacteria of specific DNA fragments would be obsolete. However, "brute force" field trials would remain necessary, as they are in classical plant breeding.

Some questions regarding the electroporation procedure remain to be solved. For example, Langridge *et al.* (85) and Shillito *et al.* (86) reported similar transformation frequencies at high voltage and low RC constant even though the former did not use PEG, found to be critical by the latter. Again, almost identical frequencies were obtained by Fromm *et al.* (83) in the absence of PEG and at a lower field strength with a very much higher RC constant. It is possible that a different mechanism of DNA uptake might take place, depending on the conditions used. Alternatively, the source of protoplasts and/or the isolation procedure might play an important role.

D. DNA Microinjection

A 30% protoplast transformation frequency by direct gene transfer is quite satisfactory. Nevertheless, this remarkably high value (let us remember that until the middle of 1985, the figures were stalled at about 10^{-4}) has been eclipsed by the 14% transformation reported by Crossway *et al.* (87). This record values was not obtained without difficulty, being the consequence of directly microinjecting DNA into plant protoplast nuclei. When DNA was microinjected into the cytoplasm, the frequency reached 6%. The vector used by these authors consisted of a bacterial sequence present between pTiA6 border fragments and cloned in a wide host-range vector. Protoplasts microinjected with this DNA were cultured in microdrops and their total DNA analyzed for the presence of integrated copies of the artificial T-DNA. No selection pressure was exercised. As noted above, a high proportion of the microinjected cells did contain integrated foreign DNA (a maximum of two copies per genome), although, as observed previously, not necessarily via the T-DNA borders.

Certainly, but perhaps temporarily, this technique yields the highest transformation frequencies obtained so far by direct gene transfer. On the other hand, given its nature, microinjection may not turn out to be the most popular procedure, especially among impatient graduate students and pressured academics.

E. *E. coli* Spheroplasts as DNA Carriers

The direct gene-transfer techniques thus bypass the cloning of foreign genes in *Agrobacterium*. However, all these procedures require purified DNA molecules. An ultimately simplified gene-transfer system would also avoid DNA purification and use *E. coli* (in which foreign DNA manipulation is performed anyway) as gene carrier. In a procedure in which cloned DNA was directly transferred from *E. coli* spheroplasts into *Brassica* protoplasts, CaMV DNA was cloned as a tandem dimer in pBR322 and transformed into *E. coli* HB101 (*88*). Spheroplasts isolated from these cells were isolated and incubated with freshly isolated *Brassica* protoplasts in the presence of polyvinyl alcohol in a high pH buffer. Such treatments are well known to stimulate fusion and endocytosis in protoplasts. Up to 8% of the treated protoplasts produced CaMV coat protein, as detected by immunofluorescence. Therefore, the CaMV genome had been transferred and expressed in these protoplasts at high frequency.

A similar approach showed that tobacco protoplasts can be transformed to hormone autotrophy and opine production (at a frequency of 2×10^{-5}) by fusion and/or endocytosis with *E. coli* spheroplasts carrying a pTiB6S3::RP4 cointegrate (*54*). The reason why the transformation frequency was very low as compared to the above (*88*) is not known. However, since such a cointegrate is expected to be present at a low copy number in *E. coli*, it might well be that the total amount of plasmid transferred was quite low.

The interactions between *E. coli* spheroplasts and *Petunia* protoplasts were maximized by Harding and Cocking (*89*) but no selectable markers were used in these experiments. It would be worthwhile to use the conditions determined by these authors to assess protoplast transformability by *E. coli* spheroplasts harboring one of the several available chimeric genes. In fact, the potential of this approach has not yet been fully explored. Its simplicity justifies more efforts in that direction.

VI. Transmission Genetics of Foreign Genes in Transgenic Plants

The classical genetic study of transgenic plants is important for at least two reasons: (1) if genetic engineering is to be used in replacement of breeding techniques to improve plants, it must be determined how the introduced foreign genes behave in the progeny of the transgenic plants, and (2) a thorough genetic analysis is still the best tool to determine linkage between foreign and endogenous traits, i.e., to determine in which plant chromosome(s) the exogenous gene(s) are integrated.

A. Mendelian Inheritance in *Agrobacterium*-Transformed Plants

The F_1 progeny of self-fertilized *N. plumbaginifolia* plants regenerated from transformed protoplasts indicate a 3 : 1 Mendelian segregation ratio of the dominant *neo* gene as well as tight linkage between the introduced chimeric gene and the nopaline synthase gene, both being present between T-DNA borders (*39*). These results can be interpreted as follows. (1) The two genes are integrated as a single unit since they do not segregate independently. This is consistant with the fact that T-DNA integrates via its border fragment, and that both marker genes are located between these borders. (2) The parental plants must be hemizygous for the *neo* and the *nos* traits, i.e., only one of two homologous chromosomes contains an integrated T-DNA.

These results were basically confirmed using the same plant material, an analogous *Agrobacterium* cocultivation technique, and markers conferring resistance to kanamycin, methotrexate, and chloramphenicol (*36*). In the case of a chloramphenicol-resistant, nopaline-producing transgenic plant, the study of the F_1 indicated that 6 seedlings out of 113 were nopaline-negative, leading the authors to conclude that the two markers could segregate independently. Furthermore, Southern blot analysis confirmed that the chloramphenicol-resistant P1 plant showed a complex pattern of T-DNA integration. The reason for this was that the engineered disabled pTi vector used to transform protoplasts contained two right T-DNA borders, one adjacent to the selectable marker (internal border) and a second (external) border containing an intact *nos* gene. Therefore, two types of T-DNA integration events were possible and were indeed detected. However, the genetic analysis provided did not allow the certainty that *cat* and *nos* did really segregate independently following a phenotypic ratio predicted by Mendel's laws. For example, it was not reported whether cat^- nos^- plants appeared in F_1, which should be the case if *nos* and *cat* were integrated at different loci.

In the same study, no kanamycin-resistant plants positive for nopaline production could be recovered. Thus, the linkage between *neo* and *nos* could not be determined. This observation also showed that in these plants, only the internal right border was used for T-DNA integration. Nevertheless, complete linkage between *nos* and MtxR could be demonstrated, and in this case of monohybrid 3 : 1 Mendelian ratio was observed. All three antibiotic resistance markers behaved as single dominant Mendelian traits.

Of course, neither of the above studies allow assignment of the dominant markers to any particular linkage group in the transformed plants. Such chromosomal localization of the foreign genes was studied using transgenic F_1 hybrids of *Petunia hybrida* VR, heterozygous for at least one endogenous marker on all seven pairs of homologs (*90*). A *nos–neo* chimeric gene was used to select transformants by the leaf disk method, while linkage analysis was based on nopaline presence or absence in test-cross progeny. Since three of the *Petunia* chromosomes contained two marker genes, three-point crosses were also performed in order to determine map distances between the marker genes and the integrated T-DNA. Southern hybridization was used to determine the patterns of T-DNA integration.

Six transgenic plants were studied. The results (*90*) can be summarized as follows: (1) three plants had T-DNA integrated in chromosome I, two plants carried it on chromosome III and one plant had the T-DNA insert in chromosome IV; (2) T-DNA was integrated in two different sites in chromosome III, both sites being distally located about 18 cM from the *prxA* (peroxidase A) gene; (3) T-DNA was present at different locations on chromosome I in these three transgenic plants, mapping at 0.7 and 20 centimorgans[5] from the *ph1* (flower color) marker; and (4) T-DNA inserted within chromosome IV was linked with the *B1* (no corolla) marker. These results suggest that T-DNA integration is a random phenomenon and show that the *nos* gene is certainly inherited as a single Mendelian trait. These observations also provide evidence that the T-DNA itself is an interesting genetic marker for mapping purposes in plants.

B. A Case of Cytoplasmic Inheritance following Agrotransformation of *N. tabacum*

As discussed above, the inheritance pattern shown by a chloramphenicol-resistant transgenic tobacco plant suggested segregation of the *nos* and *nos–cat* genes to be a possibility (*36*). This unusual segregation pattern could be explained (*91*) if the transgenic plants contained the chimeric gene in their chloroplast genome. Southern blot analysis of DNA extracted from chloroplast, mitochondrial, and nuclear fractions revealed that the chimeric *cat* gene was exclusively in chloroplast DNA. In addition, chloramphenicol acetyltransferase (EC 2.3.1.28) activity was found in the chloroplast fraction of a transgenic plant that had transmitted the *cat* gene via a maternal type of inheritance. Pollen from this plant did not transmit chloramphenicol resistance when used to fertilize a susceptible plant. Also, this trait was not maintained in the absence of chloramphenicol selection, suggesting mitotic segregation.

These experiments indicate that the *nos* promoter can also be active in chloroplasts, and the authors point out that sequences found in the *nos* promoter region could serve as prokaryotic transcription sig-

[5] One centimorgan (cM) represents a 1% recombination between two genes and is thus a measure of the distance between them.

nals suitable for expression in chloroplasts. That such signals might be used in these organelles was suggested by the following: since the expression vector carrying the *nos–cat* gene also contained a *neo* gene under prokaryotic control, the presence of APH(3′)II activity in the chloroplast fraction of this transgenic was checked. This enzyme activity was indeed found but it was not ascertained whether the *neo* gene was transcribed from its own promoter. Some of the problems, implications and potentials of these observations have been reviewed (*92*).

C. Inheritance Patterns in Transgenic Plants Obtained by Direct Gene Transfer

Two transgenic *N. tabacum* plants transformed to kanamycin resistance by fusogen-induced uptake of pABDI into protoplasts, when selfed, showed a 3 : 1 segregation pattern, while a 1 : 1 ratio was obtained in a testcross with wild-type tobacco (*76, 93*). Hence, the transgenic plants were hemizygous for the chimeric gene, which behaved as a single dominant Mendelian trait. Thus, these plants presented the same genetic characteristics as transgenic *Petunia* obtained by agrotransformation (*39, 90*). In addition, F_2 plants behaved as predicted in self- and testcrosses. Such crosses performed with transgenic kanamycin-resistant tobacco plants resulting from either calcium phosphate/polyvinyl alcohol-mediated or liposome-mediated plasmid transfer yielded essentially the same result: antibiotic resistance was transmitted as a single dominant nuclear marker (*71, 77*).

Interestingly, all groups studying direct DNA transfer into protoplasts have reported on commonly occurring tandem duplication of the vector upon integration within the host genome. However, some results (*71*) clearly showed that tandem repeats of the vector were integrated at a single site in the genome. Therefore, it was suggested that homologous recombination between vector molecules could occur prior to integration.

Finally, the segregation patterns of two different types of DNA molecules integrated within plant cells after cotransformation of protoplasts was studied (*75*). Grafted shoots regenerated from protoplasts transformed with pTi in the presence of carrier calf thymus DNA were employed. One of the regenerants contained integrated copies of the TL-DNA gene 4, the mannopine[6] (*mas*) locus of the TR-DNA as well as calf thymus DNA sequences. This regenerated shoot was testcrossed with a wild-type *N. tabacum* plant and the progeny analyzed

[6] Mannopine is (*S*)-1-deoxy-1-(N^2-glutamino)mannitol. [Eds.]

for the presence of gene 4, *mas,* and calf thymus DNA. Integrated gene 4 was transmitted as a single Mendelian trait. In addition, linkage between *mas* and calf thymus DNA sequences as well as between gene 4 and calf thymus DNA was detected. No linkage between gene 4 and *mas* was observed. These results indicated integration of foreign DNA sequences in different chromosomes (or at least at a distance of 50 cM or more if present on the same chromosome) as well as linkage between unrelated cotransformed DNA sequences. Conversely, genes initially present on the same large DNA molecule (gene 4 and *mas*) were integrated at different loci. The above experiments also showed that even though markers were originally present on the same DNA molecule, they could nevertheless integrate separately. Quite probably, this occurence depends on the physical distance between the two markers, as for example, gene 4 and *mas* are located about 14 kb apart on pTi.

Since integration of large amounts of heterologous DNA sequences might have undesirable effects on plants, it might be preferable to transform protoplasts in the absence of carrier DNA. However, this could result in decreased transformation frequencies.

In summary, it appears that even though foreign gene integration patterns are sometimes more complicated after direct DNA transfer than after agrotransformation, it seems that in the cases studied so far, the two gene transfer procedures are genetically indistinguishable. Integration at a single locus (for small vectors) in the nuclear DNA seems to be the rule.

VII. Foreign Gene Regulation in Transgenic Plants

Now that it is possible to produce transgenic plants (at least in cases where plant regeneration from callus tissue or protoplasts is feasible; see Section VIII) and transgenic plant cells routinely, some basic questions of plant molecular biology can be addressed in an entirely new fashion. Examples of these are given in this section.

A. A Novel Approach to the Study of Plant Viruses

Molecular studies on plant viruses are often hampered by factors such as limited host range and the need for insect vectors. It has been demonstrated that a full-length copy of the CaMV genome can be stably maintained and transcribed when transferred to host and nonhost plants via the T-DNA of *Agrobacterium* (*94*). Quite interestingly, gene VI and full-length (35 S) transcripts were present in very significant amounts in some nonhost cells or were found at very low concentration, depending on the plant species. Furthermore, these two promoters were used to very different extents in the same nonhost system. Therefore, this approach can be used to evaluate promoter strength in natural or artificial hosts.

Tandemly duplicated CaMV present between the borders of a disabled T-DNA harbored by pTi can escape its integration site—using an as yet undetermined mechanism—to produce systemic infection of turnip plants (*95*). As discussed in Section II T-DNA has also been used as a vector to study the replication of the tomato golden mosaic virus genome as well as its pathogenicity (*20*).

B. Construction of Promoter Probing Vehicles and Promoter Analysis

Few plant promoter regions have been isolated and characterized. An (*96*) designed a promoter-probing vehicle based on an *Agrobacterium* binary system in which the cloning vector contains a *nos–neo* chimeric gene plus a promoterless *cat* coding sequence preceded by a polylinker region. The purpose of this vector is to clone pieces of plant DNA next to the *cat* gene, transfer the recombinant molecules into cells or protoplasts, select kanamycin-resistant clones, and analyze the latter for expression of the *cat* gene. If *cat* activity can be detected in some of the transformants, there will be good evidence that the randomly cloned pieces of plant DNA present at the 5′ end of the *cat* gene do contain transcription initiation signals.

The usefulness of this system was shown by cloning the *nos* promoter in the polylinker region followed by the demonstration of *cat* activity in the kanamycin-resistant transformants. Interestingly, the transformed lines showed a wide variation in the ratio between *cat* and *neo* expression at the enzyme level. This was unexpected since both chimeric genes were under the control of a *nos* promoter and were closely linked on the vector. It is possible that epistatic effects, acting at very short range, might account for this observation. Lastly, the SV40 early promoter, the mouse metallothionein, and the herpes simplex *tk* promoters were all inactive in plant cells.

The promoter activity of DNA sequences can also be assayed rapidly by a technique now known as the "transient expression assay." This method bypasses the very long period (several months) necessary to generate sufficient material from stable transformants. The expression of chimeric *cat* genes can be rapidly detected after direct gene transfer into protoplasts, simply by analyzing extracts of the whole treated population (*82*). Similarly, APH(3′)II activity was detected for a period of at least 10 days after introduction into *Triticum monococcum* protoplasts of a chimeric corn *shrunken* promoter-*neo* gene by direct transfer (*96a*). These results strongly suggest that extrachromosomal vector molecules are correctly expressed.

C. Expression of Storage Protein Genes in Transgenic Plants or Calli

Genes coding for seed storage proteins have attracted the attention of molecular biologists for a variety of reasons, notably: (1) these genes are developmentally regulated, (2) the relative abundance of their mRNAs is high during a defined period of plant development, making specific cDNA cloning easier than in the case of other genes, and (3) since seeds are an important source of food for humans and other mammals, transfer of seed storage protein genes from one crop species to another might enhance its characteristics.

A hybrid gene between the *ocs* T-DNA promoter and the coding sequence of the phaseolin gene from bean was constructed (*97*). This chimeric gene was present in the T-DNA of a fully virulent Ti plasmid that was then transferred to sunflower via wound infection by *A. tumefaciens*. The sunflower crown-gall tumors both transcribed and translated the foreign gene. Interestingly, the introns present in the phaseolin gene were correctly spliced, as they did not appear in the mRNA. Also, in another experiment, the transferred phaseolin gene this time under the control of its own promoter, proved to be much less actively transcribed than when under the control of the *ocs* promoter. In addition, phaseolin produced in transgenic sunflower cells was actively degraded, a situation that might cause problems in applied plant genetic engineering. Further experiments demonstrated that the phaseolin gene from which the five introns had been deleted was still correctly transcribed and translated in tobacco callus (*98*).

A maize genomic clone containing a zein (maize seed storage protein) gene was transcribed from its own promoter in sunflower tissue after *Agrobacterium*-mediated transfer (*99*). However, translation products from this gene could not be detected. This indicated that the promoter from a monocotyledonous gene could at least be transcribed in dicotyledonous cells. The zein mRNA produced in the transgenic sunflower tissue was correctly translated in an *in vitro* wheat germ system, but was not translated *in vivo* for unknown reasons.

Such a problem was not encountered in the transfer of the soybean storage gene encoding the α'-subunit of β-conglycine into *Petunia* cells (*100*). The transfer strategy involved a disabled T-DNA region allowing the regeneration of fertile plants. It was shown that deposition of α'-subunit multimers occurred in the seeds of the transgenic plants, and that expression of the foreign gene was developmentally controlled, i.e., occurred maximally during embryogenesis. This latter observation might explain why zein was not found in transgenic sunflower callus cells (99).

D. Expression of Heat-Inducible Genes

The level of transcription of a soybean genomic DNA fragment containing a heat shock gene (*hs*687) into sunflower using a pTi vector was too low to be detected by Northern blot hybridization, even after heat shock (*101*). However, S1 nuclease mapping experiments showed that poly(A)-RNA isolated from transgenic sunflower callus could protect a 150-nucleotide-long DNA sequence known to be present at the correct transcription initiation site of this gene in its natural host. Such protection only occurred when the RNA was isolated from heat-shocked tissue. Therefore, the soybean gene was transcribed at very low level in the heterologous plant. Whether these transcripts can be translated has not yet been determined.

Spena *et al.* (*102*) constructed a chimeric gene containing the *neo* coding sequence under the control of the *Drosophila* heat-shock gene *hsp*70 promoter. The *ocs* terminator was used to provide a transcription termination signal and a polyadenylation site. After recombination with pGV3850, *Agrobacterium* was used to transfer the chimeric gene into tobacco protoplasts. The decision to use an animal promoter in these experiments was based on the fact that heat-shock genes contain highly conserved consensus sequences in their promoter regions. Indeed, these authors demonstrated that the chimeric *hsp–neo* gene was strongly activated when the plant cells were exposed to a temperature of 40°C, both at the level of transcription and at the level of translation. This animal promoter is so far the only nonplant promoter shown to specifically perform its function in plant cells.

E. Regulation of Light-Inducible and Organ-Specific Genes

Light plays a central role in plant development. Among the many genes that directly or indirectly respond to a light stimulus, two gene families have been extensively studied. These include the *rbc*S gene family, responsible for the synthesis of the small subunit of the enzyme ribulose-1,5-bisphosphate carboxylase (EC 4.1.1.39) and the *Cab* genes, which direct the synthesis of the light-harvesting chlorophyll *a*/*b* proteins. Both gene families are located in the nucleus, but their protein products function in chloroplasts. The light regulation of these genes has been studied in great detail, notably due to the ability to transfer natural or chimeric genes into heterologous hosts. Such experiments have been reported by essentially two groups (*103–108*) and their results have been reviewed (*109*). Basically, either genomic clones of the *rbc*S or *Cab* genes were introduced into plants by *Agrobacterium* transfer (*103, 105–107*), or chimeric genes consisting of a *cat* or *neo* coding sequence preceded by the *rbc*S or *Cab* 5′-flanking sequences were introduced into heterologous plants, also via *Agrobacterium* transfer (*104–108*). These natural or chimeric genes were shown to be light-regulated, strongly expressed in leaves, and very weakly in roots. Thus, the control regions of these genes are clearly responsible for the response to a light stimulus, and in addition, they determine organ-specific expression. Deletion mutations of these genes were produced *in vitro* by manipulating their 5′-flanking regions, and those were introduced into heterologous hosts. The aim of these experiments was to gain insight into the nature of the organ specificity and light inducibility determined by these upstream sequences. It turned out that light induction and organ specificity were both controlled by DNA sequences present within about 400 basepairs adjacent to the transcription start. However, sequences located further upstream also had an influence on the level of transcription of these genes and therefore behaved like enhancer elements (*108, 109*). The enhancer elements present in pea *rbc*S genes were also organ-specific (*109*).

Thus, it will become possible to search meaningfully for trans-acting factors recognizing these 5′-flanking sequences and regulating the expression of the gene they control.

One last example of organ-specific gene expression concerns the regulation of a chimeric soybean leghemoglobin gene in nodules of transgenic *Lotus corniculatus* (*64*); this is discussed in Section IV,D.

F. Expression of a Human Gene in Plant Cells

We have seen earlier in this section that several promoters of animal origin seem to be inactive in plant cells. Thus, if human gene products of medical importance are to be synthesized in these cells, it will probably be necessary to clone their coding sequences downstream from a plant promoter. A *nos*/human-growth-hormone chimeric gene, introduced into sunflower and tobacco cells using a Ti plasmid vector, was transcribed under the control of the *nos* promoter, but no splicing of introns was detected (*109a*). It is not known whether this would be true of all animal pre-mRNAs, but it remains that this observation raises interesting questions regarding the removal of introns in general.

VIII. Conclusions

The genetic transformation of several plant species can now be considered a rather routine task. Practical methodologies describing *Agrobacterium*-mediated or direct gene transfer have been published (*86, 110–112*). As we have seen, these procedures have already been used in the study of fundamental questions pertaining to gene regulation in plants. In the absence of transformation techniques, it would have been extremely difficult to approach these problems in the same meaningful way.

In addition, the genetic manipulation of plants can also be perceived as having a very practical, agronomical facet. Without a doubt, crop improvement via gene transfer has attracted considerable attention from commercial companies, and it is not surprising that transformation techniques are now being applied to plant species having commercial value (*113–118*). So far, the focus has mostly been on testing the feasibility of gene transfer in cultivated plant species; therefore, chimeric genes of the *nos–neo* type have been used. The next step, of course, will be to identify plant or other genes conferring desirable phenotypes and transfer them into valuable crop species. This will not be a simple task as, at least in the case of plants, such genes have only rarely been identified and usually not isolated.

Another approach to plant improvement consists in taking advantage of wild-type or mutant bacterial genes known to code for functions either nonexistant in plant cells or difficult to modify by mutation. For example, Comai *et al.* (*119*) isolated from *Salmonella* a mutant *aro*A gene whose product, 3-phosphoshikimate 1-carboxyvinyltransferase (EC 2.5.1.19), is less sensitive to the herbicide *N*-(phosphonomethyl)glycine (glyphosate) than either of the bacterial wild-type and plant enzymes. This gene was transferred into tobacco plants via a T-DNA-based vector. Regenerated transgenic tobacco plants were considerably more tolerant to the herbicide although still showing some sensitivity to it. However, these results do prove that bacteria can be used as a source of genes for plant genetic engineering.

Undoubtedly, plant transformation techniques will be used to unravel other basic problems. For example, the genetic study of plant mitochondrial functions is still in its infancy. A mitochondrial transformation system would be of great help in this regard, but so far there has been no demonstration that plant mitochondria can be transformed. In that respect, the cloning and sequencing of a small maize mitochondrial plasmid is interesting (*120*). Since this plasmid is partially homologous to maize nuclear DNA, it could be used as a vector for the transformation of both types of organelles.

As far as practical and fundamental studies with monocotyledonous plants are concerned, the establishment of monocot transformation systems either via *Agrobacterium* [in the cases of a *Liliaceae* and an *Amaryllidaceae* (*121*)], via electroporation (*83*), or via fusogen-induced DNA transfer (*79, 80*) will gain its full meaning only when efficient monocot regeneration techniques become available. This last point has so far been a major stumbling block in plant transformation, including many dicotyledonous species. However, progress is being made as, for example, a protoplast regeneration system for rice has been developed (*122*). This breakthrough will certainly be followed by serious attempts to transform this important crop genetically.

One alternative to protoplast genetic manipulation is to attempt transformation of intact cells from which plant regeneration is often considerably easier. In this regard, Morikawa *et al.* (*123*) have shown that tobacco mosaic virus RNA can be electroporated into pectinase-treated tobacco cells rather than into protoplasts (*123*). Several laboratories are undoubtedly in the process of applying this technique to monocot cells.

As for the transfer of plant nonselectable markers into other plant cells, the electroporation method, which yields such high transformation frequencies, could be used in "shotgun" experiments. Such a scenario would involve the uptake of total plant DNA possibly enriched for unique and middle repetitive sequences. Since cotransformation seems to occur at a reasonably high frequency, a *nos–neo* marker could be used to select for cells which were competent for DNA uptake. This selection would then have to be followed by the regeneration and testing of hundreds or even thousands of plantlets in order to identify those individuals expressing the desired trait. In fact, such large-scale screening is being used routinely in plant breeding schemes.

One major problem is that many desirable traits are probably of polygenic nature. Even if regenerated plantlets were shown to express such polygenes, it is likely that their constituents will be integrated at very different locations in the host chromosomes. Since segregation of cotransformed traits has been shown, sexual propagation of the transgenic plants will not be possible. Of course, vegetatively propagated plants such as potatoes are still good candidates for polygene transfer. Also, if in the future polygenes can be isolated, there is still the possibility of inserting them into closely linked chromosomal loci by homologous recombination. Such targeting would largely prevent meiotic segregation.

As a last word, it can be said that plants have definitely joined the realm of fine molecular biology; the fantastic progress made recently in the genetic transformation of plant cells is not foreign to this fact. (See Addendum, p. 253.)

IX. Glossary

A. Genetic Loci

***aphI*:** see **Tn601**.

***aphII*:** see **Tn5**.

***aphIV*:** an *E. coli* gene coding for a phosphotransferase that can phosphorylate and hence inactivate the aminocyclitol antibiotic hygromycin B.

***Cab*:** a plant nuclear gene family coding for the chlorophyll *a*/*b* binding protein found in thylakoid membranes of chloroplasts.

***cat*:** see **Tn9**.

***cos*:** a particular DNA sequence in phage λ that is required for DNA packaging into λ particles.

***dhfrMtxR*:** see **Tn7**.

***gene 4*:** a T-DNA (see Section II,B,1) gene also known as *tmr* or *roi* (see below).

***gene VI*:** a cauliflower mosaic virus (CaMV) gene specifying the synthesis of a 19-S RNA molecule coding for the virus inclusion body protein. This protein is found in the matrix bounding cytoplasmic vesicles in which CaMV particles accumulate.

***hs687*:** a cloned heat-shock gene (transcriptionally induced by heat) from soybean.

***hsp70*:** a cloned heat shock gene from *Drosophila*.

***lac*:** the *E. coli* lactose operon.

***mas*:** a TR-DNA (see Section II,B,1) locus coding for enzymes involved in the synthesis of mannopine (Section VI,C).

***neo*:** see **Tn5**.

***nos*:** a T-DNA gene coding for nopaline synthase, an enzyme involved in the synthesis of nopaline (see Section II,A).

***ocs*:** a TL-DNA (see Section II,B,1) gene coding for octopine synthase, an enzyme involved in the synthesis of octopine.

***phl*:** a *Petunia hybrida* locus affecting flower color by changing the pH of the vacuole.

***prxA*:** a *Petunia hybrida* locus determining the synthesis of a peroxidase isozyme.

***rbcS*:** a plant nuclear gene family coding for the synthesis of the small subunit of ribulose-1,5-bisphosphate carboxylase.

***tk*:** the gene coding for thymidine kinase.

***tmr* or *roi*:** a locus containing a single gene present in both the nopaline T-DNA and octopine TL-DNA of pTi. This locus codes for a function suppressing root formation in crown-gall tumors by controlling the level of the cytokinin ribosylzeatin in these cells.

***tms* or *shi*:** a locus containing two genes (gene 1 and gene 2) present in both the nopaline T-DNA and the octopine TL-DNA of pTi. This locus determines the suppression of shoot formation in crown-gall tumors through the synthesis of the auxin indole-3-acetic acid.

***vir*:** a 40-kbp region of pTi containing six distinct operons coding for functions necessary for the excision and transfer of the T-DNA from *Agrobacterium* cells to plant cells. Mutations in this regulon decrease or abolish *Agrobacterium* virulence.

B. Plasmids

1. *E. coli* Cloning Vectors

The pBR and pUC plasmids are general-purpose cloning vectors used for cloning and amplification of a particular DNA sequence in *E. coli*. These plasmids cannot replicate in *Agrobacterium* but can undergo recombination with pTi if they contain a region of homology. There is a complete description of these vectors in P. H. Pouwels, B. E. Enger-Valk and W. J. Bramm, "Cloning Vectors." Elsevier, Amsterdam, 1985.

2. *Agrobacterium* Plasmids

a. Wild-Type pTi and pRi. Virulent strains of *A. tumefaciens* contain a large tumor-inducing (Ti) plasmid (also called pTi) harboring notably a *vir* region (see under genetic loci) and T-DNA ("transferred" DNA) region. The latter is transferred from the inciting bacterial cells to the host plant cells where it becomes integrated within the nuclear genome. In addition to loci determining hormone autotrophy (*tms, tmr*) in transformed plant cells, the T-DNA also carries a *nos* or an *ocs* gene (see A above) expressed in plant cells. For example, pTiC58 and pTiT37 are *nos*-type plasmids while pTiA6 and pTiB6S3 are *ocs*-type plasmids. The T-DNA containing a *nos* gene consists of an uninterrupted DNA sequence, 24 kbp in size. The T-DNA present in *ocs*-type plasmids consists of two noncontiguous segments designated TL-DNA (for "left" T-DNA) and TR-DNA (for "right" T-DNA).

The *ocs* gene is located in the TL-DNA (14 kbp) while the *mas* locus is located in the TR-DNA (7 kbp). The TL- and TR-DNA fragments integrate independently within plant cell genomes.

Virulent strains of *Agrobacterium rhizogenes* also harbor a large plasmid called pRi (for root-inducing) equipped with a *vir* region homologous to that found in pTi as well as a T-DNA separated into TL and TR regions. The pRi T-DNA is only very partially homologous to the pTi T-DNA.

b. Disabled Ti Plasmids and Intermediate Cloning Vectors for Plant Transformation. By definition, a disabled Ti plasmid is one from which all oncogenes (*tms* and *tmr*) have been deleted but that is still equipped with either TL-DNA right and left borders (such as pGV3850) or a TL-DNA left border only (such as pTiB6S3-SE). These deleted plasmids are then recombined *in vivo* with intermediate cloning vectors containing (1) the gene to be transformed into plant cells, often cloned downstream from a *nos* promoter or a CaMV-gene-VI promoter, (2) antibiotic markers and a suitable origin of replication for selection and propagation in *E. coli*, and (3) a region of homology with the disabled pTi to allow for homologous recombination in such a way that the gene to be transferred to plant cells ends up located between T-DNA borders in the recombined pTi. Thus, disabled pTi's are always used in conjunction with an intermediate cloning vector. For example, pGV3850 is recombined with intermediate vectors of the pLGV series, whereas pTiB6S3-SE is recombined with intermediate vectors of the pMON series.

c. Binary Vectors. These vector systems are always composed of two plasmids coexisting in the same *Agrobacterium* cell. One of the plasmids carries the pTi *vir* region (for example, pAL4404) and is the provider of trans-acting functions necessary for the excision and transfer of the T-DNA, located on the second plasmid. The latter is of much smaller size than pTi and allows easy engineering of the T-DNA in *E. coli*. A plasmid like pAL4404 can be used in conjunction with any other vector able to replicate in *Agrobacterium* and containing the desired gene cloned between the right and left border of the T-DNA.

3. Plant Expression Vectors for Direct Gene Transfer.

The discovery that DNA integration within plant genomes requires neither homology between the donor and recipient DNAs nor the presence of T-DNA borders shows that, in principle, any cloning vector can be used to transform plant protoplasts. The only prerequisite is that the gene used for the selection of plant cell transformants must be expressed in the latter, i.e., that it must be flanked by appropriate control regions. Therefore, any of the intermediate cloning vectors described above (pMON's and pLGV's) can also be used in direct gene transfer experiments. Similarly, T-DNA vectors used in binary systems (like pCIB10) can be used in direct gene transfer.

Other vectors, completely devoid of T-DNA elements, have been constructed and usually contain either a CaMV 35-S promoter or a CaMV 19-S promoter upstream from the coding region to be expressed in plant cells. The 3′-ends originate from the *nos* or *ocs* genes or from the CaMV genome. Two examples of such vectors are pABDI and pCaMV-6km, both of which confer kanamycin resistance to transformed plant cells. The vector pCIB710 allows cloning of any coding sequence in a unique restriction endonuclease site located between the CaMV 35-S promoter and terminator. Direct gene transfer vectors are usually pUC or pBR replicons.

C. Transposons (Bacterial Transposable Elements)

Tn5: a DNA sequence carrying notably the *aphII* locus coding for a phosphotransferase that can phosphorylate and hence inactivate antibiotics such as neomycin and kanamycin. This locus is often referred to as *neo*.

Tn7: a DNA sequence carrying several antibiotic resistance markers and notably the *dhfrMtxR* locus coding for a mutant dihydrofolate reductase insensitive to methotrexate.

Tn9: a DNA sequence carrying notably the *cat* locus coding for an acetyltransferase that can acetylate and inactivate chloramphenicol.

Tn601: a DNA sequence carrying notably the *aphI* locus coding for a phosphotransferase functionally similar to the one encoded by *aphII*.

The *aphI*, *aphII*, *cat*, and *dhfrMtxR* coding sequences have been cloned in a variety of vectors, between promoter and terminator sequences functional in plant cells.

D. Viruses

CaMV: cauliflower mosaic virus. An icosahedral virus infecting crucifereae and containing double-stranded circular DNA. The viral genome is replicated by reverse transcription of a full-length 35-S RNA molecule. The promoter region controlling the transcription of this particular RNA molecule has been used to express foreign genes in plants. (See also *gene VI* in Section A).

CLV: cassava latent virus. A geminivirus characterized by a bipartite genome consisting of two single-stranded circular DNA molecules. Each half of the bipartite genome is encapsulated separately. Virions occur in pairs, hence the name.

SV40: simian virus 40. A virus containing double-stranded circular DNA. Its "early" promoter has been used to express foreign genes in mammalian cells.

TGMV: tomato golden mosaic virus, a geminivirus.

Acknowledgments

This work was supported in part by a grant from the State of Washington High Technology Center.

References

1. P. F. Lurquin, this series **20,** 161 (1977).
2. A. Kleinhofs and R. Behki, *ARGen* **11,** 79 (1977).
3. G. Gheysen, P. Dhaese, M. Van Montagu and J. Schell, *in* "Genetic Flux in Plants" (B. Hohn and E. S. Dennis, eds.), pp. 11. Springer-Verlag, Wien, 1985.
4. J. P. Hernalsteens, F. Van Vliet, M. De Beuckeleer, A. DePicker, G. G. Engler, M. Lemmers, M. Holsters, M. Van Montagu and J. Schell, *Nature* **287,** 654 (1980).
4a. K. A. Barton, A. N. Binns, A. J. M. Matzke and M. -D. Chilton, *Cell* **32,** 1033 (1983).
4b. C. Koncz, F. Kreuzaler, Zs. Kalman and J. Schell, *EMBO J.* **3,** 1029 (1984).
5. R. F. Barker, K. B. Idler, D. V. Thompson and J. D. Kemp, *Plant Mol. Biol.* **2,** 335 (1983).
6. J. Gielen, M. De Beuckeleer, J. Seurinck, F. Deboeck, H. De Greve, M. Lemmers, M. Van Montagu and J. Schell, *EMBO J.* **3,** 835 (1984).
7. A. DePicker, S. Stachel, P. Dhaese, P. Zambryski and H. M. Goodman, *J. Mol. Appl. Genet.* **1,** 561 (1982).
8. L. Willmitzer, P. Dhaese, P. H. Schreier, W. Schmalenbach, M. Van Montagu and J. Schell, *Cell* **32,** 1045 (1983).
9. J. A. Winter, R. L. Wright and W. B. Gurley, *NARes* **12,** 2391 (1984).
10. J. Velten, L. Velten, R. Hain and J. Schell, *EMBO J.* **3,** 2723 (1984).
11. S. B. Gelvin, S. J. Karcher and P. B. Goldsbrough, *MGG* **199,** 240 (1985).
12. R. Hull and S. N. Covey, *BioEssays* **3,** 160 (1986).
13. B. Gronenborn, R. C. Gardner, S. Schaefer and R. J. Shepherd, *Nature* **294,** 773 (1981).
14. N. Brisson, J. Paszkowski, J. R. Penswick, B. Gronenborn, I. Potrykus and T. Horn, *Nature* **310,** 511 (1984).
15. M. G. Koziel, T. L. Adams, M. A. Hazlet, D. Damm, J. Miller, D. Dahlbeck, S. Jayne and B. J. Staskawicz, *J. Mol. Appl. Genet.* **2,** 549 (1984).
16. J. T. Odell, F. Nagy and N. H. Chua, *Nature* **313,** 810 (1985).
17. E. Balazs, S. Bouzoubaa, H. Guilley, G. Jonard, J. Paszkowski and K. Richards, *Gene* **40,** 343 (1985).
18. W. D. O. Hamilton, V. E. Stein, R. H. A. Coutts and K. W. Buck, *EMBO J.* **3,** 2197 (1984).
19. R. Townsend, J. Stanley, S. J. Curson and M. N. Short, *EMBO J.* **4,** 33 (1985).
20. S. G. Rogers, D. M. Bisaro, R. B. Horsch, R. T. Fraley, N. L. Hoffman, L. Brand, J. S. Elmer and A. M. Lloyd, *Cell* **45,** 593 (1986).

21. R. Townsend, J. Watts and J. Stanley, *NARes* **14,** 1253 (1986).
22. L. Marton, G. J. Wullems, L. Molendijk and R. A. Schilperoort, *Nature* **277,** 129 (1979).
23. P. F. Lurquin and A. Kleinhofs, *BBRC* **107,** 286 (1982).
24. D. Ursic, J. D. Kemp and J. P. Helgeson, *BBRC* **101,** 1031 (1981).
25. P. J. Dix, F. Joo and P. Maliga, *MGG* **157,** 285 (1977).
26. E. Nielsen, F. Rollo, B. Parisi, R. Cella and F. Sala, *Plant Sci. Lett.* **15,** 113 (1979).
27. C. Waldron, E. B. Murphy, J. L. Roberts, G. D. Gustafson, S. L. Armour and S. K. Malcolm, *Plant Mol. Biol.* **5,** 103 (1985).
28. L. Herrera-Estrella, A. DePicker, M. Van Montagu and J. Schell, *Nature* **303,** 209 (1983).
29. L. Herrera-Estrella, M. DeBlock, E. Messens, J.-P. Hernalsteens, M. Van Montagu and J. Schell, *EMBO J.* **2,** 987 (1983).
30. R. T. Fraley, S. G. Rogers, R. B. Horsch, P. R. Sanders, J. S. Flick, S. P. Adams, M. L. Bittner, L. A. Brand, C. L. Fink, J. S. Fry, G. R. Galluppi, S. B. Goldberg, N. L. Hoffman and S. C. Woo, *PNAS* **80,** 4803 (1983).
31. M. W. Bevan, R. B. Flavell and M.-D. Chilton, *Nature* **304,** 184 (1983).
32. P. J. M. van den Elzen, J. Townsend, K. Y. Lee and J. R. Bedbrook, *Plant Mol. Biol.* **5,** 299 (1985).
33. G. Ooms, P. G. Hooykaas, G. Molenaar and R. A. Schilperoort, *Gene* **14,** 33 (1981).
35. P. Zambryski, H. Joos, C. Genetello, J. Leemans, M. Van Montagu and J. Schell, *EMBO J.* **2,** 2143 (1983).
36. M. De Block, L. Herrera-Estrella, M. Van Montagu, J. Schell and P. Zambryski, *EMBO J.* **3,** 1681 (1984).
37. J. Velten and J. Schell, *NARes* **13,** 6981 (1985).
38. R. Deblaere, B. Bytebier, H. De Greve, F. Deboeck, J. Schell, M. Van Montagu and J. Leemans, *NARes* **13,** 4777 (1985).
39. R. B. Horsch, R. T. Fraley, S. G. Rogers, P. R. Sanders, A. Lloyd and N. Hoffman, *Science* **223,** 496 (1984).
40. R. T. Fraley, S. G. Rogers, R. B. Horsch, D. A. Eichholtz, J. S. Flick, C. L. Fink, N. L. Hoffmann and P. R. Sanders, *Bio/Technology* **3,** 629 (1985).
41. S. E. Stachel, E. Messens, M. Van Montagu and J. Schell, *Nature* **318,** 624 (1985).
42. E. E. Stachel, B. Timmerman and P. Zambryski, *Nature* **322,** 706 (1986).
43. A Hoekema, P. R. Hirsch, P. J. J. Hooykaas and R. A. Schilperoort, *Nature* **303,** 179 (1983).
44. G. An, B. D. Watson, S. Stachel, M. P. Gordon and E. W. Nester, *EMBO J.* **4,** 277 (1985).
45. H. J. Klee, M. F. Yanofsky and E. W. Nester, *Bio/Technology* **3,** 637 (1985).
46. R. B. Horsch, J. E. Fry, N. L. Hoffmann, D. Eichholtz, S. G. Rogers and R. T. Fraley, *Science* **277,** 1229 (1985).
47. S. McCormick, J. Niedermeyer, J. Fry, A. Barnason, R. Horsch and R. Fraley, *Plant Cell Rep.* **5,** 81 (1986).
49. J.-F. Jia, R. D. Shillito and I. Potrykus, *Z. Pflanzenphysiol.* **112,** 1 (1983).
50. D. Hanold, *Plant Sci Lett.* **30,** 177 (1983).
51. R. T. Fraley, R. B. Horsch, A. Matzke, M.-D. Chilton, W. S. Chilton and P. R. Sanders, *Plant Mol. Biol.* **3,** 371 (1984).
52. E. L. Virts and S. B. Gelvin, *JB* **162,** 1030 (1985).
53. S. Hasezawa, T. Nagata and K. Syono, *MGG* **182,** 206 (1981).
54. R. Hain, H.-H. Steinbiss and J. Schell, *Plant Cell Rep.* **3,** 60 (1984).
55. R. A. Millman and P. F. Lurquin, *J. Plant Physiol.* **117,** 431 (1985).

56. A. Baba, S. Hasezawa and K. Syono, *Plant Cell Physiol.* **27,** 463 (1986).
57. A. Muller, T. Manzara and P. F. Lurquin, *BBRC* **123,** 458 (1984).
58. K. Pollock, D. G. Barfield, S. J. Robinson and R. Shields, *Plant Cell Rep.* **4,** 202 (1985).
59. G. An, *Plant Physiol.* **79,** 568 (1985).
60. L. Willmitzer, J. Sanchez-Serrano, E. Buschfeld and J. Schell, *MGG* **186,** 16 (1982).
61. M.-D. Chilton, D. A. Tepfer, A. Petit, C. David, F. Casse-Delbart and J. Tempe, *Nature* **295,** 432 (1982).
62. D. A. Tepfer, *in* "Genetic Engineering in Eukaryotes" (P. F. Lurquin and A. Kleinhofs, eds.), p. 153. Plenum Press, New York and London, 1983.
63. Z.-M. Wei, H. Kamada and H. Harada, *Plant Cell Rep.* **5,** 93 (1986).
64. J. S. Jensen, K. A. Marcker, L. Otten and J. Schell, *Nature* **321,** 669 (1986).
65. P. F. Lurquin and C. I. Kado, *MGG* **154,** 113 (1977).
66. S. M. Fernandez, P. F. Lurquin and C. I. Kado, *FEBS Lett.* **87,** 277 (1978).
67. M. R. Davey, E. C. Cocking, J. Freeman, N. Pearce and I. Tudor, *Plant Sci. Lett.* **18,** 307 (1980).
68. J. Draper, M. R. Davey, J. P. Freeman, E. C. Cocking and B. J. Cox, *Plant Cell Physiol.* **23,** 451 (1982).
69. P. F. Lurquin, *NARes* **6,** 3773 (1979).
70. P. F. Lurquin, *in* "Liposome Technology" (G. Gregoriadis, ed.), Vol. II, p. 187. CRC Press, Boca Raton, Florida, 1984.
71. A. Deshayes, L. Herrera-Estrella and M. Caboche, *EMBO J.* **4,** 2731 (1985).
72. M. Caboche and P. F. Lurquin, *in* "Methods in Enzymology" (L. E. Packer and R. Douce, eds.), Vol. 148, p. 39. Academic Press, Orlando, Florida, 1987.
73. F. A. Krens, L. Molendijk, G. J. Wullems and R. A. Schilperoort, *Nature* **296,** 72 (1982).
74. F. A. Krens, R. M. W. Mans, T. M. S. van Slogteren, J. H. C. Hoge, G. J. Wullems and R. A. Schilperoort, *Plant Mol. Biol.* **5,** 223 (1985).
75. R. Peerbolte, F. A. Krens, R. M. W. Mans, M. Floor, J. H. C. Hoge, G. J. Wullems and R. A. Schilperoort, *Plant Mol Biol.* **5,** 235 (1985).
76. J. Paszkowski, R. D. Shillito, M. Saul, V. Mandak, T. Hohn, B. Hohn and I. Potrykus, *EMBO J.* **3,** 2717 (1984).
77. R. Hain, P. Stabel, A. P. Czernilofsky, H.-H. Steinbiss, L. Herrera-Estrella and J. Schell, *MGG* **199,** 161 (1985).
78. P. Meyer, E. Walgenbach, K. Bussmann, G. Hombrecher and H. Saedler, *MGG* **201,** 513 (1985).
79. I. Potrykus, M. W. Saul, J. Petruska, J. Paszkowski and R. D. Shillito, *MGG* **199,** 183 (1985).
80. H. Lorz, B. Baker and J. Schell, *MGG* **199,** 178 (1985).
80a. J. Paszkowski, B. Pisan, R. D. Shillito, T. Hohn, B. Hohn and I. Potrykus, *Plant Mol. Biol.* **6,** 303 (1986).
81. E. Neumann, M. Schaefer-Ridder, Y. Wang and P. H. Hofschneider, *EMBO J.* **1,** 841 (1982).
82. M. E. Fromm, L. P. Taylor and V. Walbot, *PNAS* **82,** 5824 (1985).
83. M. E. Fromm, L. P. Taylor and V. Walbot, *Nature* **319,** (1986).
84. S. R. Ludwig, D. A. Somers, W. L. Petersen, R. F. Pohlman, M. A. Zarowitz, B. G. Gengenbach and J. Messing, *Theor. Appl. Genet.* **71,** 344 (1985).
85. W. H. R. Langridge, B. J. Li and A. A. Szalay, *Plant Cell. Rep.* **4,** 355 (1985).

86. R. D. Shillito, M. W. Saul, J. Paszkowski, M. Muller and I. Potrykus, *Bio/Technology* **3**, 1099 (1985).
87. A. Crossway, J. V. Oakes, J. M. Irvine, B. Ward, V. C. Knauf and C. K. Shewmaker, *MGG* **202**, 179 (1986).
88. N. Tanaka, M. Ikegami, T. Hohn, C. Matsui and I. Watanabe, *MGG* **195**, 378 (1984).
89. K. Harding and E. C. Cocking, *Protoplasma* **130**, 153 (1986).
90. M. Wallroth, A. G. M. Gerats, S. G. Rogers, R. T. Fraley and R. B. Horsch, *MGG* **202**, 6 (1986).
91. M. De Block, J. Schell and M. Van Montagu, *EMBO J.* **4**, 1367 (1985).
92. R. B. Flavell, *BioEssays* **3**, 177 (1986).
93. I. Potrykus, J. Paszkowski, M. W. Saul, J. Petruska and R. D. Shillito, *MGG* **199**, 169 (1985).
94. C. K. Shewmaker, J. R. Caton, C. M. Houck and R. C. Gardner, *Virology* **140**, 281 (1985).
95. N. Grimsley, B. Hohn, T. Hohn and R. Walden, *PNAS* **83**, 3282 (1986).
96. G. An, *Plant Physiol.* **81**, 86 (1986).
96a. W. Werr and H. Lorz, *MGG* **202**, 471 (1986).
97. N. Murai, D. W. Sutton, M. G. Murray, J. L. Slightom, D. J. Merlo, N. A. Reichert, C. Sengupta-Gopalan, C. A. Stock, R. F. Barker, J. D. Kemp and T. C. Hall, *Science* **222**, 476 (1983).
98. P. P. Chee, R. C. Klassy and J. L. Slightom, *Gene* **41**, 47 (1986).
99. M. A. Matzke, M. Susani, A. N. Binns, E. D. Lewis, I. Rubinstein and A. J. M. Matzke, *EMBO J.* **3**, 1525 (1984).
100. R. N. Beachy, Z.-L. Chen, R. B. Horsch, S. G. Rogers, N. J. Hoffmann and R. T. Fraley, *EMBO J.* **4**, 3047 (1985).
101. F. Schoffl and G. Baumann, *EMBO J.* **4**, 1119 (1985).
102. A Spena, R. Hain, U. Ziervogel, H. Saedler and J. Schell, *EMBO J.* **4**, 2739 (1985).
103. R. Broglie, G. Coruzzi, R. T. Fraley, S. G. Rogers, R. B. Horsch, J. G. Niedermeyer, C. L. Fink, J. S. Flick and N.-H. Chua, *Science* **224**, 838 (1984).
104. L. Herrera-Estrella, G. Van den Broeck, R. Maenhaut, M. Van Montagu, J. Schell, M. Timko and A. Cashmore, *Nature* **310**, 115 (1984).
105. F. Nagy, G. Morelli, T. T. Fraley, S. G. Rogers and N.-H. Chua, *EMBO J.* **4**, 3063 (1985).
106. G. Lamppa, F. Nagy and N.-H. Chua, *Nature* **316**, 750 (1985).
107. F. Nagy, S. A. Kay, M. Boutry, M.-Y. Hsu and N.-H. Chua, *EMBO J.* **5**, 1119 (1986).
108. J. Simpson, M. P. Timko, A. R. Cashmore, J. Schell, M. Van Montagu and L. Herrera-Estrella, *EMBO J.* **4**, 2723 (1985).
109. R. Fluhr, C. Kuhlemeier, F. Nagy and N.-H. Chua, *Science* **232**, 1106 (1986).
109a. A Barta, K. Sommergruber, D. Thompson, K. Harmuth, M. A. Matzke and A. J. M. Matzke, *Plant Mol. Biol.* **6**, 347 (1986).
110. J. B. Power, M. R. Davey, J. P. Freeman, B. J. Mulligan and E. C. Cocking, *in* "Methods in Enzymology" (A. Weissbach and H. Weissbach, eds.), Vol. 118, p. 578. Academic Press, Orlando, Florida, 1986.
111. S. G. Rogers, R. B. Horsch and R. T. Fraley, *in* "Methods in Enzymology" (A. Weissbach and H. Weissbach, eds.), Vol. 118, p. 627. Academic Press, Orlando, Florida, 1986.
112. J. Paszkowski and M. W. Saul, *in* "Methods in Enzymology" (A. Weissbach and H. Weissbach, eds.), Vol. 118, p. 668. Academic Press, Orlando, Florida, 1986.

114. G. An, B. D. Watson and C. C. Chiang, *Plant Physiol.* **81**, 301 (1986).
115. M. Deak, G. B. Kiss, C. Koncz and D. Dudits, *Plant Cell Rep.* **5**, 97 (1986).
116. T. J. Parsons, V. P. Sinkar, R. F. Stettler, E. W. Nester and M. P. Gordon, *Bio/Technology* **4**, 533 (1986).
117. J. A. Garcia, J. Hille and R. Goldbach, *Plant Sci.* **44**, 37 (1986).
118. E. J. Perkins, C. M. Stiff and P. F. Lurquin, *Weed Sci.* (in press).
119. L. Comai, D. Facciotti, W. R. Hiatt, G. Thompson, R. E. Rose and D. M. Stalker, *Nature* **317**, 741 (1985).
120. S. R. Ludwig, R. E. Pohlman, J. Vieira, A. G. Smith and J. Messing, *Gene* **38**, 131 (1985).
121. G. M. S. Hooykaas-Van Slogteren, P. J. J. Hooykaas and R. A. Schilperoort, *Nature* **311**, 763 (1984).
122. Y. Yamada, Y. Zhi-Qi and T. Ding-Tai, *Plant Cell Rep.* **5**, 85 (1986).
123. H. Morikawa, A. Iida, C. Matsui, M. Ikegami and Y. Yamada, *Gene* **41**, 121 (1986).

Epstein–Barr Virus Transformation

Samuel H. Speck and
Jack L. Strominger

Dana-Farber Cancer Institute
Harvard Medical School
Boston, Massachusetts 02114

I. Historical Perspective

The identification of Epstein–Barr virus (EBV) resulted from the pioneering epidemiologic observations of Burkitt in the late 1950s on an unusual lymphoma of children in Africa (*1, 2*). Burkitt showed that this frequent childhood lymphoma had a distribution corresponding to the malaria belt in Africa, and he therefore suggested that this disease has an infectious origin. Subseqently, endemic areas of Burkitt's lymphoma elsewhere in the world were also found to correspond to the incidence of hyperendemic malaria (*3, 4*); later, sporadic cases occurring throughout the world were identified (*5*). The ability to culture Burkitt's lymphoma tumor cells *in vitro* (*6*) greatly facilitated the identification, in 1964, of virus particles in tumor cells of African Burkitt's lymphoma patients (*7, 8*). This virus was identified as a unique herpesvirus and has since been referred to by the names of its discoverers.

Serological studies revealed a second human tumor associated with this virus, namely, nasopharyngeal carcinoma (*9–12*). Unlike Burkitt's lymphoma, this carcinoma is primarily a disease of adults, and is particularly prevalent in individuals of Chinese (Cantonese) descent (see *13*). Epidemiological data suggest that some genetic factor may be involved in its genesis, but its nature has never been

Progress in Nucleic Acid Research
and Molecular Biology, Vol. 34

clearly elucidated. A major problem is the inability to culture cells from explanted tumors, which are of epithelial origin. EBV has also been implicated in the development of at least one other humor tumor, thymic carcinoma (*14*).

Despite the fact that EBV-associated tumors display unique geographic distributions, this virus is ubiquitous among all human populations (see *15*). In the Western world, infection is primarily associated with a self-limited lymphoproliferative disorder, infectious mononucleosis (*16*). The virus is shed from the oropharynx and is thought to be transmitted through oral contact. As far as is known, infection in preadolescent children results in a benign disease, which is largely asymptomatic. However, infection after puberty can result in the debilitating, nonmalignant lymphoproliferative syndrome. The incidence of infectious mononucleosis is highest in the sanitary Western world where individuals avoid contacting the virus until late adolescence when they begin dating and kissing. In rare cases, the infection may become chronic.

In addition to Epstein–Barr virus infection, the development of Burkitt's lymphoma involves a second event, namely, the chromosomal translocation of one of the alleles of the c-*myc* gene. This translocation involves chromosome 8 (8q25 → TER) and one of the three chromosomes containing the immunoglobulin loci (chromosome 14, Ig heavy chain locus; chromosome 2, Ig κ chain locus; chromosome 22, Ig λ chain locus) (*17–20*). As opposed to African Burkitt's lymphoma, most of the sporadic cases of the disease in the Western world are not associated with EBV, are not especially prevalent in childhood, but are associated with the same chromosomal translocation (*20*, *21*). Thus, the role of the virus in the development of African Burkitt's lymphoma remained enigmatic. On the one hand, it is possible that the virus only provides a lymphoproliferative stimulus that enhances the probability of chromosomal translocation. Alternatively, it is possible that genetic information provided by the virus is essential for oncogenesis, even after the translocation. In that case, one would postulate that a mutation in the sporadic non-EBV-associated cases of Burkitt's lymphoma has provided similar genetic information.

While Epstein–Barr virus has not generally been associated with human tumors in the Western world, there has recently been a dramatic increase in the incidence of EBV-associated non-Hodgkin's lymphomas linked to immunosuppressed individuals (see *22* and *23*). Both transplant and AIDS patients have a markedly increased risk of developing EBV-associated lymphomas. The virus is also associated with a fatal lymphoproliferative disease in congenitally immunodeficient patients with Duncan's syndrome (see *24*). The correlation between EBV-associated malignancies and immunosuppression is further substantiated by the finding that EBV-specific T lymphocyte control is impaired during acute malaria induced by *Plasmodium falciparium* (*25*).

In addition to Burkitt's lymphoma cell lines, the study of Epstein–Barr virus has relied heavily on the ability to infect human B lymphocytes *in vitro*. This infection leads to two phenomena, latency and proliferation, that are pertinent to the pathology induced by the virus. "Latency" is defined as the continued presence of virus in infected cells as autonomously replicating episomes without production of viral particles. The presence of EBV genetic information also results in a growth transformation (immortalization) of the infected B lymphocytes. However, these cells are *not* oncogenically transformed and do *not* carry the chromosomal translocation associated with Burkitt's lymphoma. Cell lines transformed *in vitro* (lymphoblastoid cell lines), as well as cell lines derived from Burkitt's lymphoma patients, have been invaluable in the study of viral gene expression in EBV-immortalized lymphocytes (see *15*).

The limited range of cell types that Epstein–Barr virus can infect (B lymphocytes and some epithelial cells in the nasopharynx) appears to be restricted by the expression of the receptor for the virus. It has recently been determined that the virus receptor is one of the complement receptors, C3dR (also called CR2) (*26, 27*). Interestingly, homologies between the EBV glycoprotein, gp350, and C3d have now been identified (*28*).

The biology of Epstein–Barr virus poses a special problem to investigators because its mode of existence (latent infection) does not lend itself to the classic analysis of viral function by mutagenesis coupled with plaque assays, an approach that has been so successfully employed in the analysis of lytic viruses. Furthermore, its cell and host range are effectively limited to human B lymphocytes (and those of a few higher primates), a cell type generally not ammenable to culture *in vitro* in the absence of transforming virus. In addition, the absence of a fully permissive cell type (one in which the virus lytically replicates) has severely limited the ability to produce large amounts of the virus. Thus, studies of this virus have "come of age" only with the advent of recombinant DNA technology, which has permitted the genome to be cloned (*29, 30*) and sequenced (*31*). The extensive information provided by the DNA sequence has now allowed relatively rapid progress in the study of the molecular biology of EBV and the associated growth transformation of human B lymphocytes.

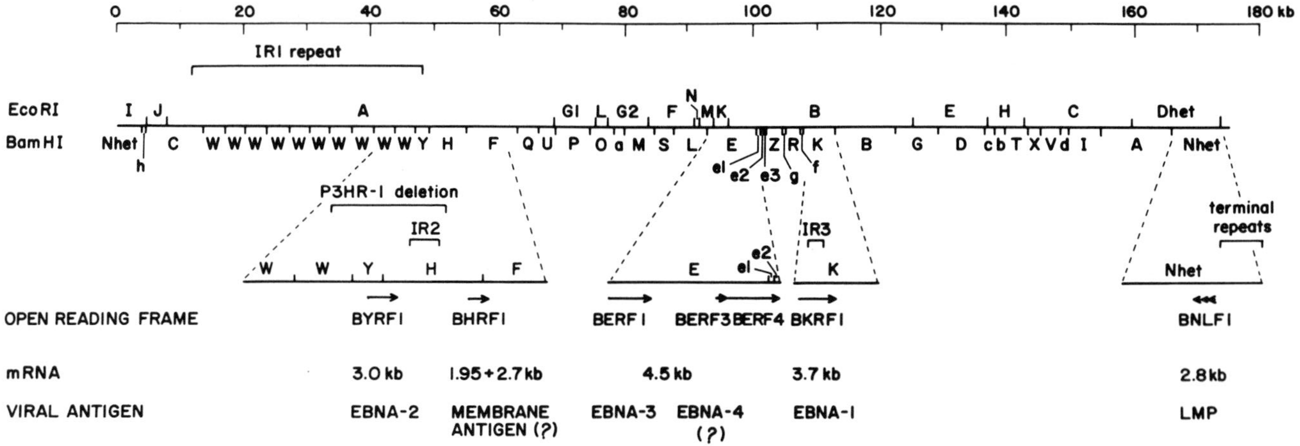

FIG. 1. *Bam*HI and *Eco*RI restriction endonuclease maps of the B95.8 strain of EBV. Regions of the viral genome that are transcribed in latently infected lymphocytes are shown. The relevant open reading frames and the viral proteins they encode are indicated. Abbreviations:

EBV, Epstein–Barr virus.
EBNA, Epstein–Barr nuclear antigen.
BERF, *Bam*HI E rightward open reading frame.
BHRF, *Bam*HI H rightward open reading frame.
BKRF, *Bam*HI K rightward open reading frame.
BYRF, *Bam*HI Y rightward open reading frame.
BMLF, *Bam*HI M leftward open reading frame.
BNLF, *Bam*HI N leftward open reading frame.
BZLF, *Bam*HI Z leftward open reading frame.

Epstein–Barr virus is a large DNA virus with a genome approximately 170×10^3 bases long (*29–31*). The genome exists as a linear molecule in the packaged virus. However, upon infection the genome circularizes and resides largely as a free episome in the nucleus of infected cells (*32*). It was initially cloned and sequenced as *Bam*HI and *Eco*RI restriction fragments, and the restriction map for the B95-8 strain of EBV is shown in Fig. 1. The viral genome is composed of both unique and repetitive regions. The ends of the genome contain a number of direct terminal repeats involved in circularization of the virus. The major internal repeat element, IR-1, near the left hand end of the genome, is composed of a variable number (about 8–12 in different virus strains) of 3.2 kb direct repeats. Three other smaller repeat regions have been identified. The complete DNA sequence (published only in part, but available on computer tape) has provided a wealth of information about the locations of possible promoters, open-reading frames, and polyadenylation signals (*31*). However, as is the major focus of this review, the structures of the transcripts encoding viral antigens associated with the latent virus life cycle are extremely complex. Thus, it has only been through the cloning and sequencing of viral transcripts, as cDNAs, that an understanding of viral transcription in latently transformed cells has begun to be elucidated.

II. Viral Antigens Expressed during Latent Infection

A. Nuclear Antigens

Historically, the only viral antigen detectable in latently infected lymphoblastoid cell lines by immunofluorescence employing patient sera was a nuclear antigen (Epstein–Barr nuclear antigen, EBNA) (*33*). Subsequent evidence suggested that there might be more than one EBNA expressed in latently infected cells (*34, 35*). Indeed, it has been determined that several are expressed, and to date five distinct EBNAs have been identified (*36–46*). Of these, EBNA-1 is the best characterized. It was first identified by gene transfer coupled with immunofluorescent staining using human sera with high anti-EBNA titers (*36*). EBNA-1 is encoded in a 2-kb open reading frame mapping at the left end of the viral *Bam*HI K fragment (*Bam*HI *K* *R*ight reading *F*rame *1*: BKRF-1) and contains the IR-3 repeat region (*37, 47*) (Fig. 1). This gives rise to a polypeptide of approximately 70 kDa (the size of EBNA-1 varies among different strains of EBV due to variation in the length of IR-3), approximately 200 residues of which are composed of an irregular copolymer of glycine and alanine encoded by IR-3 while the remaining portion of the protein is rich in arginine and proline (*37*).

EBNA-1 is present in all lymphoblastoid and Burkitt's lymphoma lines that have been examined. A major insight into the function of this viral antigen came from efforts to identify the viral origin of replication responsible for the maintenance of viral genomes in latently infected cells (*48–50*). A cloned region of the *Bam*HI C fragment functions as an origin (*ori*-P), but only in eukaryotic cells that express EBNA-1. Thus, EBNA-1 appears in this respect to be similar to the SV40 large T antigen–an analogy further strengthened by the subsequent observation that there is an EBNA-1-dependent enhancer element associated with *ori*-P (*51*).

EBNA-2 is encoded, at least in part, by a 1.5 kb open reading frame (BYRF-1) that spans the junction between the viral *Bam*HI Y and H fragments (*39, 40, 52*) (Fig. 1). Like EBNA-1, EBNA-2 varies in size in different cell lines (86 kDa in the B95-8 marmoset cell line). Furthermore, there are two serologically distinct forms of EBNA-2 owing to divergent viral genomic sequences in BYRF-1 (*39*). In addition, the naturally occurring mutant virus P3HR-1, which cannot transform cord blood lymphocytes (*53, 54*), contains a deletion that spans the region encoding EBNA-2 (*55, 56*). Superinfection of an EBV-positive Burkitt's lymphoma cell line (Raji) with P3HR-1 virus, and recovery of recombinant immortalizing viruses, demonstrated that this deletion is responsible for the mutant phenotype of the P3HR-1 virus (*57*). However, as is evident from the discussion of viral transcription in latently infected cells (see Section III), the inability of P3HR-1 virus to immortalize B lymphocytes cannot unambiguously be assigned to the deletion of BYRF-1. Unlike EBNA-1, which is expressed in all EBV-positive cell lines, there are several EBV-positive Burkitt's lymphoma lines that do not express EBNA-2 (Daudi, P3HR-1, and Naliaka all have deleted EBNA-2 coding sequences).

The other EBNAs have only recently been identified, and there are as yet no insights into their functions. EBNA-3 and probably EBNA-4 are encodes by long open reading frames encoded in the viral *Bam*HI E fragment (Fig. 1). BERF-1 encodes EBNA-3 (*42*) and, based on cloning of viral transcripts, the open reading frames BERF-3 an BERF-4, together appear to encode EBNA-4 (*58*). EBNA-5 is encoded almost entirely by small exons from the major internal repeat, IR-1 (*44, 45*). Both EBNA-3 and EBNA-4 are large viral antigens ranging in size from 140 to 160 and 150 to 180 kDa, respectively (*42, 43*). EBNA-5 is actually a family of small nuclear antigens that varies from

22 to 70 kDa (*44, 45*). As is discussed below, this variation in size arises from differences in the number of IR-1 encoded repeat exons used to generate the coding sequence. All Burkitt and lymphoblastoid cell lines tested express both EBNA-3 and EBNA-4 (*43*). EBNA-5 is expressed in all lymphoblastoid cell lines examined, but may not be expressed in some Burkitt's lymphoma lines (*44, 45*). Clearly, the observation that some of the EBV-coded EBNAs are expressed in all lymphoblastoid, but not in all Burkitt's lymphoma cell lines, suggests that the other events responsible for the oncogenic transformation in the latter (i.e., c-*myc* translocation) have supplanted the need for EBV to supply those (or similar) functions for continued cell proliferation.

B. Membrane Antigens

Although initial studies on EBV-transformed cell lines did not identify serologically any viral membrane antigens, the demonstration of cytotoxic T lymphocytes specific for EBV-transformed B cells in infected individuals suggested the existence of such an antigen (the "lymphocyte-detected-membrane antigen," or LYDMA) (*59–61*). A potential candidate for this antigen was identified from the analysis of transcriptionally active regions of the viral genome in latently infected cells. It was predicted that an open reading frame in the *Bam*HI Nhet fragment at the right hand end of the viral genome (BNLF-1) encoded a membrane antigen associated with the latent viral life cycle (*62, 63*). This was confirmed by subsequent studies in which antibodies generated to portions of this antigen were used to demonstrate its presence on lymphoblastoid and Burkitt cell lines; this antigen is currently referred to as the latent membrane protein (*64, 65*). This region actually encodes two related membrane antigens, resulting from two promoters with differential activity and a single 3′ end (*66, 67*). The shorter of the two proteins lacks 138 residues at the amino terminus. In the deduced amino-acid sequence of this protein there are a number of interesting features: (1) the amino terminus is highly charged (in the longer form of the protein), and does not appear to contain a signal sequence; (2) the hydrophobic domain appears to contain six membrane-spanning regions connected by short hydrophilic linker sequences of 5 to 7 residues (similar to several transport proteins); and (3) the carboxyl terminus is long (200 residues), very acidic (the overall charge of the protein is −41 at neutral pH), and also proline-rich (*63, 64, 67*).

The predicted structure of the latent-membrane protein suggests that it plays an important role in EBV-induced growth transformation of B lymphocytes. Indeed, in a recent study, its expression in an established rodent fibroblast cell line, "Rat-1," led to loss of contact inhibition, anchorage independence, and tumorigenicity in nude mice (*68*). The latter point is perhaps surprising, given that EBV-transformed B lymphocytes are not tumorigenic in nude mice. This may, however, reflect differences in the cell types and their prior state of "activation." The role of the latent membrane-protein in EBV-infected B lymphocytes has yet to be elucidated.

The question of whether this viral antigen is the target for cytotoxic T lymphocyte killing of EBV-infected B cells is still unresolved. Recent data indicate that, in some Burkitt's lymphoma cell lines, only the truncated form of this protein is expressed (*67*). In addition, some Burkitt's lymphoma cell lines do not express the membrane antigen (*69*). However, the correlation between expression of the latent membrane protein and the presence of the antigen remains unclear.

We have recently identified another region of the EBV genome that is transcriptionally active in latently infected lymphoblastoid cell lines (*70, 71*). This region contains an open reading frame that spans the *Bam*HI H and F junction (BHRF-1), and an analysis of the deduced amino-acid sequence indicates that it encodes a membrane antigen (*70*). This putative antigen is potentially very exciting, because it shares homology with the *bcl-2* gene product (*72*). Analysis of chromosomal rearrangements in human follicular lymphomas revealed that the t(14;18) translocation, observed in greater than 90% of these lymphomas, joins the *bcl-2* gene to the IgH gene (*73, 74*). It has been proposed that this rearrangement leads to an activation of the transcription of *bcl-2*, analogous to that of the c-*myc* gene in Burkitt's lymphoma. This is the first example of a potential oncogene (or any cellular gene) that shares homology with an EBV-encoded antigen. The direct implication of the *bcl-2* gene product in the transformation of B lymphocytes in follicular lymphoma makes it very attractive to speculate that the BHRF-1 gene product serves a similar function during EBV infection. Expression of the BHRF-1 encoded polypeptide on lymphoblastoid cell lines has not been investigated, and awaits the generation of specific antibodies.

III. Viral Transcription in Latently Infected Lymphocytes

A. Long Primary Transcriptional Units

To date, seven transcriptionally active regions of the EBV genome have been identified in latently infected cells (see Fig. 1) (*41, 62, 75*). The most abundant transcripts are two small Pol-III products encoded at the left end of the *Bam*HI C fragment ("Epstein–Barr-encoded RNAs," EBERs) (*63, 76*). Although they are not homologous in nucleotide sequence to the adenovirus-encoded VA-1 and VA-2 RNAs, they can complement mutants that lack these small RNAs (*77, 78*). Their role during latent infection is unclear. However, they have recently been located in the nuclei of infected cells by *in situ* hybridization (*79*).

All of the viral messages, with the single exception of the transcripts encoding the two forms of the latent membrane protein, are transcribed from left to right (see Fig. 1) (*62*). A striking feature of these transcripts is their unusually large size relative to the size of the coding sequence. Indeed, EBNA-1 is encoded on a 3.7 kb message mapped at the left end of the *Bam*HI K fragment. However, based on the size of EBNA-1, and also from transfection data, only about 2 kb of coding sequence is required, leaving an additional 1.7 kb of transcript. Furthermore, S1 nuclease protection experiments identified only a single 2 kb exon from this region of the genome (*47*). Subsequent sequence analyses clearly identified a 1912 bp open reading frame (BKRF-1), but there did not appear to be a consensus eukaryotic promoter in the 5′-flanking sequences, although there is a polyadenylation signal 62 bp downstream from the termination codon (*31*). It therefore seemed likely that the exon that encodes EBNA-1 is spliced to a region relatively far upstream from the *Bam*HI K fragment.

To address the question of the origins of the 5′-flanking sequences, we decided to clone the transcript that encodes EBNA-1 (*80*). This turned out to be very difficult, most likely because of (1) the abundance of the EBNA transcripts is very low (only approximately 1–3 copies per cell) (*63*), (2) the regions of the viral genome encoding the EBNAs are extremely rich in G and C, and (3) the large size of these viral transcripts. As a result, EBNA-1 cDNA clones were underrepresented in libraries generated from poly(A)$^+$ RNA isolated from latently infected lymphoblastoid cell lines (only present at a level of 1 per 100,000 to 200,000 recombinants, or at about one-tenth the expected abundance), and no clones longer than 1.5 kb were generated (all the isolated clones terminated in the IR-3 repeat region of the EBNA-1 exon, a region which is approximately 75% G+C). However, these clones did allow us to verify that the polyadenylation signal at the 3′ end of the EBNA-1 exon is indeed utilized (*76*). To circumvent the problem of obtaining clones containing 5′-flanking sequences, large cDNA libraries were generated using as primers for reverse transcription both oligo(dT) and a specific oligonucleotide homologous to a region of the EBNA-1 exon approximately 250 bp downstream from the putative initiation codon. From 10^6 recombinants, one primer-extended EBNA-1 cDNA clone was obtained, JY-K2. Its exon map is shown in Fig. 2 (*80*).

Comparison of the DNA sequence determined for the JY-K2 cDNA clone to the genomic sequence of the virus revealed that this clone has seven exons. Two exons (W1 and W2) are encoded within the *Bam*HI W direct repeats (IR-1) of 63 and 131 bp, respectively, separated by an 82 bp intron. There are also two exons from the unique region of the viral *Bam*HI Y fragment (Y1 and Y2) of 31 and 121 bp, respectively, separated by an 85 bp intron. The W2 and Y1 exons are separated by a 2200-bp intron. The Y2 exon is spliced approximately 19,500 bp downstream to a 172-bp exon encoded in the

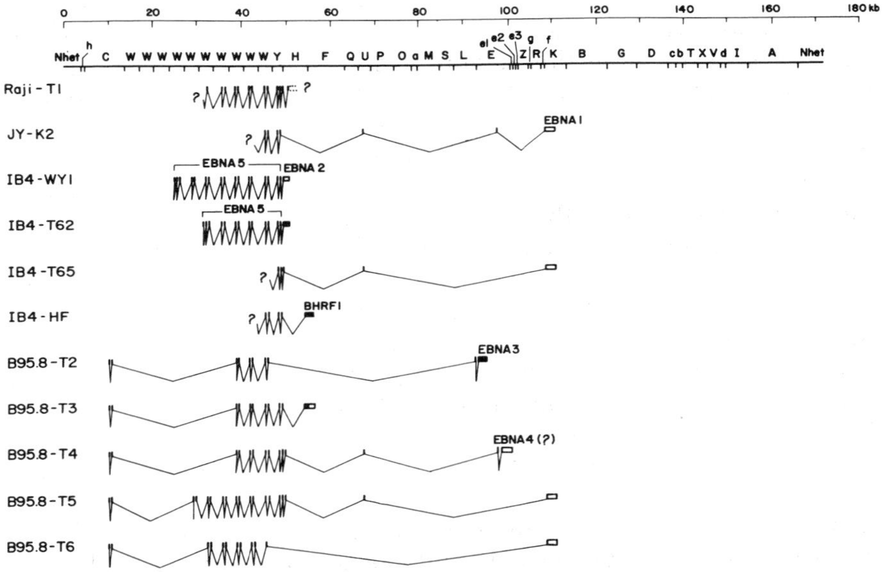

FIG. 2. Schematic exon maps of rightward viral transcripts are shown with respect to the *Bam*HI restriction endonuclease map of the B95.8 strain of EBV. Filled boxes represent sequences present in each cDNA clone, and open boxes represent proposed structures. The coding sequences for the known viral antigens present in latently infected lymphocytes are indicated. The cDNA clones are named according to the cell line from which they were derived (Raji, a latently infected Burkitt's lymphoma cell line; JY and IB4, latently infected lymphoblastoid cell lines; B95.8, productively infected marmoset lymphoblastoid cell line). Exon structures of the cDNA clones are from the following references: Raji-T1 (*82*), JY-K2 (*80*), IB4-WY1 (*86*), IB4-T62, -T65 (*85*), IB4-HF (*71*), B95.8-T2 (*83*), B95.8-T3, -T4, -T5, -T6 (*84*).

*Bam*HI U fragment. This exon in turn is spliced 38,600 bp downstream to a 367-bp-exon mapped in the *Bam*HI E fragment, which is spliced to the EBNA-1 exon 9000 bp away. Thus, the transcript that encodes EBNA-1 was found to be composed of exons spread over at least 70 kb of the viral genome and, since this clone is incomplete at the 5′ end, it is likely to be more than 10^5 bases.

In addition to the startling revelation that the primary transcriptional unit encoding EBNA-1 contains nearly one-half of the EBV genome, there were two other intriguing aspects of this clone. First, the W1, W2, Y1, and Y2 exons had also been found in another viral transcript from the EBV latently infected Burkitt's lymphoma cell line, Raji (Raji-T1, see Fig. 2) (*82*). In the Raji-T1 clone there are three and a half copies of the W1–W2 repeat exons, and also a third exon from the unique region of the *Bam*HI Y fragment (Y3) and an exon from the *Bam*HI H fragment (this clone splices around BYRF-1 which encodes EBNA-2). Neither the JY-K2 nor Raji-T1 cDNA clones contain the complete 5′ (or 3′) end of their respective transcripts, leaving unresolved the identification of the promoter(s) for these messages. The finding of two transcripts with common 5′ exons and distinct 3′ exons raised the possibility that a single or limited number of viral promoters might be involved in generating the rightward transcribed messages present in latently infected cells.

Second, an examination of the 5′ region of the EBNA-1 transcript in JY-K2 revealed three other open reading frames in addition to the one encoding EBNA-1. The first extends through exons W1, W2, and Y1 and nearly to the end of Y2. There are also two short ones encoded in the *Bam*HI E exon that use alternate reading frames. The larger of the two begins 9 bp from the 5′ end of the exon and terminates 48 bp within the *Bam*HI K exon. It would encode a protein of 135 residues (15 kDa). The shorter starts 57 nucleotides downstream from the initiation codon for the longer, terminates within the *Bam*HI E exon, and would encode a protein 72 residues long (8.4 kDa). Interestingly, there are three potential initiation codons within 37 nucleotides clustered at the start of the shorter open reading frame. The significance of these additional open reading frames is unclear. However, the potentially polycistronic nature of some of the viral transcripts present in latently infected cells is addressed in more detail at the end of this section.

Since the initial reports on the Raji-T1 (*82*) and JY-K2 (*80*) cDNA clones, several other viral transcripts have been partially cloned and characterized (see Fig. 2). Five very interesting clones have been obtained from the B95-8 (marmoset) cell line (B95.8-T2, -T3, -T4, -T5, and -T-6; see Fig. 2) (*83, 84*). These clones share a number of structural similarities to the two clones described above. Unfortunately, they are from the B95.8 cell line, which introduces two major problems in assessing their relevance: (1) B95-8 is a virus-producer cell line in which a relatively high percentage of the cells in culture is lytically replicating virus, so it is impossible to assign these clones unambiguously to transcripts associated with the latent life cycle of the virus; and (2) B95-8 is a marmoset B lymphocyte cell line, and the pathology of EBV in marmosets is known to be distinctly different from that in humans, raising the question of whether it is relevant to study control of viral transcription in this system.

Notwithstanding these reservations, these clones provide potentially interesting insights. The 5′ end of all of the clones map in the viral *Bam*HI C fragment 30 to 90 nucleotides downstream from a CCAAT-N_{37}-TACAAAA sequence, suggesting that they may represent closely the 5′ ends of these transcripts. Furthermore, all the clones contain a variable number of the W1–W2 repeat exons and three of the clones have either two or all three of the *Bam*HI Y encoded exons. The clones all have distinct 3′ structures, with the exception of T5 and T6, both of which contain the 5′ end of the EBNA-1 exon. Attempts in our laboratory to obtain cDNA clones from latently infected human lymphoblastoid cell lines that contain exons from the *Bam*HI C fragment have so far been unsuccessful. Although the use of a promoter in *Bam*HI C in close proximity to *ori*-P is very attractive, the use of such a promoter in human cells remains to be established.

While the B95-8 clones implicate the use of a promoter in the viral *Bam*HI C fragment during latent infection, three clones (IB4-WY1, IB4-T62, and IB4-T65) encoding EBNA-5 and possibly EBNA-2 (recovered from the tightly latent lymphoblastoid cell line IB4) provide strong evidence for the use of the consensus eukaryotic promoter present in the *Bam*HI W repeats (IR-1) (*85, 86*). The 5′ end of the IB4-WY1 cDNA clone, which was isolated in our laboratory, contains one copy of a 26-nucleotide exon (W0) that is 28 nucleotides downstream from the promoter element (CAATT-N_{34}-TATAAA) (*86*). Whether this promoter is utilized in the marmoset cell line, B95-8, is unclear since the clones that have been characterized were selected by screening with the *Bam*HI C fragment (*83, 84*) which would not hybridize to any transcripts initiating in IR-1.

Interestingly, the splice between the W0 exon and the first repeat exon (W1′) generates an AUG initiation codon, and is the only potential translation initiation codon in the entire EBNA-5 open reading frame. The IB4-T65 cDNA clone contains the entire EBNA-2 open reading frame in addition to that of EBNA-5, again raising the question of the significance of more than one in a single transcript. While the IB4-WY1 clone has seven copies of the W1–W2 repeat exons, both the IB4-T62 and IB4-T65 clones have only three copies of these repeat exons, indicating either that (1) more than one of the *Bam*HI W promoters are active in latently infected cells, or that (2) some of the W1–W2 repeat exons can be eliminated by splicing. However, immunoblots using affinity-purified anti-EBNA-5 antibodies do indicate that predominantly only one or two EBNA-5 antigens are synthesized in any given lymphoblastoid cell line (although in some cell lines as many as five or six different EBNA-5 polypeptides can be detected) (*44, 45*).

B. Alternative Splicing

In addition to the long transcriptional units necessary to generate the viral transcripts present in latently infected cells, considerable diversity is also generated by alternative splicing. Table I summarizes the splice donor and acceptor pairs identified to date. While several exons seem to splice to only one specific acceptor (C1–C2, C2–W1, and Y1–Y2), a number of exons are capable of splicing to two or three different splice acceptor sites. In two cases the splice donor signal is ignored (see Fig. 2; clones IB4-WY1, IB4-T62, IB4-T65, and IB4-W2′), one in which the transcript reads through the W2 splice junction (IB4-W2′) (*87*), and the other in which the transcript reads through the Y3 splice junction (IB4WY-1, IB4-T62, and IB4-T65) (*85, 86*).

Alternative splicing may also offer a trivial explanation for the puzzling existence of apparently polycistronic viral transcripts. This is most apparent in the case of the recently characterized EBNA-1 cDNA clones, all of which lack the small *Bam*HI E exon (E2) present in the JY-K2 cDNA clone (*84, 85*). This suggests that there is heterogeneity among transcripts containing the EBNA-1 open reading frame, only some of which may actually be translated to produce EBNA-1. Furthermore, the B95.8-T4 clone is similar to the 3′ region of the JY-K2 clone, except that the E2 exon splices to the large E3 exon instead of to the EBNA-1 exon. In fact the BERF-3, encoded in the E2 exon, is in frame with the BERF-4 (proposed to encode EBNA-4) and may encode a portion of EBNA-4. With regard to the putative polycistronic EBNA-2 transcript, the observation that the W0 splice to the first repeat exon does not always generate an AUG (two different splice

TABLE I

SPLICE DONOR AND ACCEPTOR PAIRS[a]

Exon border	Splice donor	Splice acceptor	Reference
Consensus	MAG/GTRAGT	$(Y)_n$ NYAG/G	*95*
C1/C2	ACC/GTAAGT	(CTTCCCT)CTAG/G	*83, 84*
C2/W1	CAT/GTATCT	(CCGCATC)CAAG/C	*83, 84*
W0/W1′	AAT/GTAAGA	(TCCAAGC)CTAG/G	*85, 86*
W0/W1	AAT/GTAAGA	(CGCCATC)CAAG/C	*87*
W1/W2	GAG/GTAAGT	(ACCCGTC)TCAG/G	*71, 80, 82–86*
W1/L	GAG/GTAAGT	(TTTTGTT)GCAG/A	*83*
W1/K	GAG/GTAAGT	(ATCTCTT)TTAG/T	*84*
W2/W1	GGG/GTAAGT	(CGCCATC)CAAG/C	*71, 80, 82–86*
W2/Y1	GGG/GTAAGT	(TTACAAC)CAAG/C	*71, 80, 82–86*
Y1/Y2	CGG/GTAAGT	(TTCCAAT)GTAG/T	*71, 80, 82–86*
Y2/Y3 and Y3′	CAG/GTGATT	(TCCACCC)GCAG/T	*82–86*
Y2/U	CAG/GTGATT	(AATTTCT)GCAG/G	*80*
Y2/HF	CAG/GTGATT	(TGGTTTT)CTAG/T	*71, 84*
Y3/H	CAG/GTACAT	(TCCATCT)ATAG/A	*82*
Y3/U	CAG/GTACAT	(AATTTCT)GCAG/G	*84, 85*
U/E2	AAG/GTGCTG	(CATATTT)TCAG/A	*80, 85*
U/K	AAG/GTGCTG	(ATCTCTT)TTAG/T	*84, 85*
L/E1	ACG/GTGAGC	(GTTGGTT)TCAG/C	*83*
E2/E3	AAG/GTGAGT	(ATTAATT)TTAG/C	*84*
E2/K	AAG/GTGAGT	(ATCTCTT)TTAG/T	*80*

[a] The exons are named according to the viral genomic *Bam*HI restriction endonuclease fragments in which they are encoded. In the cases where more than one exon present in the latent transcripts is encoded by a single *Bam*HI fragment, the exons are numbered with respect to their position in the fragment (5′ → 3′). Exons containing major open reading frames are as follows: Y3′, BYRF-1; HF, BHRF-1; E1, BERF-1; E2, BERF-3; E3, BERF-4; K, BKRF-1.

acceptor sites, separated by five nucleotides) (*86, 87*), raises the likely possibility that EBNA-2 transcripts lacking a translatable EBNA-5 open reading frame exist.

IV. Concluding Remarks

At present we only have a partial picture of the structure of the transcripts present in latently infected B lymphocytes. As such, it is perhaps premature to speculate about control of viral transcription in latently infected cells. However, given the consensus use of exons at the left end of the viral genome in all the rightward transcribed messages, it is difficult to avoid concluding that control is exerted by the utilization of only a very limited number of viral promoters (perhaps only one or two) encoded near the left end of the genome.

A number of central issues remain unanswered and need to be resolved before a clear understanding of viral transcription in latently infected cells can be achieved. (1) Is there a promoter(s) in the *Bam*HI C region of the viral genome that is utilized during latent infection? (2) If there is more than one promoter utilized (i.e., *Bam*HI C and IR-1 promoters; since the consensus promoter present in each of the IR-1 repeats is identical, in terms of a discussion of control of viral transcription we will consider them as a single promoter), is there a selective use of a given promoter for the expression of any given viral antigen? (3) What, if any, is the role of the EBERs in latent transcription? (4) Are some of the viral transcripts polycistronic (i.e., more than one viral protein translated from a single transcript), or are the apparently polycistronic messages that have been characterized merely byproducts of semirandom complex splicing of long primary transcripts? Clearly, most of these questions will be resolved only after a sufficiently large number of viral transcripts have been characterized.

It is attractive to speculate that EBNA-1 may play a pivotal role in controlling viral transcription. As discussed above, the observation that EBNA-1 is necessary for replication of *ori*-P and, in addition, can function to transactivate an enhancer associated with this element supports such a role for EBNA-1. However, the low level of viral transcription in latently infected cells also suggests that transcription may be down-regulated. If EBNA-1 serves this function, its being encoded by the exon most distal to the region of transcriptional initiation may ensure the proper expression of the other viral antigens associated with the latent life cycle. In the case of bacteriophage λ, the λ repressor can function to both up-regulate and down-regulate its own expression (see *88*).

The transcription of the latent membrane protein (*60*) is clearly an exception to the overall picture of viral transcription in latently infected cells. This may reflect a need for this protein to be regulated by a completely different control mechanism. However, it should be noted that the 5′ end of its transcripts may actually not be very far removed from elements controlling the rightward transcripts, as most of the Epstein–Barr virus genomes in infected cells exist as episomes in which the terminal repeats are covalently attached and thus bring the right and left ends of the genome in close proximity.

One other form of control that the virus most likely exerts is the regulation of lytic gene expression. Two viral gene products that appear to be intimately associated with induction of the lytic cycle have recently been identified (*89–91*). Both genes are oriented from left to right on the viral genome (encoded by the open reading frames BZLF-1 and BSLF-2/BMLF-1) (see Fig. 1) such that their transcription is in the opposite direction to that of the latent transcripts (with the exception of latent membrane protein). This may be very important since both genes are contained in the left half of the viral genome from which the long primary transcripts are generated. Expression of BZLF-1 in latently infected cells can disrupt latency (*89*), and the BSLF-2/BMLF-1-encoded polypeptide can function as a promiscuous transactivator of a number of EBV and non-EBV viral promoters (*90, 91*). Whether there is a virally encoded repressor that keeps the lytic switch off during latent infection is not known, but it is attractive to speculate that one of the EBNAs might serve such a function.

The last point we wish to consider is the requirements of latent infection. At least at a superficial level, the aim of latency is the maintainance of the viral genome without the production of viral particles. This is achieved by regulating viral gene expression and adapting a mechanism for propagation of the viral genome. The data at hand, for viruses known to infect cells latently, suggest that the major constraint is imposed by the cell type in which the virus sets up a latent infection. Perhaps a reasonable model, and certainly the best characterized, is bacteriophage λ lysogeny (see *88*). λ ensures its propagation to subsequent generations of lysogenically infected *E. coli* by integrating into the host chromosome, whereby it is passively replicated. Expression of lytic functions are actively repressed by constitutive (or semiconstitutive) expression of the λ repressor. Since *E. coli* is essentially "immortal," there is no requirement for λ to provide a growth stimulus to the infected cell.

The herpesviruses, Herpes Simplex virus, and Varicella Zoster virus both infect latently the trigeminal ganglion of neurons (see *92* and *93*). In these cases, it is likely that both viruses need only provide a mechanism for suppression of lytic gene functions, since the host cells are terminally differentiated and do not divide. At the other extreme, Epstein–Barr virus and *Herpesvirus saimiri* (*H. saimiri*), a T lymphotrophic virus that causes a rapidly progressing malignant lymphoma in some New World primates (see *94*), latently infect lymphocytes. These cells most likely have only a limited life span *in vivo* (although **memory** B and T cells may be long lived). Thus, at least in the case of Epstein–Barr virus (little is known about *H. saimiri* latent infection), the virus ensures its continued maintenance by inducing the infected lymphocytes to proliferate. This requires that (1) the viral genome must be efficiently replicated, (2) expression of lytic gene functions must be suppressed, and (3) the virus must provide a proliferative stimulus to the infected cells. It is therefore not surprising that the latent life cycle of the Epstein–Barr virus is complex and involves the expression of a number of viral antigens.

Central to understanding Epstein–Barr virus latency is the careful characterization of the patterns of viral transcription in latently infected cells. An important tool for dissecting the role of this virus in Burkitt's lymphoma will be the careful comparison of regions of the viral genome essential for latent infection in lymphoblastoid compared to Burkitt's lymphoma cell lines. This may also, at least partially, distinguish between viral functions involved in replication and repression of lytic gene expression from those involved in growth transformation.

References

1. D. Burkitt, *Br. J. Surg.* **46,** 218 (1958).
2. D. Burkitt, Nature **194,** 232 (1962).
3. K. Booth, D. P. Burkitt, D. J. Bassett, R. A. Cooke and R. J. Biddulph, *Br. J. Cancer* **21,** 657 (1967).
4. D. P. Burkitt, *in* "Burkitt's Lymphoma" (D. P. Burkitt and D. H. Wright, eds.), p. 186. Livingston, Edinburgh, 1970.
5. D. P. Burkitt, *Int. J. Cancer* **2,** 562 (1967).
6. M. A. Epstein and Y. M. Barr, *Lancet* **I,** 252 (1964).
7. M. A. Epstein, B. G. Achong and Y. M. Barr, *Lancet* **I,** 702 (1964).
8. M. A. Epstein, G. Henle, B. G. Achong and Y. M. Barr, *J. Exp. Med.* **121,** 761 (1965).
9. L. J. Old, E. A. Boyse, H. F. Wettgen, E. deHarven, G. Geering, B. Williamson and P. Cufford, *PNAS* **56,** 1699 (1966).
10. A. DeSchryver, S. Friberg, G. Klein, G. Henle, W. Henle, G. de-Thé, P. Cufford and C. H. Ho, *Clin. Exp. Immunol.* **5,** 443 (1969).
11. G. de-Thé, *Pathobiol. Annu.* **2,** 235 (1972).
12. G. Klein, *PNAS* **69,** 1056 (1972).
13. G. Klein, *in* "The Epstein–Barr Virus" (M. A. Epstein and B. G. Achong, eds.), p. 340. Springer-Verlag, Berlin, 1979.
14. S. Leyvraz, W. Henle, A. P. Chahinian, C. Perlmann, G. Klein, R. E. Gordon, M. Rosenblum and J. F. Holland, *N.E.J. Med.* **312,** 1296 (1985).
15. "The Epstein–Barr Virus" (M. A. Epstein and B. G. Achong, eds.). Springer-Verlag, Berlin, 1979.
16. W. Henle and G. Henle, *in* "Oncogenesis and Herpesviruses" (P. M. Biggs, G. de-Thé and L. N. Payne, eds.), IARC, Lyon, 1972.
17. G. Manolov and Y. Monolova, *Nature* **237,** 33 (1972).
18. L. Zech, V. Haglund, K. Nilsson and G. Klein, *Int. J. Cancer* **17,** 47 (1976).
19. G. Klein, *Nature* **294,** 313 (1981).
20. P. Leder, J. Batey, G. Lenoir, C. Moulding, W. Murphy, H. Potter, T. Stewart and R. Taub, *Nature* **222,** 765 (1983).
21. R. B. Mann, E. S. Jafe, R. C. Braylan, K., Nanba, M. M. Frank, J. L. Ziegler and C. W. Benard, *N.E.J. Med.* **295,** 685 (1976).
22. M. L. Cleary, R. F. Dorfman and J. Sklar, *in* "The Epstein–Barr Virus: Recent Advances" (M. A. Epstein and A. G. Achong, eds.). John Wiley, New York, 1986.

23. D. L. Birx, R. R. Redfield and G. Tosato, *N.E.J. Med.* **314,** 874 (1986)
24. D. D. Weisenburgen and D. T. Purtilo, *in* "The Epstein–Barr Virus: Recent Advances" (M. A. Epstein and A. G. Achong, eds.). John Wiley, New York, 1986.
25. H. C. Whittle, J. Brown, and K. Marsh, *Nature* **312,** 449 (1984).
26. J. D. Fingeroth, J. J. Weis, T. F. Tedder, J. L. Strominger, P. A. Biro and D. T. Feron, *PNAS* **81,** 4510 (1984).
27. R. Frade, M. Barel, B. Ehlin-Henriksson and G. Klein, *PNAS* **82,** 1490 (1985).
28. G. R. Nemerow, C. Mold, V. K. Schwend, V. Tollefson and N. R. Copper, *J. Virol.* **61,** 1416 (1987).
29. T. Dambaugh, C. Beisel and M. Hummel, *PNAS* **77,** 2999 (1980).
30. J. Skare and J. L. Strominger, *PNAS* **77,** 3860 (1980).
31. B. Baer, A. Bankier, M. Biggin, P. L. Deninger, P. J. Farrell, T. J. Gibson, G. Hatful, G. Hudson, S. C. Satchwell, C. Séguin, P. S. Tuffnell and B. G. Barrell, *Nature* **311,** 207 (1984).
32. T. Lindahl, A. Adams, G. Bjursell, G. W. Bornkamm, G. Kaschka-Dierich and V. Jehn, *JMB* **102,** 511 (1976).
33. B. M. Reedman and G. Klein, *Int. J. Cancer* **11,** 499 (1973).
34. K. O. Fresen, B. Merkt, G. W. Bornkamm and H. zur Hausen, *Int. J. Cancer* **19,** 317 (1977).
35. B. C. Strnad, T. C. Schuster, R. F. Hopkins, R. H. Neubauer and H. Rabin, *J. Virol.* **38,** 996 (1981).
36. W. Summers, E. Grogan, D. Shedd, M. Robert, C. Lui and G. Miller, *PNAS* **79,** 5688 (1982).
37. K. Hennessy and E. Kieff, *PNAS* **80,** 5665 (1983).
38. D. K. Fischer, M. F. Robert, D. Shedd, W. P. Summers, J. E. Robinson, J. Wolak, J. E. Stefano and G. Miller, *PNAS* **81,** 43 (1984).
39. T. Dambaugh, K. Hennessy, L. Chamnankit and E. Kieff, *PNAS* **81,** 7632 (1984).
40. K. Hennessy and E. Kieff, *Science* **227,** 1238 (1985).
41. K. Hennessy, S. Fennewald and E. Kieff, *PNAS* **82,** 5944 (1985).
42. K. Hennessy, F. Wang, E. W. Bushman and E. Kieff, *PNAS* **83,** 5693 (1986).
43. B. Kallin, J. Dillner, I. Ernberg, B. Ehlin-Henriksson, A. Rosen, W. Henle, G. Henle and G. Klein, *PNAS* **83,** 1499 (1986).
44. J. Dillner, B. Kallin, H. Alexander, I. Ernberg, M. Uno, Y. Ono, G. Klein and R. A. Lerner, *PNAS* **83,** 6641 (1986).
45. S. H. Speck and J. L. Strominger, in preparation.
46. D. T. Rowe, P. J. Farrell and G. Miller, *Virology* **156,** 153 (1987).
47. K. Hennessy, M. Heller, V. van Santen and E. Kieff, *Science* **220,** 1396 (1983).
48. J. L. Yates, N. Warren, D. Reisman and B. Sugden, *J. Virol.* **81,** 3806 (1984).
49. J. L. Yates, N. Warren and B. Sugden, *Nature* **313,** 812 (1985).
50. S. Lupton and A. J. Levine, *MCBiol.* **5,** 2533 (1985).
51. D. Reisman and B. Sugden, *MCBiol.* **6,** 3838 (1986).
52. T. Dambaugh, F. Wang, K. Hennessy, E. Woodland, A. Rickinson and E. Kieff, *J. Virol.* **59,** 453 (1986).
53. Y. Hinuma, M. Konn, J. Yamaguchi, D. J. Wudarski, J. R. Blakeslee and J. T. Grace, Jr., *J. Virol.* **1,** 1045 (1967).
54. M. Rabson, L. Gradoville, L. Heston and G. Miller, *J. Virol.* **44,** 834 (1982).
55. G. W. Bornkamm, J. Hudewent, V. K. Freese and V. Zimber, *J. Virol.* **43,** 952 (1982).
56. S. D. Hayward, S. G. Lozardwitz and G. S. Hayward, *J. Virol.* **43** 201 (1982).
57. J. Skare, J. Farley, J. L. Strominger, K. O. Fresen, M. S. Cho and H. zurHausen, *J. Virol.* **55,** 286 (1985).

58. M. Bodescot and M. Perricaudet, *NARes.* **14,** 7103 (1986).
59. D. J. Moss, A. B. Rickinson, L. E. Wallace and M. A. Epstein, *Nature* **291,** 664 (1981).
60. A. B. Rickinson, D. J. Moss and L. E. Wallace, *Cancer Res.* **41,** 4216 91981).
61. A. B. Rickenson, *in* "The Epstein–Barr Virus: Recent Advances" (M. A. Epstein and B. G. Achong, eds.). John Wiley, New York, 1986.
62. V. van Santen, A. Cheung and E. Kieff, *PNAS* **78,** 1930 (1981).
63. S. Fennewald, V. van Santen and E. Kieff, *J. Virol.* **51,** 411 (1984).
64. K. Hennessy, S. Fennwald, M. Hummel, T. Cole and E. Kieff, *PNAS* **81,** 7207 (1984).
65. K. P. Mann, D. Staunton and D. A. Thorley-Lawson, *J. Virol.* **55,** 710 (1985).
66. G. S. Hudson, P. J. Farrell and B. G. Barrell, *J. Virol.* **53,** 528 (1985).
67. S. Modrow and H. Wolf, *PNAS* **83,** 5703 (1986).
68. D. Wang, D. Liebowitz and E. Kieff, *Cell* **43,** 831 (1985).
69. D. T. Rowe, M. Rowe, G. I. Evan, L. E. Wallace, P. J. Farrell and A. B. Rickinson, *EMBO J.* **5,** 2599 (1986).
70. A. Pfitzner, E. Tsai, J. L. Strominger and S. H. Speck, *J. Virol.* (in press).
71. P. Austin, J. L. Strominger and S. H. Speck, (unpublished).
72. M. L. Cleary, S. D. Smith and J. Sklar, *Cell* **47,** 19 (1986).
73. S. Fukahara, J. D. Rowley, D. Variakojis and H. M. Golomb, *Cancer Res.* **39,** 3119 (1979).
74. J. J. Yunis, M. M. Oken, M. E. Kaplan, K. M. Ensrud, R. R. Howe and A. Thedlogides, *N.E.J. Med.* **307,** 1231 (1982).
75. J. Sample, A. Tanaka, G. Lancz and M. Nonoyama, *Virology* **139,** 1 (1984).
76. M. D. Rosa, E. Gottlieb, M. R. Lerner and J. A. Steitz, *MCBiol.* **1,** 785 (1981).
77. R. A. Bhat and B. Thimmappaya, *PNAS* **80,** 4789 (1983).
78. R. A. Bhat and B. Thimmappaya, *J. Virol.* **56,** 750 (1985).
79. J. G. Howe and J. A. Steitz, *PNAS* **83,** 9006 (1986).
80. S. H. Speck and J. L. Strominger, *PNAS* **82,** 8305 (1985).
81. S. H. Speck, unpublished.
82. M. Bodescot, B. Chambraud, P. Farrell and M. Perricaudet, *EMBO J.* **3,** (1984).
83. M. Bodescot, O. Brison and M. Perricaudet, *NARes.* **14,** 2611 (1986).
84. M. Bodescot and M. Perricaudet, *NARes.* **14,** 7103 (1986).
85. J. Sample, M. Hummel, D. Braun, M. Birkenbach and E. Kieff, *PNAS* **83,** 5096 (1986).
86. S. H. Speck, A. Pfitzner and J. L. Strominger, *PNAS* **83,** 9298 (1986).
87. R. Rogers, J. L. Strominger and S. H. Speck, (unpublished).
88. M. Ptashne, *in* "A Genetic Switch: Gene Control and Phage Lambda." Cell Press, Cambridge, MA 1986.
89. J. Countryman and G. Miller, *PNAS* **82,** 4085 (1985).
90. P. M. Lieberman, P. O'Hare, G. S. Hayward and S. D. Hayward, *J. Virol.* **60,** 140 (1986).
91. K. M. Wong and A. J. Levine, *J. Virol.* **60,** 149 (1986).
92. T. J. Hill, *in* "The Herpesviruses" (B. Roizman, ed.), Vol. 3. Plenum Press, New York, 1985.
93. R. W. Hyman, *in* "The Herpesviruses" (B. Roizman, ed.), Vol. 1. Plenum Press, New York, 1982.
94. B. Fleckenstein and R. C. Desrosiers, *in* "The Herpesviruses" (B. Roizman, ed.), Vol. 1. Plenum Press, New York, 1982.
95. S. Mount, *NARes* **10,** 459 (1982).

Proteins Covalently Linked to Viral Genomes

ANDREI B. VARTAPETIAN
AND
ALEXEI A. BOGDANOV

A. N. Belozersky Laboratory of Molecular Biology and Bioorganic Chemistry
Moscow State University
Moscow, USSR

Covalently linked DNA– and RNA–protein complexes were discovered in the late 1950s (see *1* for references). Since then, many attempts have been made to elucidate their structures and functions, and the results of these studies have been systematically reviewed in this series (*2–4*). However, early investigators dealt with preparations of total cellular RNA and DNA; proteins covalently bound to nucleic acids were not characterized.

Several hypotheses have been proposed to explain the possible functional role of the covalent binding of proteins to nucleic acids (*2*), but all of them failed in experimental testing. The most important results of this early period were the development of a general strategy of the experimental proof of the existence of a covalent bond between nucleic acid and protein (*1*), and the extensive study of different types of nucleotide–peptide bonds using numerous synthetic models (*4*).

Progress in Nucleic Acid Research and Molecular Biology, Vol. 34

The last decade saw a dramatic increase of interest in covalently linked nucleic acids and proteins. It was found that the 5′ termini of many viral DNAs and RNAs are covalently bound to specific proteins. Moreover, the attachment of these proteins to DNA or RNA was shown to be an important step in viral genome replication. This review summarizes the achievements and trends of research in this expanding field.

I. How Genome-Linked Proteins May Be Identified

The fact that proteins can be covalently linked to the nucleic acids of well-investigated viruses came as a surprise. This phenomenon was detected during an investigation of viral nucleic acids isolated without proteolytic treatment, a procedure frequently employed for eliminating protein impurities from nucleic acid preparations. It turned out that the viral DNAs thus obtained possess certain properties making them different from the protease-treated DNAs. Such DNAs acquire anomalous electrophoretic mobilities in polyacrylamide gel and in agarose (*5–7*); they have lower buoyant densities (*8, 9*); they can bind tightly to some sorbents, such as BND-cellulose (10), GF/C, and nitrocellulose filters (*11, 12*); and they form circular and oligomeric structures detectable by electron microscopy (*13, 14*). The normal behavior of these DNAs could in all cases be restored by protease treatment. The 5′ termini of the DNAs must have been blocked by some agent, and this made it impossible to carry out a 5′-terminal radioactive labeling or a 5′ → 3′ exonuclease DNA hydrolysis (*15–17*), though 3′ → 5′ exonucleases readily hydrolyze such DNAs. In the case of eukaryotic RNA viruses, the RNAs, isolated from the viral particle (virion), likewise had the blocked 5′ end instead of the characteristic cap-structure or the 5′-terminal nucleoside mono- (di-, tri-) phosphate (*18–21*). In some cases, the infectivity of protease-treated DNAs and RNAs proved to be much lower than of those not treated with protease (*5, 8, 22*). These results suggested the existence of a blocking group, possibly of protein nature, at or close to the 5′ terminus of viral nucleic acids, that is not removed by standard deproteinization procedures.

Thus, there is reason to suspect that some protein must be tightly linked to the nucleic acid. If so, two questions arise: how may one identify this protein? and what can be done to prove that it is covalently linked to the nucleic acid? To tackle the first problem, the following approach is possible. A viral particle contains, besides the genome nucleic acid which may be represented by a molecule(s) of DNA and RNA, many copies of structural protein(s) that are not covalently linked to the nucleic acid. These proteins may be fully separated from the nucleic acid by repeated phenol deproteinization in the presence of sodium dodecyl sulfate. This procedure detaches the noncovalently bound proteins from the nucleic acids and transfers the protein to the phenol phase. If some protein was indeed bound to the nucleic acid covalently, then it should be retained in the bound state in the aqueous phase and be subject to analysis. It would be reasonable to suggest (and such is the case) that one molecule of protein is bound to each polynucleotide chain (if it is bound at all), and thus the quantity of the protein at the researcher's disposal is rather limited, usually at the picomole level. Thus, radioactive labeling becomes necessary for the identification of this protein.

The protein may be identified in a variety of ways. The most direct and rigorous way involves a nucleic acid labeled *in vivo* by ^{32}P. Hydrolysis of such a nucleic acid with nucleases yields mononucleotides. If the protein was covalently linked to the nucleic acid, the nucleotide participating in the formation of this linkage and carrying the ^{32}P label remains attached to the protein in the form of a protein–nucleotide complex; DNases and RNases do not hydrolyze such a bond (*23–25*). The task is therefore to separate the nucleotidylated protein from free nucleotides. This poses a problem since only one nucleotide residue out of several thousand is bound to the protein and thus the compound contains approximately 0.01% of the initial radioactivity. This problem may be solved by taking advantage of the positive charge of the protein at acidic pH (paper electrophoresis at pH 3.5, ion-exchange chromatography) (*23, 26*) or the size of the protein (electrophoresis of nucleic acid hydrolyzate) (*25, 27, 28*). In the latter, most convenient procedure, the protein covalently linked to the nucleic acid is detected by autoradiography, since it carries a covalently bound ^{32}P nucleotide. The detected zone is sensitive to protease treatment, and this testifies to the presence of a protein component (Fig. 1). This method allows one to assess the molecular weight of the investigated protein and its homogeneity.

Another way to identify a protein covalently linked to nucleic acid is to specifically label *in vitro* the preparation of the investigated viral nucleic acid, thoroughly purified from the noncovalently bound protein, by labeling the protein but not the nucleic acid. A convenient reagent for this purpose may be ^{125}I Bolton–Hunter reagent (*29*). If the viral nucleic acid contains a covalently linked protein, the radioactive label is visibly incorporated into the preparation. This label comigrates with the nucleic acid during subsequent fractionation and may be removed by protease treatment. After the treatment of the iodi-

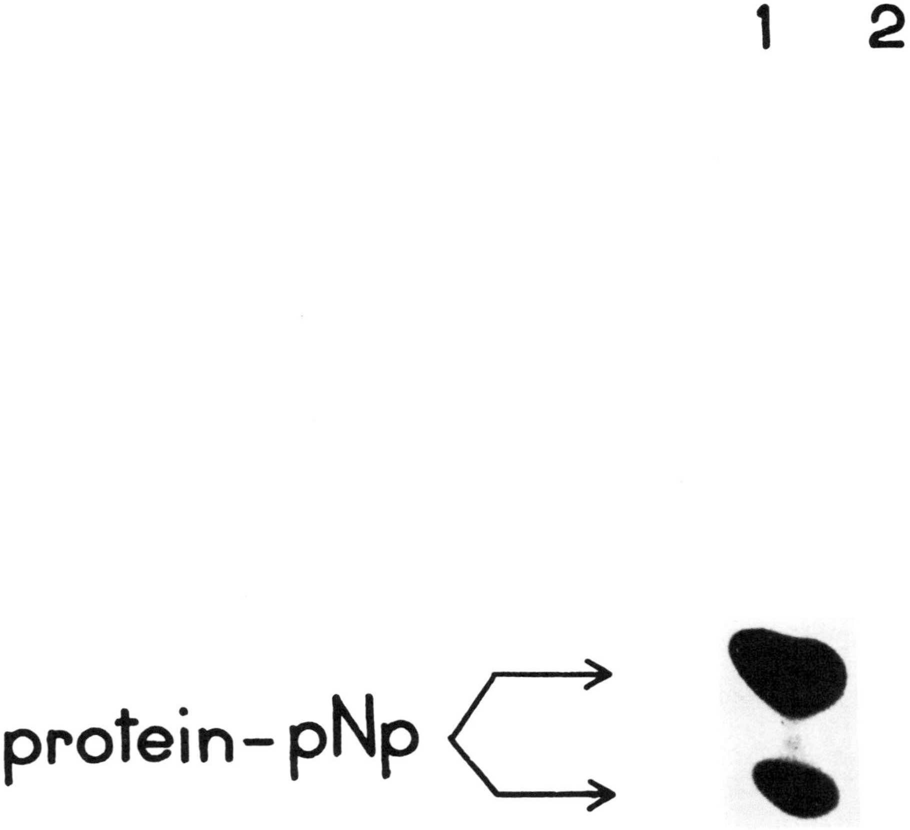

FIG. 1. Autoradiograph of the 12% polyacrylamide gel/dodecyl sulfate electrophoresis, demonstrating the presence of two different forms of protein, covalently bound to the virion RNA of encephalomyocarditis virus. Encephalomyocarditis virus virion RNA was labeled with ^{32}P *in vivo*, isolated, and hydrolyzed with RNases. The hydrolyzate was subjected to electrophoresis prior (lane 1) or after (lane 2) protease treatment.

nated preparation of nucleic acid with nucleases, the nucleotide-linked protein may be isolated and characterized, e.g., by electrophoresis, as described above. Although this approach is simpler than one involving ^{32}P labeling *in vivo*, the result thus obtained is not as unambiguous, for the conclusion that the identified protein is linked to the nucleic acid is predicated on the supposition that the noncovalently linked proteins were completely eliminated.

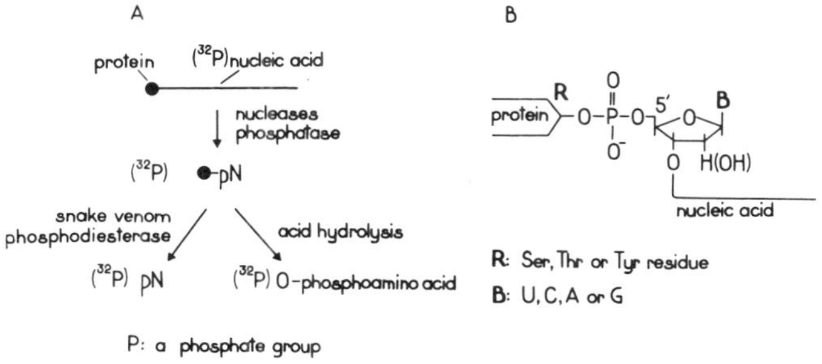

FIG. 2. A scheme analyzing the covalent linkage between protein and nucleic acid (A) and the deduced structure of this linkage (B).

The data obtained as a result of such experiments support the supposition that the identified protein was linked to the nucleic acid. Next comes the second problem: to prove that the protein was linked covalently. To give an definitive answer to this question, one must determine the type of a bond between the protein and nucleic acid molecules and find out what nucleotide and what amino acid are involved in its formation. This has been done for some DNA and RNA viruses. In all the cases studied, the protein proved to be linked by a phosphodiester bond connecting the phosphate group of the 5′-terminal nucleotide of the nucleic acid and the hydroxy group of a hydroxyamino acid residue of the protein, i.e., Ser, Thr, or Tyr (*29–34*). A scheme analyzing the linkage and its structure is given in Fig. 2. This scheme illustrates the use of the radioactive phosphate label in the nucleic acid preparation to identify the covalently linked protein and determine the nucleotide of the nucleic acid and the amino acid of the protein involved in bond formation.

In some cases, the amino acid-pNp compound was isolated as a result of the nuclease and acid treatment of the starting compound. The treatment of this compound by micrococcal nuclease cleaves the phosphodiester bond, with two radioactive phosphate-labeled products being formed: amino acid-P and Np, which can easily be identified (*31*).

It is noteworthy that a bond formed by the hydroxyamino acid of the protein with the phosphate or nucleotide is energy-rich, with the free energy of hydrolysis about 10 kcal/mol, which is comparable to the free energy of hydrolysis of the interphosphate bond in ATP (*35, 36*).

Thus, the above analysis allows us to infer unambiguously that the nucleic acids of some viruses are covalently linked to proteins. Of considerable interest in this respect are the following questions: What kind of proteins are these? What are their functions? What are the mechanisms of the formation and degradation of these compounds? These problems will be discussed as we examine specific examples in Sections II and III.

II. Genome-Linked Proteins of RNA Viruses

A. "VPg," a Protein Linked to the RNA Genome of Picornaviruses

Picornaviruses are replicated in the cytoplasm of infected cells and have a single-stranded polyadenylated RNA genome, about 7500 nucleotides long (*37*) and with positive polarity, i.e., the RNA has the same sequence as the viral mRNA. The genome RNA of picornaviruses, isolated from virions, contains a covalently linked virus-specific protein, designated VPg for "*V*iral *P*rotein *G*enome-Linked" (*23, 26, 31, 38–43*).

1. Structural Aspects

VPg was localized at the 5′ end of the RNA by means of enzymatic and chemical treatments (*26, 31, 44*) and shown to be bound to the phosphate group of the terminal uridylic acid of RNA by a phosphodiester bond to the Tyr hydroxy-group (Fig. 2B) (*29–31*). However, the viral mRNA contains no such protein and has a free pU at its 5′ terminus (*45*), which is the only difference between these two types of RNA (*20, 46, 47*). This circumstance distinguishes viral mRNA from capped cellular mRNAs, and may serve as a basis for their discrimination in the process of translation in a virus-infected cell.

Virus-infected cells contain two other types of virus-specific RNA that are related to RNA replication. One of them, the replicative intermediate (RI) RNA, is composed of a minus-strand template RNA with several nascent plus strands (*48*). This is a metabolically active structure (*49*) that supplies viral plus strand RNA for translation and virion assembly. The other is a by-product of replication termed replicative form (RF) RNA. This is a double-stranded RNA molecule composed of full-length plus and minus strands (*50*). Both the plus and the minus strands of RI and RF RNAs contain VPg on the 5′ termini (*28, 44, 51*). Since the picornaviral RNA has a genetically coded 3′-poly(A) sequence, a uridylic acid residue acts as a 5′-terminal nucleotide, linked to VPg, in the cases of minus strands as well (*52*); thus, the structure VPg (Tyr-pU) RNA is common both for plus and minus strands of RNA.

VPg may be obtained by exhaustive hydrolysis of virion ^{32}P-labeled RNA by RNases, in the form of the nucleotide derivative [^{32}P]VPg-pUp and then characterized (*23, 26, 31*). Initially its molecular weight was considerably overestimated (10,000–12,000) (*30, 31*), because the VPg exhibits anomalous mobility in electrophoresis and in gel filtration, a feature characteristic of some small proteins. VPg displays a number of other unusual properties. Since it sorbs readily on various surfaces, working with it involves major difficulties. This nonspecific sorption may be prevented by dodecyl sulfate (*23*). However, in a complex with dodecyl sulfate, VPg is not precipitated under standard assay conditions (e.g., by acetic and trichloroacetic acids; *53*). The latter poses a significant difficulty for VPg identification, for VPg is eluted from polyacrylamide gels during staining.

Nevertheless, polio VPg was isolated, and its amino acid sequence was partly determined (*54*). Now that the complete primary structure of polio RNA is known (*37*) (as well as that of the RNA of many other picornaviruses), it has become possible to locate the VPg gene and deduce the full amino acid sequence of the protein (*53*). By using the similarities of picornavirus genome structures, the gene for VPg was located and the corresponding amino acid sequence was deduced for the VPg of other picornaviruses (*55–61*), even before the VPg itself was identified. VPg proved to be a basic protein (or peptide), consisting of only 20–24 amino acids (2.5–3 kDa). Sequences of VPg's for different picornaviruses are shown in Table I. The VPg's of different picornaviruses show significant similarity; thus their sequences can be aligned. There are a number of fairly well-conserved amino acid residues, located predominantly in the amino-terminal and central parts of the molecule. These include the only Tyr residue, the third from the N-terminus, which is involved in formation of a phosphodiester bond with RNA. This Tyr residue seems to be located on the polypeptide chain bend caused by the adjacent Pro residue. The central and carboxy-terminal portions of the VPg molecule are markedly rich in basic amino acids, which might be involved in the interaction with the RNA. Hydrophobicity patterns of amino acid sequences of different VPg's seem to be similar (*43*).

It should be remarked that although one gene of VPg was found for most picornaviruses, an analysis of the protein bound to the virion RNA of EMC (*31*) and polioviruses (*62*) reveals two electrophoretically different forms of VPg (Fig. 1). It appears that the heterogeneity of poliovirus VPg is related to the procedure of isolating viral RNA and VPg (*62*). The genome of the foot-and-mouth disease virus, containing a tandem repeat of three similar but actually different VPg

TABLE I

A TENTATIVE VPg ALIGNMENT[a]

	1		3		5		7		9		11		13		15		17		19		21		23		References
Poliovirus type 1	G	A	Y	T	G	L	P	–	N	K	K	P	–	N	V	P	T	I	R	T	A	K	V	Q	53
ECHO virus type 9	G	A	Y	T	G	L	P	–	N	K	K	P	–	K	V	P	T	L	R	Q	A	K	V	Q	61
Human rhinovirus type 14	G	P	Y	S	G	N	P	P	H	N	K	L	–	K	A	P	T	L	R	P	V	V	V	Q	58
Human rhinovirus type 1a	G	P	Y	S	G	E	P	–	K	P	K	T	–	K	V	P	E	R	R	–	V	V	A	Q	61
Human rhinovirus type 2	G	P	Y	S	G	E	P	–	K	P	K	T	–	K	I	P	E	R	R	–	V	V	T	Q	59
Foot-and-mouth disease virus																									
Gene 1	G	P	Y	S	G	P	L	E	R	Q	K	P	L	K	V	R	A	K	K	P	–	Q	Q	E	55
Gene 2	G	P	Y	A	G	P	M	E	R	Q	K	P	L	K	V	K	V	K	A	P	V	V	K	E	55
Gene 3	G	P	Y	E	G	P	V	K	K	P	V	A	L	K	V	K	A	R	N	L	I	V	T	E	55
Encephalomyocarditis virus	G	P	Y	N	E	T	A	–	R	V	K	P	–	K	T	K	Q	L	L	–	–	D	I	Q	57
Hepatitis A virus	G	V	Y	H	G	V	T	–	K	P	K	Q	–	V	I	K	L	D	A	D	–	P	V	E	43

[a] VPg, genome-linked viral protein. Letters are symbols for the α-amino acids (*J. Biol. Chem.* **243**, 3557, 1968; **250**, 14, 1985).

genes (*55*), makes an interesting exception. The products of these three genes are active, judging by the fact that RNA in the virion may contain any of these proteins (*63*).

Like all picornaviral proteins, VPg is a product of the proteolytic processing of polyprotein, formed during translation of viral mRNA (*64*). This polyprotein is subjected to posttranslational cleavage to give a broad spectrum of intermediate precursor proteins, which ultimately go to form mature viral proteins. With the aid of antibodies raised against poliovirus synthetic VPg (or its N- and C-terminal fragments), yet another six precursor proteins with the VPg amino acid sequence were detected in poliovirus-infected cells (*65*) [apart from the VPg bound to the virion RNA and the free VPg (*66*)]. One of them, a protein of 12 kDa, is present in considerable amounts in the membrane structures of infected cells (*67*) where replication of picornaviral RNA is known to occur. These anti-VPg antibodies have been used in the study of the functions of VPg, a subject considered in Section II,A,2.

2. Functional Aspects

Nucleic acid–protein interactions constitute a very important and multifarious stage in the process of gene expression. Practically all essential requirements known have been satisfied by means of noncovalent interactions of proteins and nucleic acids. The variety of such nucleoproteins is infinite indeed. One may ask: why then should a protein be bound to nucleic acid covalently? Since all the covalent RNA–protein complexes investigated are of viral origin, this question may also be worded: at what stage of the virus reproduction cycle does the need arise for the presence of a RNA-linked protein, and how is this compound formed?

Picornaviral RNA enters the cell and is translated with the resultant formation of structural and nonstructural viral proteins. The latter include VPg and the RNA-dependent RNA polymerase, the appearance of which makes possible the replication of virion RNA. This process includes formation of the complementary (minus) strand of RNA which, in turn, is used as a template for the synthesis of new plus strands of the RNA. These plus strands are destined for translation, replication, or association with structural proteins to produce new virions. This set of processes may be triggered by the VPg-linked virion RNA of picornaviruses. Both the viral mRNA without VPg and the virion RNA subjected to proteolytic treatment retain their infectivity (*26*). It thus becomes clear that viral RNA does not need VPg for penetrating the cell and subsequent translation.

It is now believed that picornaviral RNA, formed in the process of replication, has VPg at the 5′ terminus (see below). The presence of viral mRNA without VPg in infected cells may be indicative of the activity capable of unlinking VPg from RNA, i.e., of specific hydrolysis of the Tyr-pU bond. Such unlinking activity was indeed found both in picornavirus-infected cells and in some noninfected eukaryotic cells (*68–71*). This activity is ascribed to a protein of 27 kDa, detected in the cytoplasm and in the nuclear salt wash (*69*). The possible significance of the discovery of such an enzyme, going beyond the viral VPg–RNA compounds, is discussed in Section II,C.

The capacity of the VPg-linked RNA for translation is of some interest. It might be that the VPg-bound RNA, unlike the viral mRNA, cannot be translated, even though it appears to be capable of forming an initiator complex of translation *in vitro* (*72*). If its translation *in vivo* does not take place at all, then it becomes possible to regulate the quantity of newly synthesized viral plus strand RNA dispatched to translation (without VPg) and to virion assembly (with VPg) at the stage of VPg detachment from RNA (*73*).

One of the possible functions of VPg could be its involvement in virion formation (*73*). Indeed, RNA within the virion is linked to VPg, while the VPg-free viral mRNA, present in large amounts in infected cells, is not incorporated into virions. Unfortunately, this idea has not been verified experimentally.

The possible VPg participation in the replication of viral RNAs has come to be the focus of attention (*73–75*). Localization of the polio VPg at the 5′ end of virion RNA has lent ground for an original suggestion that VPg may act as a primer in the initiation of viral RNA synthesis (*23, 26*). There is good reason for the priming suggestion. Indeed, picornaviral RNA, with a 5′-terminal nucleoside tri- or diphosphate, or a cap structure, has never been identified; instead, the nascent strands of viral RI RNA were already linked to VPg (*28*), which indicates that the attachment of VPg to RNA occurs at an early stage of its synthesis. Furthermore, the purified polioviral RNA-dependent RNA polymerase is capable of synthesizing a full-size copy of virion 3′-polyadenylated RNA, but only when oligo(U) is added to the reaction *in vitro* as a primer (*76, 77*). This points to the problem of RNA synthesis initiation, which is proposed to be solved with the participation of VPg (see Fig. 3A). The model suggests that VPg or one of its precursors binds a uridylate residue to the Tyr hydroxy group and forms the primer, VPg-pU (*74*). This process may or may not depend on template RNA. VPg-pU is specifically located at the 3′ end of the viral RNA; the 3′-OH group of the VPg-bound uridylate then acts as a

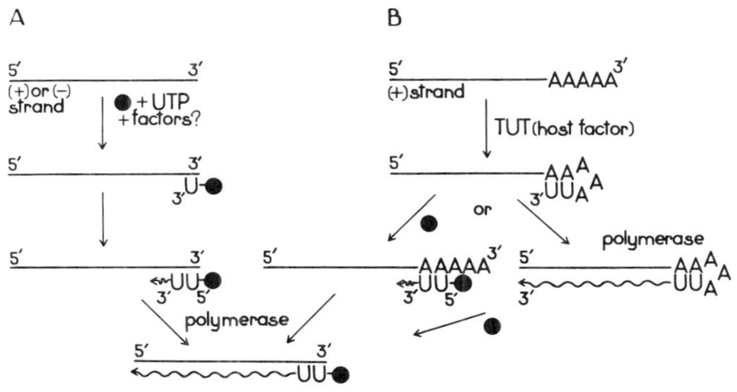

FIG. 3. Replication of the RNA genome of picornaviruses with participation of genome-linked viral protein (VPg). (A) VPg-pU priming model for initiating plus- and minus-strand synthesis; (B) hairpin priming model for initiation of minus-strand synthesis. The solid circle denotes VPg or pre-VPg.

growing point for the complementary RNA strand synthesis accomplished by RNA polymerase. The RNA product will carry a covalently linked VPg as a "memory" about the mechanism of initiation of its synthesis. This model is attractive since it is fit for the initiation of the synthesis of both plus and minus strands of RNA that have the identical 5′-terminal structure of VPg-pUpU ....

This model gained experimental support when it turned out that the initiation of picornaviral RNA synthesis could be studied in a cell-free system, a crude membrane fraction of the cytoplasm of virus-infected cells (*78*). It has long been known that such an *in vitro* system is capable, by using endogenous RNA as a template, of carrying out elongation of viral RNA strands with the formation of RNA of positive polarity, i.e., of RNA replication (*79*). It is in such a cell-free system, obtained from the encephalomyocarditis virus-infected cells, that the postulated primer VPg-pU and the more elongated forms were detected (*78, 80*). The nucleotide component of these compounds is linked to VPg via a Tyr-pU bond, as is in the case of VPg linked to the virion RNA (*80*). Cell membranes play a tangible role in VPg nucleotidylation, as their destruction by detergents precludes VPg uridylation (*78*). Formation of the VPg-bound dinucleotide pUpU was then demonstrated for the poliovirus in an analogous *in vitro* system (*81*), and more importantly, *in vivo* in poliovirus-infected cells (*66*). Polio VPg-pU may be chased to VPg-pUpU, and possibly to longer VPg-linked RNA species (*82*). Yet this is a low-efficiency process, probably because of the abortive elongation, or the absence of an essential component in the *in vitro* system (*83*). VPg-pU elongation resulting in formation of full-length VPg-linked RNA has not been documented thus far.

At present yet another, membrane-free *in vitro* system is being studied (*84*); it consists of purified polioviral RNA-polymerase (53 kDa) and virion RNA of the poliovirus. The RNA synthesis in this system is limited to the copying of the virion RNA, i.e., to the formation of a complementary minus strand. Strictly speaking, we cannot regard this process as RNA replication. Polymerase is the only protein necessary for elongation. As noted above, this process requires an oligo(U) primer; in the absence of nucleotide primer, this requirement may be met by the addition of cellular protein [designated "host factor," HF (*85*)]. Although VPg was not detected in polymerase preparations, the synthesis of complementary RNA in such a system, containing RNA template (VPg-bound or unbound), polymerase and HF, is inhibited by antibodies to synthetic VPg (*86–88*). Furthermore, part of the RNA product may be immunoprecipitated by these antibodies and contains a protein linked to the RNA via a Tyr-pU bridge; this protein is a putative VPg precursor. ATP is needed for this product to be formed (*89*). At the same time, the antibodies exert no effect on RNA synthesis in the oligo(U)-primed system, thus indicating that the antibodies do not block elongation (*86, 87*). These results may serve as corroborative evidence for VPg involvement in the initiation of viral RNA synthesis.

Since the process requires HF participation, its role in the initiation of minus strand RNA synthesis is of major interest. This protein, isolated from the cytoplasm of HeLa cells and partly associated with ribosomes, has a molecular weight of 67,000 (*90*) and is capable of binding to viral polymerase (*91*). The activity of terminal uridylyltransferase (RNA uridylyltransferase, EC 2.7.7.52) was shown to copurify with HF (*92*). This enzyme adds a short oligo(U) sequence to the 3′ terminus of various synthetic polyribonucleotides and, possibly, to the 3′ terminus of polioviral RNA (*92–94*). This result made it possible to formulate an alternative model of initiation of the synthesis of the viral RNA minus strand. According to this model (see Fig. 3B), the transferase activity may be involved in RNA synthesis *in vitro* by adding several U-residues to the 3′-terminal poly(A) sequence of the poliovirus RNA template. The oligo(U) tail thus synthesized could fold back and form a hairpin thus providing a 3′-hydroxyl group for priming RNA synthesis (*92, 95*). The product of this reaction *in vitro* should be a double-stranded RNA with the strands linked covalently at one end, which, as a result of denaturation, would be twice as long as the RNA template. Such covalently linked RF molecules have indeed been found in picornavirus-infected cells (96), and were formed *in vitro* in the presence of polio RNA template, viral RNA polymerase and HF (95). Formation of a full-length minus strand RNA molecule is thought to require the cleavage of the oligo(U) loop at the template RNA 3′ end by VPg or its precursor; VPg might concomitantly become attached to the 5′ terminus of the newly synthesized RNA strand.

Formation of the VPg-linked viral RNA (a possible criterion that the process is "native") has not yet been demonstrated for the membrane-free system *in vitro*. Thus, no evidence is available to link the formation of dimeric RNA by the hairpin model *in vitro* to picornaviral RNA replication *in vivo*. Besides, the supposition that VPg may specifically cleave the dimeric RNA and become covalently linked to the 5′ terminus at the nick lacks experimental proof. It should be noted that the fact that VPg per se is capable of carrying out this process is not immediately obvious, for VPg is of minute size (20–24 amino acids).

Another research group detected protein kinase activity in the preparation of purified HF capable of phosphorylating HF itself, and also the translation initiation factor, eIF-2. This phosphorylation is stimulated by double-stranded RNAs (97). The presence of protein kinase activity agrees with the need for ATP for initiation of minus strand RNA synthesis in the presence of HF (89). However, it is not yet clear how the protein kinase activity could be related to the synthesis of viral RNA.

Thus, two ways seem possible for VPg-bound viral RNA to be formed; either as a result of protein priming, or a result of the polynucleotide chain cleavage in the RNA precursor with the simultaneous linkage of VPg to the newly formed 5′ end of the nascent RNA molecule (nicking-linking). There are data supporting both models. It is clear that further investigations are needed to choose between the two, or to demonstrate that both ways are relevant *in vivo*. It must be noted that both mechanisms of initiation of RNA synthesis have no precedent in RNA biosynthesis. These two principles of protein–nucleic acid covalent complex formation, protein priming and nicking-linking, apply in the case of DNA viruses and are examined in Section III.

B. VPg Linked to the Virion RNAs of Some Other Viruses

The detection of VPg covalently linked to the RNA of picornaviruses stimulated a search for similar compounds in other RNA viruses. This search was successful, and demonstrated that picornaviruses are no exception concerning the existence of the genome-linked protein (VPg). At present the viral protein, covalently linked to the virion RNA, has been found in a number of eukaryotic RNA viruses with the positive genome that belong to the picorna- and caliciviruses (*98*) (animal viruses), como- (*99, 100*), nepo- (*101*), poty- (*102*), luteo- (*103*), sobemo- (*104*), and pea enation mosaic (*105*) viruses (plant viruses) and possibly to birnaviruses with a double-stranded RNA genome (*106*) (see Table II).

Although, by and large, these complexes are less well characterized than those of picornaviruses, certain common characters are detectable. The genome-linked proteins have low molecular weights and are heterogeneous in some cases. Their sizes may differ from the data given in Table II and based on VPg electrophoretic mobility, for the unusual behavior of VPg was noted for picornaviruses. A protein linked to a double-stranded genome of the infectious pancreatic necrosis virus (110 kDa) may be a possible exception (*106*). VPg, as has been suggested (or demonstrated), is located at the 5′ terminus of genomic RNAs and is bound to a uridylate residue, as a rule. The VPg bond to RNA in the case of comoviruses is a phosphodiester, as in the case of picornaviruses, though a hydroxyl group of the serine, and not tyrosine, residue is involved (*112*).

There seems to be no correlation between the presence of VPg and the number of RNA components of the viral genome: thus, the genome of picornaviruses and some other viruses is monopartite, and that of como-, nepo-, and pea enation mosaic viruses is bipartite. If the viral genome is represented by two molecules of RNA (and not one), both RNAs are linked to VPg (*99, 114*). VPg may be detected in replicative double-stranded forms of RNA, although in comoviruses not all RF molecules contain VPg (*123*). In some instances, when the strategy of viral RNA translation involves the formation of subgenomic RNAs, VPg is linked to the subgenomic RNA as well (*121*).

The presence of satellite RNA in infected cells is a characteristic feature of some strains of plant nepoviruses. This RNA is not homologous with the viral genome and depends on genome RNAs in replication and encapsidation. The replication of these satellite RNAs is likely to be effected by viral RNA polymerase according to the same mechanism as the synthesis of the nepoviral RNAs. This is indicated by the presence of VPg at the 5′ terminus of the satellite RNA of tomato black ring virus (*117*). At the same time, the satellite RNAs of some other nepoviruses contain no VPg; they are formed by autocatalytic cleavage of a multimeric RNA precursor synthesized, possibly, according to the rolling-circle model (*124*). This pathway is starkly different from the one suggested for the replication of nepoviral RNA (picornaviral RNA, too); yet some analogy with the hypothetical scheme of picornaviral RNA replication, including the cleavage of dimeric viral RNA by the VPg, is observable.

A startling difference of VPg-containing viral RNAs was found when the infectivity of protease-treated RNA preparations was being investigated. As in the case of picornaviruses, this treatment did not diminish the infectivity of the RNAs of como- (*107*), poty- (*102*), and luteoviruses (*103*). However, the infectivity of the RNAs of nepo- (22, 116), calici- (98), and sobemo- (*104*) viruses dropped after corresponding treatment (Table II). The suppositions about the function of the VPg of the viruses under study depend in many respects on analogy with more-investigated complexes of picornaviruses; it is assumed that VPg is involved in the replication of viral RNAs and in viral morphogenesis. In this connection, the need of intact VPg for the infectivity of RNAs of some viruses may reveal yet another function of VPg.

As noted above, the genomes of viruses that have VPg linked to RNA are of positive polarity (with the exception of a double-stranded RNA of birnaviruses). They may or may not have a 3′-terminal poly(A) sequence (*103, 104*). Proteolytic treatment of the VPg-containing RNAs in all the investigated cases does not annihilate the RNA translation capability (*26, 99, 116*) (Table II). The most frequent translation strategy for these RNAs is that of polyprotein formation and its subsequent cleavage to mature viral proteins (*37, 111, 125–127*). The most important thing is that various viruses that have VPg and utilize this strategy—such as picornaviruses of animals and como- and, possibly, potyviruses of plants—possess a surprisingly similar genome organization (*126, 128, 129*), suggesting a common ancestry. Consequently, a deeper investigation into the structural and genetic organization of viruses that have a genome-linked protein will make it possible to reveal more profound analogies between these viruses. The presence of VPg is a marker, one of the essential manifestations of this similarity. The common characters detected in VPg-RNA compounds of different viruses give hope that such a peculiarity as the presence of the protein covalently linked to the RNA genome reflects a general fundamental mechanism in the processes of replication of these viruses.

C. Do Cellular Analogues of the Covalent Complexes of Viral RNA and Protein Exist?

Protein covalently linked to the 5′ terminus of the RNA of the viral genome was found in viruses of plants and animals belonging to nine

TABLE II

RNA Viruses Possessing Genome-Linked Proteins (VPg)

Virus families	Representatives	Genome	MW of VPg	Linkage	Effect of proteolytic degradation of VPg on the RNA's		Presence of 3′-poly(A)	References
					Infectivity	Translation capability		
Picorna	Poliovirus, foot-and-mouth disease virus, encephalomyocarditis virus, human rhinovirus, etc.	ss[a] RNA, plus strand, 7500–8000 nt	20–24 residues, 3–12 kDa[c], heterogeneous	Tyr-O-5′pU	–[b]	–[b]	+	23, 26, 29–31, 37–43, 54
Como	Cowpea mosaic virus, squash mosaic virus, etc.	Two ssRNAs, plus strands, 3500 and 5000 nt	28 (?) residues, 4–5 kDa[c]	Ser-O-5′pU	–	–	+	99, 100, 107–113
Nepo	Tobacco ringspot virus, tomato black ring virus, etc.	The same, 4000 and 7000 nt, satellite RNA 1400 nt	4–6 kDa[c], heterogeneous	X-5′pU	+	–	+	22, 101, 114–117

Poty	Tobacco etch virus, tobacco vein mottling virus	ss RNA, plus strand, 10,000 nt	6–24 kDa[c]	X-5′pA	–	?	+	102, 118
Luteo	Potato leafroll virus	The same, 6000 nt	7 kDa[c]	?	–	?	–	103
Calici	Vesicular exanthema virus, San Miguel sea lion virus	The same, 8000 nt	10–15 kDa[c]	?	+	?	+	98, 119, 120
Sobemo	Southern bean mosaic virus	The same, 4500 nt	10–14 kDa[c]	5′(?)	+	?	–	104, 121, 122
Pea enation mosaic virus (unclassified)		two–three ssRNAs, plus strands 4500, 3500 and ±600 nt	17.5 kDa[c]	5′(?)	?	?	–	105
Birna	Infectious pancreatic necrosis virus	Two dsRNAs, 3100 and 3600 bp	110 kDa[c]	5′(?)	?	?	–	106

[a] ss, single-stranded; ds, double-stranded.
[b] minus, no effect; plus, decreased infectivity.
[c] MW was estimated by polyacrylamide gel/dodecyl sulfate electrophoresis.

different groups. Thus, this phenomenon is rather widespread in the viral world. Viruses are like a proving ground for different kinds of structures and mechanisms employed by the host cell. A question arises as to whether covalent complexes of RNA and protein exist in normal noninfected cells. This question is all the more legitimate as the unlinking activity, capable of selective hydrolysis of a phosphodiester bond between picornaviral RNA and VPg, appears not only in picornaviruse-infected, but also in uninfected cells (*68*). With polio VPg-RNA as a substrate, this activity was detected in a number of plant and animal cells; possibly, it is absent in bacteria (*E. coli*) (*69*). The enzyme is not specific to the protein component of the complex, from the fact that peptidyl-RNA, remaining after the protease treatment of VPg, is a good substrate. However, the RNA component is important for the effect of the unlinking activity: VPg, linked to the 5′-terminal nonanucleotide, is unlinked significantly less well than is VPg-RNA; VPg-pUp is not hydrolyzed by the enzyme at all (*69*). The specificity of this activity and its function(s) in uninfected cells are still obscure, but its very existence suggests that the cells contain RNAs covalently bound to proteins and that viruses become attuned to a corresponding cellular mechanism.

No such stable complexes have been found thus far. According to some data, aminoacyl-tRNA synthetases form a short-lived adduct with tRNA by binding the enzyme's nucleophilic group to the C-6 of the conserved uridine at position 8 in a tRNA molecule (*130*). The formation of these intermediate compounds must have little in common with the stable VPg-RNA complexes that have been dealt with. Some evidence on the possible existence of stable covalent complexes has been obtained, strange as it may seem, in investigations of proteins and not RNAs. These are glutamine synthetase I from *Drosophila* (*131*), a nucleolar protein of 125 kDa from Novikoff hepatoma cells (*132*), and the product of a cellular oncogene, protein p53 (*133*). In all cases, some oligoribonucleotide material was linked to the purified protein. These short oligonucleotides, it is believed, may be the products of RNA degradation, with the RNA being covalently bound to the proteins listed above. Yet, as far as we know, these RNAs have not been found and characterized.

Thus, one of the intriguing problems related to the studies of covalent viral RNA–protein compounds is the following: do analogous cellular compounds exist, and if they do, what are their structure and function?

III. Proteins Linked to the Genomes of DNA Viruses

A. Linear Double-Stranded DNA Genomes with Terminal Proteins

Linear double-stranded genomes of several viruses, i.e., adenoviruses (Ad) (*10–12, 14, 15*), small bacteriophages of *Bacillus subtilis* (ϕ29 and related phages) (*7, 13, 16, 17, 134*), *Steptococcus pneumonia* (Cp-1 and related phages) (*135, 135a, 136*), and broad-host-range bacteriophage PRD 1 (*137*) have proteins covalently attached to the 5′ ends of DNA. The proteins bound to both ends are identical and are about 30–55 kDa (*16, 135, 137, 137a, 138*). They are linked to the phosphate group of the 5′-terminal nucleotide via a phosphodiester bond to a serine (*32, 33*), threonine (*34*), or tyrosine (*139*) residue of the respective protein (see Table III and Fig. 2B). Because of their location on the DNA molecule, these proteins have been called *T*erminal *P*roteins (TP). Another characteristic feature of the genomes of these viruses is the presence of the inverted terminal repetitions (ITR) of up to several hundred base-pairs (for phage PRD 1, DNA sequence data are not available) (*140–143*), which means that both ends of duplex DNA, bearing the TPs, are identical (see Table III).

TPs are coded by "early" viral genes (*139, 144, 145*). Within each group of viruses, TPs are well-conserved judging by their peptide maps (*141, 143, 146*), but no sequence homology has been found between Ad and ϕ29 TPs (*147*). Adenovirus TP is synthesized in the 80-kDa precursor form (compare with VPg) called pTP (*144*), which is probably not the case for ϕ29 TP. pTP but not TP can be detected in the infected cells both free and linked to the newly synthesized DNA strands (*25, 148, 149*). Adenovirus pTP is processed to the mature 55-kDa TP during the morphogenesis of virion late in the infection. This process is accomplished by the Ad-encoded protease (*148, 150*). The

TABLE III
CHARACTERISTICS OF VIRAL DNA-TP COMPLEXES

Virus	Genome (kb)	ITR (bp)	TP (kDa)	Linkage
Adenovirus	36	100–200	80/55	Ser-5′p(dC)
Bacteriophage ϕ29	18	6	31	Ser-5′p(dA)
Bacteriophage Cp-1	18	240–350	28	Thr-5′p(dA)
Bacteriophage PRD 1	15	?	30	Tyr-5′p(dG)

serine residues participating in the formation of the covalent bond with Ad and ϕ29 DNA were localized in the carboxyl-terminal portion of the TP molecule (*151, 152*). The linking sites are located in β-turns preceded by α-helical regions. They are probably located on the surface of the protein molecule (*152*).

TPs at both ends of the viral genomes have affinity for each other and/or for the terminal regions of DNA which results in the formation of circular and oligomeric forms of DNA, visualized by electron microscopy (*13, 14, 153*). The presence of TPs markedly changes the electrophoretic mobility of viral DNAs and their terminal restriction fragments (*17, 135, 154*), as well as conferring affinity of the DNA-TP complexes for GF/C and nitrocellulose filters and BND-cellulose (*10, 11, 155*) (see Section I). This is not well understood, yet these properties permit detection and easy isolation of such complexes.

The most intriguing question concerning the TPs is their function. Proteolytic degradation of TP linked to viral DNA, resulting in the formation of a DNA molecule with attached short terminal peptides, drastically diminishes the infectivity of DNA (*5, 8, 135, 135a*). This was not because of the inability of the viral DNA to penetrate the cell. Studies both *in vivo* and *in vitro* with mutated or degraded TP show that the synthesis of viral DNA is impaired (*145, 156–158*), which suggests that the parental (template-bound) TP plays a role in the replication of viral DNA. The fact that short nascent viral DNA chains in the replicative intermediates are already covalently linked to (p)TP (*10, 12, 159*), and other data obtained with TP mutants (*160, 161*), indicates that free (not template-bound) TP is involved in DNA synthesis as well, probably at the initiation stage. Antibodies raised against TP inhibit viral DNA replication *in vitro* (*162, 163*).

Initiation of Ad and ϕ29 duplex DNA replication *in vivo* occurs at either end of the DNA and proceeds in the 5′-to-3′ direction by a strand displacement mechanism (*157, 164, 165*). The displaced strand is then replicated independently. Initiation of DNA synthesis on linear replicons poses a problem. Indeed, all known DNA polymerases need a 3′-hydroxyl group of the primer to elongate DNA chains. Such a 3′-hydroxyl group could be supplied by an RNA primer, yet its subsequent removal should result in the formation of a gap at the 5′ terminus of the DNA molecule that could not be filled since DNA synthesis proceeds only in the 5′-to-3′ direction (*166*). In some cases, this problem can be overcome by making use of direct terminal repeats or cohesive ends of linear DNA molecules. This would allow formation of circular or concatemeric forms of DNA and thus fill the gap. In case of self-complementary terminal structures of DNA, a 3′-

terminal hairpin of the template could provide the 3′-hydroxyl group for DNA polymerase to synthesize a new DNA chain. Neither of these models can be applied to initiate DNA synthesis on the viral DNAs described as they do not possess the necessary structures (*167*). Instead, they have two peculiar terminal structures: identical nucleotide sequences at both DNA termini (ITRs) and 5′-linked TPs. To resolve the problem of initiation and explain the presence of 5′ terminal TPs, Rekosh *et al.* (*137a*) proposed that the newly synthesized molecules of the TP serve as primers for viral DNA synthesis. According to this model, the initial reaction is the interaction of free (p)TP with dNTP destined to become the 5′-terminal nucleotide of the synthesized strand. This results in the formation of the covalent TP-p(dN) complex, the primer itself. The 3′-hydroxyl group of the TP-bound nucleotide is then utilized by the DNA polymerase to further elongate the polynucleotide chain (see Fig. 4; compare with VPg-pU priming model of RNA replication, Fig. 3A). New DNA strands thus synthesized would possess TP linked to the 5′-terminal nucleotide.

This model gained experimental support when cell-free systems capable of initiating and elongating Ad and ϕ29 DNA synthesis on exogenous template DNA were developed (*168, 169*). These systems permitted investigation of the mechanism of initiation of viral DNA

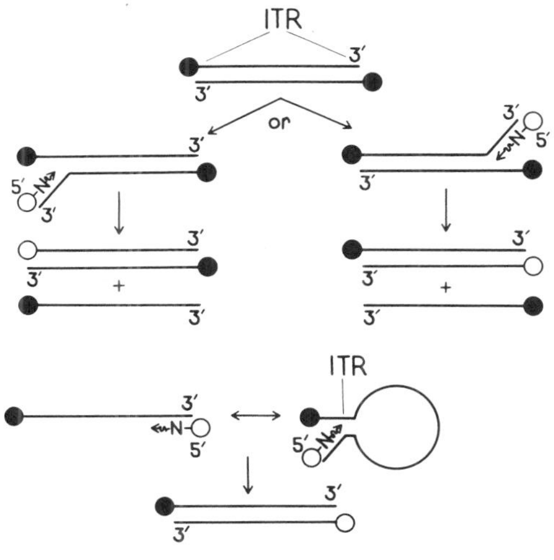

FIG. 4. Protein-primed initiation of replication of linear double-stranded viral genomes. The solid circle denotes the parental terminal protein (TP); the open circle denotes the daughter TP or pTP (pre-TP).

synthesis, template requirements for initiation, and characterize proteins participating in the replication of viral DNA. Using Ad or ϕ29 DNA-TP or Ad DNA-pTP as template, these systems catalyzed the formation of Ad pTP- or ϕ29 TP-p(dN) primer from free (p)TP and dNTP (*25, 170–176*). Mutations, especially in the carboxyl terminus of ϕ29 TP, drastically decrease the effectivity of TP-p(dA) formation (*177, 178*). Upon the addition of appropriate proteins this primer is elongated to the full-sized DNA molecule (*170, 179*). Formation of TP-p(dN) primers *in vitro* has been also documented for PRD 1 (*137, 139*) and Cp-1 phages (*34*).

Formation of (p)TP-p(dN) appears to occur on the template, and the primer remains associated with the template DNA (*180*). The best template for initiation (primer formation) is the DNA-TP complex. Viral DNA with denatured TP or with residual peptides obtained via protease treatment of DNA-TP appears to be an inefficient template for DNA replication and primer formation (*170–172, 176*) unless, in the later case, a cellular protein ("pL") is added to the *in vitro* system (181). In contrast, viral DNA with unblocked termini obtained by piperidine treatment of DNA-TP can substitute for the DNA-TP template (*182, 183*), although it is far less efficient in TP-p(dN) formation. This defect can be partially restored by pL factor *in vitro* (*181*). These results show the requirement for the parental TP for the initiation of DNA replication. Since this requirement is not absolute, TP-deficient viral ITR cloned into plasmid is capable of supporting initiation and elongation of DNA replication. For this process to occur *in vitro*, it was necessary to linearize the plasmid DNA in such a way that the Ad ITR is located very close to the end of the molecule (*182, 184*). Even in this case DNA synthesis is not initiated strictly from the correct nucleotide of template (*182, 184*). On the contrary, circular DNA constructions possessing two Ad ITRs appear to replicate quite well *in vivo* (*185–187*) (in the presence of helper adenovirus) and give rise to the protein-bound double-standard linear DNA molecules (*187*). These data are in agreement with the fact that covalently closed Ad DNA molecules with covalently joined ITRs (*188*), found in Ad-infected cells, are infectious (*189*). The progeny DNA molecules were reported to be linear. The above results demonstrate that it is not the DNA terminus or TP per se that is required for the protein-primed initiation of viral DNA replication, but a specific origin sequence.

The *in vivo* origin of Ad replication is at the terminal 18 to 30–45 bp of the ITRs (*190–192*). Notably, this region contains a 10 bp sequence (nucleotides 9 to 18 of human Ad DNA sequence) which is perfectly conserved among human adenoviruses and is well conserved in animal and avian adenoviruses (*193–195*). Studies *in vitro* located the "minimal" origin required for TP-p(dN) formation within the first 18 bp of the Ad genome (*196–198*) and within the first 10 bp of the ϕ29 genome (*199*). Yet the adjacent region, up to the 40th bp, is frequently necessary for restoration of a completely active Ad origin (*196, 200, 201*). This region contains a binding site for the host-coded protein engaged in Ad DNA replication (see below). Mutations within the conserved Ad sequence impair formation of the primer, while those outside the conserved sequence are tolerated (*186, 197, 198*). Nevertheless, bp 1 to 8 of the Ad DNA sequence cannot be eliminated without loss of the origin function and the ability to nucleotidylate pTP (*201*). It was gratifying to find that pTP can bind specifically to the conserved origin region (bp 9–22 of the Ad DNA sequence) of the circular or linearized template DNA. The presence of the template-bound TP seems to have no effect on this binding (*202*). TP was shown to possess higher affinity to the denatured Ad DNA terminus than to the native one (*203*).

Curiously, different single-stranded DNAs, including synthetic oligonucleotides, serve as templates for Ad pTP-p(dC) formation, though less efficiently than the Ad DNA-TP complex (*182, 197, 204*). Ad-encoded single-stranded DNA binding protein inhibits primer formation (*204*). dCTP is the preferred nucleotide in the reaction of pTP nucleotidylation; yet in the absence of complementary dG in the single-stranded template, e.g., oligo(dT), pTP-p(dA) is formed (*198*). Thus, any nucleotide that in principle can hydrogen-bond to the template can be linked to pTP.

(p)TP by itself appears to have no enzymatic activity (*160, 205, 206*). Nucleotidylation of (p)TP depends upon virus-coded DNA polymerase (*205, 207–210*), which is tightly associated with TP both before and after primer formation (*173, 207a, 211*). Ad DNA polymerase is driven to the origin in the form of a pTP–polymerase complex; this specific binding to the 9–22 bp sequence is mediated by pTP (*202*). Virus-coded DNA polymerase is engaged not only in the formation of the primer, but in its subsequent elongation as well (*205, 212, 213*). Some data suggest that the daughter TP is involved also in the elongation step (*162*) and can protect nascent DNA strands from exonuclease digestion (*214*). Nevertheless, viral polymerases retain enzymatic activity in the absence of TP (*205*) and they can utilize oligonucleotide primers as well (*212, 213*). There appear to be two active sites in these DNA polymerases, one for TP-p(dN) formation, and the other for elongation (*213, 215*). Ad and ϕ29 DNA polymerases appear to contain $3' \rightarrow 5'$ exonuclease activity on the single-stranded DNA, which is thought to be implicated in proofreading (*213, 216*). This property appears to be an intrinsic activity of the described polymerases. DNA polymerases of adenovirus and ϕ29 phage, belonging to two quite different groups and yet adopting the same strategy for replicating their DNA, show striking homology with each other and with Epstein–Barr virus-encoded DNA polymerase (*147*).

Only two ϕ29-encoded proteins, TP and DNA polymerase, are sufficient for the synthesis of full-length ϕ29 DNA *in vitro*, using ϕ29 DNA-TP as template (*179*). To make this process more efficient, phage-encoded protein 6, which stimulates initiation and elongation (*217*) and some host-coded proteins might be required. To replicate *in vitro* Ad genome, which is twice as large as that of ϕ29 phage DNA, several additional proteins are needed (for review, see *218*). One of them, a host-coded protein from the nucleus of uninfected HeLa cells (called nuclear factor I), is a sequence-specific DNA binding protein (*203*) stimulating the formation of pTP-p(dC) with the double-stranded Ad template, but not with single-stranded template. It becomes indispensable for primer formation in the presence of single-stranded DNA binding protein (*219*). Nuclear factor I binds Ad DNA within the ITR, immediately adjacent to the pTP-polymerase binding site (bp 19–43 of the Ad DNA sequence) (*200–203*). Bearing these facts in mind, it might be suggested that the binding of nuclear factor I causes a partial unwinding of the pTP-polymerase binding site and thus enables effective pTP-polymerase association with the single-stranded region of the template. Template-bound TP might be required for proper positioning of pTP to initiate DNA synthesis from the very first nucleotide of the template, some 10–15 bp from the pTP-polymerase binding site. Associated Ad polymerase forms the primer via nucleotidylation of pTP and further elongates the polynucleotide chain.

Binding activities similar to nuclear factor I have been detected in a variety of cells (*220*), their function being obscure. It has been shown, however, that the "TGGCA" DNA binding protein is functionally related to nuclear factor I (*221*). It should be noted that the role of nuclear factor I in Ad DNA replication seems somewhat subordinate since some adenoviruses (e.g., human Ad type 4, 9, CELO) lack a binding site for nuclear factor I and no stimulatory effect was observed in these cases (*192*). Nevertheless, these adenoviruses are quite infectious. A number of other host factors affect the initiation of Ad DNA synthesis *in vitro*, but their function seems to be auxillary (*222, 223*).

Thus, a small number of proteins are needed to replicate Ad and ϕ29 DNA *in vitro* and, presumably, *in vivo*. This is probably because of the unique protein-priming mechanism employed by these viruses to initiate DNA synthesis: the preformed "pre-primer" (p)TP capable of specific binding to the viral origin by itself, and the virus-encoded DNA polymerase capable of both formation of the primer (p)TP-p(dN) and its subsequent elongation.

One more function of TP linked to viral DNA might be connected with encapsidation of the viral genome. It is at this stage that processing of Ad DNA-bound pTP to TP occurs, though the requirement for this process is obscure. Experiments with the phage ϕ29 DNA-TP complex demonstrate that TP has an essential role in DNA packaging *in vivo* and *in vitro* (*161, 224*).

By analogy with the VPg-RNA complexes of picornaviruses, it might be asked if viral DNA genomes known to possess TPs could exist in the protein-free form. Infectious covalently closed circular Ad DNA molecules with joined ITRs appear to be formed in the Ad-infected cells (*189*). Whether such circles could be intermediates in integration of viral DNA into the host chromosome remains to be seen. It is interesting to note in this regard that the transforming and oncogenic activities of Ad DNA devoid of TP by protease treatment are much higher than those of the Ad DNA-TP complex (*225*).

Another intriguing question is whether analogous covalent complexes of linear DNA with TP exist in uninfected cells. Linear double-stranded plasmid DNAs from *Streptomyces rochei* (*226*), the DNA killer plasmids pGKL1 and pGKL2 from yeast (*227–229*), plasmid-like mitochondrial DNAs S1 and S2 from maize (*230*), and N1 and N2 from sorghum (*231*) possess TPs attached to the 5′ termini of DNA. Significantly, the cellular DNAs with a TP, in which the nucleotide sequences at the ends have been determined, possess an ITR of variable length (*232–235*). Another feature in common between viral and cellular DNA-TP complexes is the ability to replicate autonomously. Although in the case of cellular complexes the mechanism of initiation of replication is unknown, the structural similarity of the termini of their DNA to the described viral genomes suggests that they may replicate their DNA in an analogous fashion. More generally, detection of the ITRs in the linear DNA might be indicative of the presence of covalently linked TPs and of the characteristic mechanism of DNA replication (protein priming, strand displacement).

B. Interaction of the Bacteriophage-Encoded Gene A Initiator Protein with the DNA of Isometric Single-Stranded Bacteriophages

Small isometric bacteriophages of *E. coli* (ϕX174 and related phages) possess single-stranded circular DNA genomes 5000–6000 nucleotides long (*236, 237*). Studies on the replication of these genomes provided an insight into the fundamental mechanisms of DNA replication. The main point considered in this section is the covalent interaction between the phage-encoded gene-A protein, an initiator of DNA replication, and the DNA of isometric bacteriophages in the course of DNA replication. Structural and functional aspects of the covalent bond formation are likewise discussed.

In the phage DNA replication cycle the single-stranded DNA genome is first converted to a double-stranded replicative form (RF) DNA by the synthesis of a complementary DNA strand. This is accomplished with the aid of host proteins only. Then the rolling circle mode of replication is employed for the RF DNA replication (formation of progeny RF DNA) and single-stranded circular viral DNA synthesis (for recent review, see *238*). This process starts, for the best-studied phage ϕX174, with the introduction of a single nick in the viral strand of the superhelical RF (RF-1) DNA by the ϕX174 gene-A protein (*239, 240*). The cleavage site is located within the structural gene of A-protein itself, in the 30 bp origin region (*241*), which is highly conserved among several isometric phages (*242–244*) (nucleotides 4299–4328 of the ϕX174 DNA sequence). During the cleavage, the A-protein becomes covalently attached to the 5′ end of the nicked strand. This results in the formation of the RF DNA with a discontinuity in the viral strand (RF-II DNA) linked to the A-protein, an RF-II DNA·A complex (*245*). The free 3′-hydroxyl group of the nicked strand is used then as a primer for viral strand DNA synthesis (*240*). This is accomplished by the combined action of the DNA-linked A-protein and three host-coded proteins: *rep* protein (helicase), which unwinds the double-stranded RF-II DNA, single-stranded DNA (ssDNA) binding protein (*246*), and DNA polymerase III holoenzyme, which synthesizes the new viral strand.

During replication, the A-protein moves along with the replication fork owing to the interaction with the *rep* protein (*245*) and, as a result, a looped rolling-circle is formed (*247*) (see Fig. 5). Unless the displaced viral strand is immediately encapsidated (*248, 249*), it is covered with ssDNA binding protein and is used then as a template for the complementary strand synthesis. When a round of replication is completed, the A-protein linked to the 5′-end cleaves the regenerated origin sequence and binds covalently to the 5′ end of the cleavage site of the newly synthesized viral strand (reinitiation of replication) (*247, 250*). Concomitantly with its transfer, the A-protein circularizes the displaced viral strand (termination of replication) (*247, 251*). This complex phosphodiester bond rearrangement results in (1) production

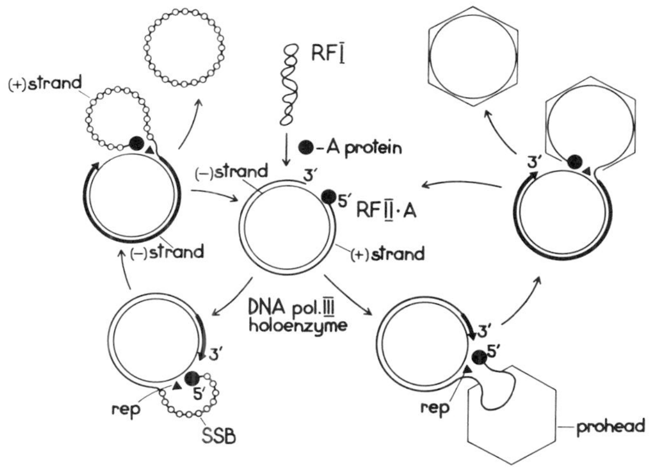

FIG. 5. The bacteriophage ϕX174 gene-A initiator protein is a multifunctional enzyme in phage RF DNA replication and single-stranded DNA synthesis. The process of virion formation is not detailed.

of single-stranded circular progeny DNA molecule, and (2) regeneration of the covalent RF-II DNA·A complex, capable of reinitiating the DNA synthesis and participating in multiple rounds of replication (*247*). Thus, the gene-A protein of bacteriophage ϕX174 is a multifunctional enzyme. It participates in the initiation, reinitiation/termination of DNA synthesis, and in chain elongation in the rolling-circle replication of phage DNA (*247*). Many properties of the A-protein required to accomplish these functions, originally postulated from indirect considerations, have been reproduced *in vitro* with purified host- and phage-encoded proteins. These studies have contributed greatly to the understanding of molecular mechanisms of interaction of the A-protein with DNA and have given experimental support for the above model.

The ϕX A-protein (59 kDa, *252*) possesses both nicking (endonuclease) (*239*, *240*, *253*) and ligase activities (*254*). Cleavage of the specific site within the target DNA molecule is coupled with the covalent attachment of the A-protein to the 5′ end of the cleaved DNA strand. The linkage is a phosphodiester bond between a tyrosine moiety of the A-protein and the 5′ phosphate of the dAMP residue of phage DNA in position 4306 [Tyr-p(dA) linkage, (*250*, *255*, *256*), see Fig. 2B]. This phosphodiester bond retains the energy of the hydrolyzed internucleotide phosphodiester bond, which permits the covalently linked A-protein to ligate the broken strand later on without an additional energy supply. Two closely located and probably juxtaposed tyrosine residues in the carboxyl-terminal half of the A-protein molecule were implicated in DNA cleavage, either of which can function as the acceptor of the DNA chain (*257*). This offers a mechanism for the A-protein-catalyzed cleavage and ligation of DNA. The hydroxyl groups of these two tyrosine residues are supposed to participate in an alternating manner in successive cleavage and ligation reactions, which take place during the initiation and reinitiation/termination of the DNA replication (*247, 250, 257*).

Multimeric forms of the A-protein are probably required for the cleavage reaction (*245, 253*). However, only one molecule of the A-protein remains covalently linked to the DNA (*252*). An *in vivo* cis-action of the A-protein on the ϕX DNA has been reported (*253, 258*). However, both *in vitro* with phage DNA (*245*) and *in vivo* with plasmid DNA containing the ϕX origin (*259*), the A-protein can act in trans, initiating the rolling-circle mode of replication. There is no clear explanation for this discrepancy.

The A-protein turns out to be a site-specific endonuclease. It cleaves RF-I DNA of bacteriophage ϕX174 and related G4, St-1, α3, U3, and G14 phages at the conservative origin region only and produces a covalent RF-II DNA·A complex (*242–244, 260*). Supercoiling is essential for the A-protein cleavage of double-stranded circular DNA (*240, 261, 262*), probably because it makes for the unwinding of the cleavage site. The A-protein cleaves single-stranded viral ϕX174 DNA at the origin as well, and at one additional site (*263–265*). However, in the presence of ssDNA binding protein, cleavage is restricted to the origin only (*265*). The A-protein binds covalently to the 5′ end of the cleaved DNA (*255, 256*). These observations strongly suggest that the A-protein is an endonuclease specific for single-stranded DNA. Only the first 10 nucleotides of the origin (nucleotides 4299–4308) are required for this cleavage reaction to occur when single-stranded oligonucleotides are used as substrate (*266*). However, the presence of this sequence in a supercoiled double-stranded DNA is not sufficient for the cleavage by the A-protein (*267*); nor is it sufficient for the cleavage of single-stranded DNA in the presence of ssDNA binding protein (*265*). In both cases a sequence corresponding to the first 27 nucleotides of the origin region is required (*265, 268*). Though the sequence requirements for the DNA replication initiated by A-protein appear to be somewhat more stringent than those for DNA cleavage per se (*269*), this 27 nucleotide region meets both.

Likewise, reinitiation and termination reactions were supposed to require the first 25 (but more than 18) nucleotides of the origin (*251, 270*). The last three nucleotides of the ϕX origin region (4326–4328) are implicated in the packaging of the newly synthesized DNA (*271*).

Gene A of bacteriophage ϕX174 and related phages codes for another product called A*-protein which is translated from an internal in-frame start codon within gene A. The A*-protein (35 kDa) thus lacks the N-terminal part of the A-protein (*272*). A*-protein cannot substitute for A-protein in ϕX DNA replication. Nevertheless, it retains some of the enzymatic activities of A-protein. A*-protein is an endonuclease specific for single-stranded DNA. It cleaves ϕX viral strand DNA at the origin and at many other sites (*254, 263, 273*). Evidently, site specificity in the cleavage reaction of A*-protein is highly relaxed compared to that of the A-protein, and it is not restored by the addition of ssDNA binding protein (*265*). Each of the two tyrosine residues in the A*-protein active site can form a covalent linkage with the 5′ end of the cleaved DNA (*257, 274*). Covalently linked A* protein can ligate DNA (*275*). The A*-protein can be isolated from ϕX174 infected cells covalently linked to oligonucleotides via a Tyr-p(dA) bond (*276, 277*). These oligonucleotides can be transferred to the acceptor oligonucleotide with the origin sequence via the A*-protein-mediated cleavage/ligation (*276*). These oligonucleotides probably arise from A*-protein cleavage of ϕX174 DNA at the additional A-protein cleavage site, followed by degradation of the covalently bound DNA (*265, 278*). This cleavage might permit switching of DNA synthesis to the production of virions (*265*). Most importantly, the A*-protein does not cleave supercoiled ϕX RF-I DNA (*254, 275*). Thus, the ability to bind the ϕX origin specifically and/or to unwind the superhelical RF-I DNA seems to residue in the N-terminal part of the A-protein, which is absent in the A*-protein. Instead, the A*-protein binds nonspecifically to the double-stranded DNAs (*279*). This property may be related to the inhibitory effect of the A*-protein of host DNA replication (*279, 280*).

The available data on the specificity of the A-protein action and mutational analysis of the ϕX origin region permit formulation of a model for A-protein participation in the initiation of DNA replication (*267*). The 30 bp ϕX origin is thought to contain two functional domains: a recognition sequence and a binding sequence, which are separated by an (A + T)-rich region. The recognition sequence is located at the 5′ end of the conserved region and is essentially the same as that of the single-stranded decamer, which can be cleaved by the A- and A*-proteins (*266*). The binding sequence is supposed to provide a specific binding site for the A-protein. Mutations within these regions drastically decrease the efficiency of the initiation reaction (*269, 281*). The (A + T)-rich spacer sequence is about eight nucleotides long, and within this sequence many nucleotide substitutions are tolerated (*238*). Initiation of ϕX DNA replication on the RF-I DNA template appears to involve a noncovalent binding of the A-protein to the binding site. This binding brings the protein into a proper orientation toward the recognition site and induces local unwinding of the recognition sequence, which now can be cleaved by the ssDNA-specific endonuclease activity of the A-protein.

It should be noted that the described model of interaction of the initiator protein with the ϕX origin of DNA replication resembles that at the adenovirus origin in several respects. However, the mechanisms of primer formation in these two instances seem to be completely different. Interestingly, each of these mechanisms, protein priming and nicking-linking, has analogy with the two possible pathways of VPg participation in picornavirus RNA replication (see Fig. 3).

A comparison of the nicking-closing mode of action of the A-protein during ϕX174 DNA replication with that of some cellular enzymes reveals important analogies. Nicking-closing activity is a well-known property of topoisomerases, and the analogy of the A-protein with bacterial type-I topoisomerases in particular is straightforward. These enzymes nick one strand of the superhelical DNA and form covalent protein–DNA complexes, in which the protein is linked to the 5′-phosphate of the DNA via a tyrosine residue (*282*). The energy of the hydrolyzed internucleotide bond is thus conserved in the phosphotyrosine bond and is subsequently utilized for a ligation step. However, the ϕX A-protein is a sequence-specific nicking-closing enzyme with a protracted coupling period.

Another example is a relaxation protein of relaxation complexes of bacterial plasmids, e.g., Col-E1 plasmid. This protein nicks one strand of superhelical Col-E1 DNA (*283*) specifically at the origin of conjugative transfer (*284*), and binds covalently to the 5′-phosphate of this strand (*285*) via a Ser-p(dC) linkage (*286*). The existing model of the Col-E1 DNA conjugative transfer includes the rolling circle mode of DNA replication in the donor cell and circularization of the transferred strand by the DNA-linked protein within the recipient (*284*). A different functional role is played by the RepC protein encoded by *S. aureus* plasmid pT181. It was claimed that RepC protein generates a single-stranded nick in the pT181 DNA and binds, probably covalently, to the 5′ end of the DNA at the cleavage site (*287*), presumably at the origin of replication (*288*). This might mean that unlike most other plasmids, pT181 replicates by a rolling-circle mechanism.

C. Proteins Linked to the Replicative Form DNA of Autonomous Parvoviruses

The autonomous parvoviruses of vertebrates possess a linear single-stranded DNA genome about 5000 nucleotides long with short terminal hairpins (*289*). Following infection, single-stranded virion DNA is converted, via a self-priming mechanism, to a linear double-stranded RF DNA, which, in turn, serves as a template for the synthesis of progeny single-stranded viral DNA. The RF-DNA molecules of rodent parvoviruses, i.e., the H-1 virus (*9*), the Kilham rat virus (KRV) (*290*), and the minute virus of mice (MVM) (*291, 292*), are covalently associated with proteins. The proteins appear to be covalently attached to both 5′ ends of RF-DNA (*9, 290, 292*). While almost all of the H-1 virus RF-DNA molecules are protein-bound (*9*), the protein is associated with only a fraction of the MVM RF-DNA (*292*). A variable fraction of MVM single-stranded virion DNA has the same protein bound to its 5′ terminus as well (*292*), but for other parvovirus genomes, this is not established.

The proteins linked to both ends of RF DNA of H-1 virus and MVM have masses of 60 kDa (*9, 291*); two proteins, with masses of 90 and 40 kDa, were reported to be attached to RF-DNA of KRV [the latter seems to be the product of the proteolytic degradation of the former (290)]. The protein linked to the MVM RF-DNA was identified in the free form in the cytoplasm of the MVM-infected cells (*292*), making the existence of the precursor form unlikely. The genetic origin of the parvoviral RF-DNA-linked proteins is unresolved. They are immunologically unrelated to the capsid or known nonstructural proteins of parvoviruses (*292*), and thus, unlike other genome-linked proteins, might be host-coded.

It is not known at what stage during the synthesis of parvovirus RF-DNA the proteins become attached. Since the covalently bound proteins are known to participate in the replication of the viral genomes (see above) and since the existing models of parvoviral DNA synthesis include site-specific nicking of the oligomeric replication intermediates to synthesize a progeny viral strand (*293*), it was implied that parvoviral DNA-linked proteins participate in this cleavage reaction. Concomitantly with the cleavage, the protein is supposed to bind covalently to the 5′ DNA end and produce a free 3′-hydroxyl group for the DNA chain elongation (*293*). The bacteriophage ϕX174-encoded gene-A protein might provide some analogy on this point.

D. A Protein Attached to the 5′ Terminus of the Minus Strand of the Hepadnavirus Genome

The hepatitis B-like viruses (hepadnaviruses) have a small circular partially double-stranded DNA genome of about 3.2 kb. Both strands of this DNA molecule, however, are linear, the circular conformation of the genome being maintained by base-pairing between the cohesive 5′ ends of the two strands. The long minus strand is of fixed length (about 3200 nucleotides). The plus strand is incomplete and heterogeneous in length due to the variable 3′ end location (see Fig. 6, and *294* for a review). A protein is covalently bound to the 5′ terminus of the minus strand of virion DNA. This protein might be virus-encoded; its mass was estimated to be about 100 kDa (*295*). Unless treated with protease, this protein causes partition of the associated DNA into the organic phase upon phenol extraction (*295, 296*). Full-length single-stranded viral minus-strand species detected in the infected liver (*296*), as well as short nascent minus strands isolated from replication complexes (*297, 298*), are also protein-linked. Thus the protein appears to become attached to the 5′ terminus of the minus strand early in its synthesis. On this basis, the protein was supposed to function as a primer for minus strand DNA synthesis and to remain covalently bound to DNA throughout virus morphogenesis (*297*), with the obvious analogy to the terminal proteins of adenoviruses, etc. Yet the genome-linked protein of hepadnaviruses is bound to the minus strand DNA only, the plus strand being primed by oligoribonucleotide (*299*).

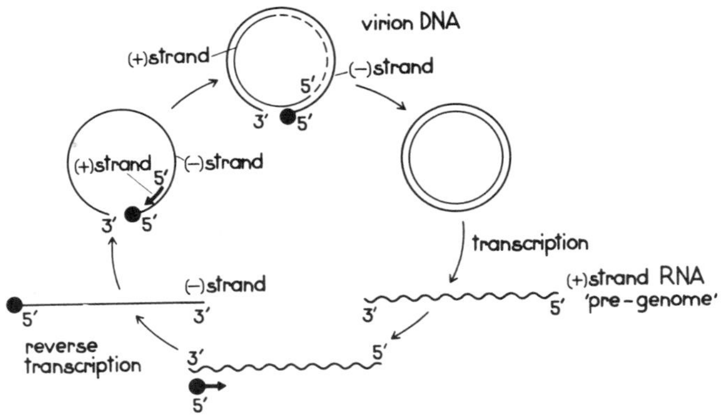

FIG. 6. A schematic representation of hepadnavirus genome structure and replication. The solid circle denotes a protein covalently bound to the minus strand of viral DNA.

The mechanism by which hepadnaviruses replicate their genomes differs strikingly from that of other DNA viruses. It involves a reverse transcription step (*300*), accomplished presumably by the virus-encoded enzyme. The RNA template for reverse transcription of slightly more than the genome length, termed pre-genome, is produced by copying the minus strand DNA in the closed circular double-stranded viral DNA (*300, 301*), to which the virion DNA is converted (see Fig. 6). The details of the formation of the covalently closed double-stranded DNA molecule are not well understood; yet this process seems to require a splitting-off of the template-bound protein. The pre-genome is then used as a template for reverse transcription to synthesize a full-length viral minus-strand DNA (*300*). It is at this stage where the protein molecule is supposed to be employed to initiate the process via a protein-priming mechanism. The protein-linked 5′ terminal sequence of the minus strand of virion DNA corresponds to the origin of reverse transcription (*302, 303*), this result being consistent with the protein-priming hypothesis. The origin is located in one of the direct repeat sequences (*302*) conserved in the genomes of all hepadnaviruses. The minus strand generated in this way is further used as a template for DNA-dependent plus-strand DNA synthesis (*301*). Complete synthesis of plus strand is not required for the formation of mature virus particles.

Thus, the replication cycle of hepadnaviruses resembles that of retroviruses (for discussion, see *303*); but unlike the retroviruses, hepatitis B-like viruses appear to use a protein rather than a tRNA as a primer for reverse transcription. The grounds for this difference remain obscure.

IV. Concluding Remarks

The overview presented permits the conclusion that covalent nucleic acid–protein complexes are abundant in the virus world. Both RNA and DNA viruses possess genome-linked proteins, which are specified, where known, by the viruses themselves.

From the structural point of view, it is of major importance that the proteins are found linked exclusively to the 5′ termini of nucleic acids, which has major consequences for protein functioning. The nucleic acid–protein covalent linkage in all the cases studied is a phosphodiester bond, imitating that of the internucleotide bond. This probably arises from two causes. In the case of protein-primers (adenoviruses, phage ϕ29), the linkage is formed by polymerases, which form just such bonds, thus eliminating the need for supplementary activities at the initiation step. In the case of nicking-linking proteins capable of forming the linkage themselves, the nucleic acid–protein bond accumulates the energy of the hydrolyzed internucleotide bond of the same type for its subsequent resealing.

The genome-linked proteins play a fundamental role in the replication of viral genomes, whatever this process might be: DNA-dependent DNA synthesis, RNA-dependent RNA synthesis, and presumably reverse transcription. The template-bound (parental) protein may be absent (isometric bacteriophages and, probably, hepadnaviruses) or, if present, may have no role in replication (i.e., picornaviruses). In other cases, the DNA-linked protein seems to play an important (judging by protease-sensitive infectivity of adenoviral and other complexes), though somewhat auxiliary role: to provide the proper orientation of its free (daughter) counterpart toward the origin of replication. Nevertheless, the problem of specific binding of the daughter protein to the origin region is resolved by the incoming daughter protein itself. Thus, the specific nucleic acid sequence (origin of replication) is probably of greater significance for the daughter protein functioning.

Participation of the free, "prelinked" protein in the initiation of replication may be achieved in two different ways, depending on the structure of template and the properties of the protein. If the double-stranded viral genome is circular or oligomeric, and if the protein possesses a site-specific nicking activity, the protein may become attached to the 5′ terminus of the displaced strand of the template DNA molecule at the cleavage site. The free 3′-hydroxyl group then serves as a primer for nucleic acid synthesis. The bacteriophage ϕX174 gene-A initiator protein is an example of this kind. It demonstrates, furthermore, that the DNA-bound protein may participate not only in the initiation of DNA replication, but also at the elongation and termination/reinitiation steps. The same molecule of the protein thus takes part in multiple rounds of DNA replication.

If the genome is linear and the protein lacks enzymatic activity, its participation in the initiation event may follow a different pathway. This approach requires an ability of the protein to interact with an additional factor, a DNA (or RNA?) polymerase, to form protein-primer via nucleotidylation of the protein. The 3′-hydroxyl group of the protein-bound nucleotide then serves as a primer for the same polymerase to elongate the chain. In this case the protein becomes covalently bound to the 5′ end of the nascent polynucleotide strand. This permits an accurate copying of the complete viral genome, including its 5′-terminal regions. To conclude our comparison of these two strategies of initiation of nucleic acid replication, it is worth noting that in spite of their significant differences, either of them might be suspected to be employed in the replication of the same genome (*304*). The two possible modes of picornavirus genome replication demonstrate this point fairly well.

Another function of the genome-linked proteins might be their participation in virus morphogenesis. Data supporting this notion have been reported. This bifunctional role seems to be the rule for protein-primers.

The genome-linked proteins, free and bound to nucleic acids, thus play an important role in the life cycle of diverse viruses. The existence of DNA–protein covalent complexes in uninfected cells, resembling those of the viruses described, as well as the ability of cellular enzymes to unlink VPg from the picornaviral genomes, testifies to the ubiquity of such structures (and corresponding apparatus) in bacterial and eukaryotic cells.

Acknowledgments

We thank Drs. V. Agol, M. Balakireva, and E. Koonin for a critical reading of the manuscript and helpful discussions, and Dr. A. Gorbalenya for help in aligning the sequences of viral-linked proteins (VPg).

References

1. A. A. Bogdanov, *Uspekhy Sovrem. Biokhimii* (*Russ.*) **55**, 321 (1963).
2. A. Bendich and H. S. Rosenkranz, this series **1**, 219 (1963).
3. H. G. Zachau and H. Feldman, this series **4**, 217 (1965).
4. Z. A. Shabarova, this series **10**, 145 (1970).
5. P. A. Sharp, C. Moore and J. L. Haverty, *Virology* **75**, 442 (1976).
6. N. E. Harding and J. Ito, *Virology* **73**, 389 (1976).
7. M. R. Inciarte, J. M. Lazaro, M. Salas and E. Vinuela, *Virology* **74**, 314 (1976).
8. H. Hirokawa, *PNAS* **69**, 1555 (1972).
9. D. Revie, B. Y. Tseng, R. H. Grafstrom and M. Goulian, *PNAS* **76**, 5539 (1979).
10. B. W. Stillman and A. J. D. Bellet, *Virology* **93**, 69 (1979).
11. D. H. Coombs and G. D. Pearson, *PNAS* **75**, 5291 (1978).
12. A. J. Robinson, J. W. Bodnar, D. H. Coombs and G. D. Pearson, *Virology* **96**, 143 (1979).
13. J. Ortin, E. Vinuela, M. Salas and C. Vasquez, *Nature NB* **234**, 275 (1971).
14. A. J. Robinson, H. B. Younghusband and A. J. D. Bellett, *Virology* **56**, 54 (1973).
15. E. A. Carusi, *Virology* **76**, 380 (1977).
16. C. O. Yehle, *J. Virol.* **27**, 776 (1978).
17. J. Ito, *J. Virol.* **28**, 895 (1978).
18. D. Frisby, M. Eaton and P. Fellner, *NARes* **3**, 2771 (1976).
19. Y. F. Lee, A. Nomoto and E. Wimmer, This series **19**, 89 (1976).
20. R. Fernandez-Munoz and U. Lavi, *J. Virol.* **21**, 820 (1977).
21. J. Klootwijk, I. Klein, P. Zabel and A. van Kammen, *Cell* **11**, 73 (1977).

22. B. D. Harrison and H. Barker, *J. Gen. Virol.* **40,** 711 (1978).
23. Y. F. Lee, A. Nomoto, B. Detjen and E. Wimmer, *PNAS* **74,** 59 (1977).
24. Yu. F. Drygin, A. B. Vartapetian and K. M. Chumakov, *Mol. Biol.* (Russ.) **13,** 777 (1979).
25. M. D. Challberg, S. V. Desiderio and T. J. Kelly, Jr., *PNAS* **77,** 5105 (1980).
26. J. B. Flanegan, R. F. Pettersson, V. Ambros, M. J. Hewlett and D. Baltimore, *PNAS* **74,** 961 (1977).
27. A. B. Vartapetian, Yu. F. Drygin and K. M. Chumakov, *Bioorgan. Khim.* (Russ.) **5,** 1876 (1979).
28. R. F. Pettersson, V. Ambros and D. Baltimore, *J. Virol.* **27,** 357 (1978).
29. P. G. Rothberg, T. J. R. Harris, A. Nomoto and E. Wimmer, *PNAS* **75,** 4868 (1978).
30. V. Ambros and D. Baltimore, *JBC* **253,** 5263 (1978).
31. A. B. Vartapetian, Yu. F. Drygin, K. M. Chumakov and A. A. Bogdanov, *NARes* **8,** 3729 (1980).
32. S. V. Desiderio and T. J. Kelly, Jr., *JMB* **145,** 319 (1981).
33. J. M. Hermoso and M. Salas, *PNAS* **77,** 6425 (1980).
34. P. Garcia, M. Hermoso, J. A. Garcia, E. Garcia, R. Lopez and M. Salas, *J. Virol.* **58,** 31 (1986).
35. H. Holzer and R. Wohlhueter, *Adv. Enzyme Reg.* **10,** 121 (1972).
36. Y. Fukami and F. Lipman, *PNAS* **80,** 1872 (1983).
37. N. Kitamura, B. L. Semler, P. G. Rothberg, G. R. Larsen, C. J. Adler, A. J. Dorner, E. A. Emini, R. Hanecak, Y. F. Lee, S. van der Werf, C. W. Anderson and E. Wimmer, *Nature* **291,** 547 (1981).
38. D. V. Sangar, D. J. Rowlands, T. J. R. Harris and F. Brown, *Nature* **268,** 648 (1977).
39. F. Golini, A. Nomoto and E. Wimmer, *Virology* **89,** 112 (1978).
40. D. E. Hruby and W. K. Roberts, *J. Virol.* **25,** 413 (1978).
41. R. Perez-Bercoff and M. Gander, *FEBS Lett.* **96,** 306 (1978).
42. Y. Lorch, M. Kotler and A. Friedmann, *J. Virol.* **45,** 1150 (1983).
43. M. Weitz, B. M. Baroudy, W. L. Maloy, J. R. Ticehurst and R. H. Purcell, *J. Virol.* **60,** 124 (1986).
44. A. Nomoto, B. Detjen, R. Pozzatti and E. Wimmer, *Nature* **268,** 208 (1977).
45. M. J. Hewlett, J. K. Rose and D. Baltimore, *PNAS* **73,** 327 (1976).
46. R. F. Pettersson, J. B. Flanegan, J. K. Rose and D. Baltimore, *Nature* **268,** 270 (1977).
47. A. Nomoto, N. Kitamura, F. Golini and E. Wimmer, *PNAS* **74,** 5345 (1977).
48. O. C. Richards, S. C. Martin, H. G. Jense and E. Ehrenfeld, *JMB* **173,** 325 (1984).
49. M. Girard, *J. Virol.* **3,** 376 (1969).
50. G. R. Larsen, A. J. Dorner, T. J. R. Harris and E. Wimmer, *NARes* **8,** 1217 (1980).
51. M. Wu, N. Davidson and E. Wimmer, *NARes* **5,** 4711 (1978).
52. Y. Yogo, M. Teng and E. Wimmer, *BBRC* **61,** 1101 (1974).
53. C. J. Adler, M. Elzinga and E. Wimmer, *J. Gen. Virol.* **64,** 349 (1983).
54. N. Kitamura, C. J. Adler, P. G. Rothberg, J. Martinko, S. G. Nathenson and E. Wimmer, *Cell* **21,** 295 (1980).
55. S. Forss and H. Schaller, *NARes* **10,** 6441 (1982).
56. A. R. Carroll, D. J. Rowlands and B. E. Clarke, *NARes* **12,** 2461 (1984).
57. A. C. Palmenberg, E. M. Kirby, M. R. Janda, N. L. Drake, G. M. Duke, K. F. Potratz and M. C. Collett, *NARes* **12,** 2969 (1984).
58. G. Stanway, P. J. Hughes, R. C. Mountford, P. D. Minor and J. W. Almond, *NARes* **12,** 7859 (1984).

59. T. Skern, W. Sommergruber, D. Blaas, P. Gruendler, F. Fraudorfer, C. Pieler, I. Fogy and E. Kuechler, *NARes* **13**, 2111 (1985).
60. R. Najarian, D. Caput, W. Gee, S. J. Potter, A. Renard, J. Merryweather, G. van Nest and D. Dina, *PNAS* **82**, 2627 (1985).
61. G. Werner, B. Rosenwirth, E. Bauer, J.-M. Scifert, F.-J. Werner and J. Besemer, *J. Virol.* **57**, 1084 (1986).
62. O. C. Richards, K. Morton, S. C. Martin and E. Ehrenfeld, *Virology* **136**, 453 (1984).
63. A. M. Q. King, D. V. Sangar, T. J. R. Harris and F. Brown, *J. Virol.* **34**, 627 (1980).
64. R. Hanecak, B. L. Semler, C. W. Anderson and E. Wimmer, *PNAS* **79**, 3973 (1982).
65. M. A. Pallansch, O. M. Kew, B. L. Semler, D. R. Omilianowski, C. W. Anderson, E. Wimmer and R. Rueckert, *J. Virol.* **49**, 873 (1984).
66. N. M. Crawford and D. Baltimore, *PNAS* **80**, 7452 (1983).
67. B. L. Semler, C. W. Anderson, R. Hanecak, L. F. Dorner and E. Wimmer, *Cell* **28**, 405 (1982).
68. V. Amrbos, R. F. Pettersson and D. Baltimore, *Cell* **15**, 1439 (1978).
69. V. Ambros and D. Baltimore, *JBC* **255**, 6739 (1980).
70. D. V. Sangar, J. Bryant, T. J. R. Harris, F. Brown and D. J. Rowlands, *J. Virol.* **39**, 67 (1981).
71. A. J. Dorner, P. G. Rothberg and E. Wimmer, *FEBS Lett.* **132**, 219 (1981).
72. F. Golini, B. L. Semler, A. J. Dorner and E. Wimmer, *Nature* **287**, 600 (1980).
73. E. Wimmer, *in* "The Molecular Biology of Picornaviruses" (R. Perez-Bercoff, ed.), p. 175. Plenum, New York, 1979.
74. E. Wimmer, *Cell* **28**, 199 (1982).
75. A. B. Vartapetian, *Bioorgan. Khim.* (Russ.) **8**, 1581 (1982).
76. J. B. Flanegan and D. Baltimore, *PNAS* **74**, 3677 (1977).
77. M. H. Baron and D. Baltimore, *JBC* **257**, 12359 (1982).
78. A. B. Vartapetian, E. V. Koonin, K. M. Chumakov, A. A. Bogdanov and V. I. Agol, *Dokl. Akad. Nauk SSSR* **267**, 963 (1982).
79. D. Etchison and E. Ehrenfeld, *Virology* **111**, 33 (1981).
80. A. B. Vartapetian, E. V. Koonin, V. I. Agol and A. A. Bogdanov, *EMBO J.* **3**, 2593 (1984).
81. T. Takegami, R. J. Kuhn, C. A. Anderson and E. Wimmer, *PNAS* **80**, 7447 (1983).
82. N. Takeda, R. J. Kuhn, C.-F. Yang, T. Takegami and E. Wimmer, *J. Virol.* **60**, 43 (1986).
83. T. M. Dmitrieva, M. V. Shcheglova and V. I. Agol, *Virology* **92**, 271 (1979).
84. T. A. Van Dyke and J. B. Flanegan, *J. Virol.* **35**, 732 (1980).
85. A. Dasgupta, P. Zabel and D. Baltimore, *Cell* **19**, 423 (1980).
86. M. H. Baron and D. Baltimore, *Cell* **30**, 745 (1982).
87. C. D. Morrow and A. Dasgupta, *J. Virol.* **48**, 429 (1983).
88. D. Baltimore, *in* "The Microbe 1984: P. I. Viruses" (B. W. J. Mahy, ed.), p. 109. Cambridge University Press, Cambridge, 1984.
89. C. D. Morrow, J. Hocko, M. Navab and A. Dasgupta, *J. Virol.* **50**, 515 (1984).
90. M. H. Baron and D. Baltimore, *JBC* **257**, 12351 (1982).
91. A. Dasgupta, *Virology* **128**, 252 (1983).
92. N. C. Andrews, D. Levin and D. Baltimore, *JBC* **260**, 7628 (1985).
93. N. C. Andrews and D. Baltimore, *PNAS* **83**, 221 (1986).
94. N. C. Andrews and D. Baltimore, *J. Virol.* **58**, 212 (1986).
95. D. C. Young, D. M. Tuschall and J. B. Flanegan, *J. Virol.* **54**, 256 (1985).

96. T. G. Senkevich, I. M. Chumakov, G. Y. Lipskaya and V. I. Agol, *Virology* **102**, 339 (1980).
97. C. D. Morrow, G. F. Gibbons and A. Dasgupta, *Cell* **40**, 913 (1985).
98. J. N. Burroughs and F. Brown, *J. Gen. Virol.* **41**, 443 (1978).
99. J. Stanley, P. Rottier, J. W. Davies, P. Zabel and A. van Kammen, NARes **5**, 4505 (1978).
100. S. D. Daubert, G. Bruening and R. C. Najarian, *EJB* **92**, 45 (1978).
101. M. A. Mayo, H. Barker and B. D. Harrison, *J. Gen. Virol.* **43**, 735 (1979).
102. V. Hari, *Virology* **112**, 391 (1981).
103. M. A. Mayo, H. Barker, D. J. Robinson, T. Tamada and B. D. Harrison, *J. Gen. Virol.* **59**, 163 (1982).
104. A. Ghosh, R. Dasgupta, T. Salerno-Rife, T. Rutgers and P. Kaesberg, *NARes* **7**, 2137 (1979).
105. D. Reisman and G. A. de Zoeten, *J. Gen Virol.* **62**, 187 (1982).
106. R. H. Persson and R. D. Macdonald, *J. Virol.* **44**, 437 (1982).
107. S. D. Daubert and G. Bruening, *Virology* **98**, 246 (1979).
108. R. Goldbach, G. Rezelman, P. Zabel and A. van Kammen, *J. Virol.* **42**, 630 (1982).
109. M. V. Sapotskij and Yu. F. Drygin, *Biokhimia* (Russ.) **48**, 1838 (1983).
110. P. Zabel, M. Moerman, G. Lomonossoff, M. Shanks and K. Beyreuther, *EMBO J.* **3**, 1629 (1984).
111. J. Wellink, G. Rezelman, R. Goldbach and K. Beyreuther, *J. Virol.* **59**, 50 (1986).
112. M. Jaegle, J. Wellink, R. Goldbach and A. van Kammen, *in* "Molecular Plant Virology" (EMBO Workshop), p. 23. Wageningen, 1986.
113. G. P. Lomonossoff and M. Shanks, *EMBO J.* **2**, 2253 (1983).
114. M. A. Mayo, H. Barker and B. D. Harrison, *J. Gen. Virol.* **59**, 149 (1982).
115. I. Koenig and C. Fritsch, *J. Gen. Virol.* **60**, 343 (1982).
116. P. W. G. Chu, G. Boccardo and R. I. B. Francki, *Virology* **109**, 428 (1981).
117. M. Meyer, O. Hemmer and C. Fritsch, *J. Gen. Virol.* **65**, 1575 (1984).
118. M. F. E. Siaw, M. Shahabuddin, S. Ballard, J. G. Shaw and R. E. Rhoads, *Virology* **142**, 134 (1985).
119. D. N. Black, J. N. Burroughs, T. J. R. Harris and F. Brown, *Nature* **274**, 614 (1978).
120. F. L. Schaffer, D. W. Ehresmann, M. K. Fretz and M. E. Soergel, *J. Gen. Virol.* **47**, 215 (1980).
121. A. Ghosh, T. Rutgers, M. Ke-giang and P. Kaesberg, *J. Virol.* **39**, 87 (1981).
122. K. Mang, A. Ghosh and P. Kaesberg, *Virology* **116**, 264 (1982).
123. G. P. Lomonossoff, M. Shanks and D. Evans, *Virology* **144**, 351 (1985).
124. G. A. Prody, J. T. Bakos, J. M. Buzayan, I. R. Schneider and G. Bruening, *Science* **231**, 1577 (1986).
125. S. A. Jobling and K. R. Wood, *J. Gen. Virol.* **66**, 2589 (1985).
126. R. Allison, R. E. Johnston and W. G. Dougherty, *Virology* **154**, 9 (1986).
127. R. Duncan and P. Dobos, *NARes* **14**, 5934 (1986).
128. P. Argos, G. Kamer, M. J. H. Nicklin and E. Wimmer, *NARes* **12**, 7251 (1984).
129. H. Franssen, J. Leunissen, R. Goldbach, G. Lomonossoff and D. Zimmern, *EMBO J.* **3**, 855 (1984).
130. R. M. Starzyk, S. W. Koontz and P. Schimmel, *Nature* **298**, 136 (1982).
131. R. Caizzi and F. Ritossa, *Biochem. Genet.* **21**, 267 (1983).
132. Y. S. Ahn, Y. C. Choi, I. L. Goldknopf and H. Busch, *Bchem* **24**, 7296 (1985).
133. A. Samad, C. W. Anderson and R. B. Carroll, *PNAS* **83**, 897 (1986).
134. F. Arwert and G. Venema, *J. Virol.* **13**, 584 (1974).
135. E. Garcia, A. Gomez, C. Ronda, C. Escarmis and R. Lopez, *Virology* **128**, 92 (1983).

135a. C. Ronda, R. Lopez, A. Gomez and E. Garcia, *J. Virol.* **48,** 721 (1983).
136. R. Lopez, C. Ronda, P. Garcia, C. Escarmis and E. Garcia, *MGG* **197,** 67 (1984).
137. D. Bamford, T. McGraw, G. MacKenzie and L. Mindich, *J. Virol.* **47,** 311 (1983).
137a. D. M. K. Rekosh, W. C. Russell, A. J. D. Bellett and A. J. Robinson, *Cell,* **11,** 283 (1977).
138. N. E. Harding, J. Ito and G. S. David, *Virology* **84,** 279 (1978).
139. D. H. Bamford and L. Mindich, *J. Virol.* **50,** 309 (1984).
140. C. F. Garon, K. W. Berry and J. A. Rose, *PNAS* **69,** 2391 (1972).
141. H. Yoshikawa and J. Ito, *PNAS* **78,** 2596 (1981).
142. C. Escarmis, A. Gomez, E. Garcia, C. Ronda, R. Lopez and M. Salas, *Virology* **133,** 166 (1984).
143. C. Escarmis, P. Garcia, E. Mendez, R. Lopez, M. Salas and E. Garcia, *Gene* **36,** 341 (1985).
144. B. W. Stillman, J. B. Lewis, L. T. Chow, M. B. Mathews and J. E. Smart, *Cell* **23,** 497 (1981).
145. M. Salas, R. P. Mellado, E. Vinuela and J. M. Sogo, *JMB* **119,** 269 (1978).
146. M. Green, K. Brackmann, W. S. M. Wold, M. Cartas, H. Thornton and J. H. Elder, *PNAS* **76,** 4380 (1979).
147. P. Argos, A. D. Tucker and L. Philipson, *Virology* **149,** 208 (1986).
148. M. D. Challberg and T. J. Kelly, Jr., *J. Virol.* **38,** 272 (1981).
149. M. Green, J. Symington, K. H. Brackmann, M. A. Cartas, H. Thornton and L. Young, *J. Virol.* **40,** 541 (1981).
150. M. L. Tremblay, C. V. Dery, B. G. Talbot and J. Weber, *BBA* **743,** 239 (1983).
151. J. E. Smart and B. W. Stillman, *JBC* **257,** 13499 (1982).
152. J. M. Hermoso, E. Mendez, F. Soriano and M. Salas, *NARes* **13,** 7715 (1985).
153. A. C. Arnberg and F. Arwert, *J. Virol.* **18,** 783 (1976).
154. M. Girard, J.-P. Bouche, L. Marty, B. Revet and N. Berthelot, *Virology* **83,** 34 (1977).
155. R. Padmanabhan and R. V. Padmanabhan, *BBRC* **75,** 955 (1977).
156. S. Yanofsky, F. Kawamura and J. Ito, *Nature* **259,** 60 (1976).
157. M. D. Challberg and T. J. Kelly, Jr., *JMB* **135,** 999 (1979).
158. R. P. Mellado, M. A. Penalva, M. R. Inciarte and M. Salas, *Virology* **104,** 84 (1980).
159. B. W. Stillman, *J. Virol.* **37,** 139 (1981).
160. I. Prieto, J. M. Lazaro, J. A. Garcia, J. M. Hermoso and M. Salas, *PNAS* **81,** 1639 (1984).
161. M. A. Bjornsti, B. E. Reilly and D. L. Anderson, *J. Virol.* **41,** 508 (1982).
162. A. W. M. Rijnders, B. G. M. van Bergen, P. C. van der Vliet and J. S. Sussenbach, *Virology* **131,** 287 (1983).
163. M. Shih, K. Watabe, H. Yoshikawa and J. Ito, *Virology* **133,** 56 (1984).
164. R. L. Lechner and T. J. Kelly, Jr., *Cell* **12,** 1007 (1977).
165. M. R. Inciarte, M. Salas and J. M. Sogo, *J. Virol.* **34,** 187 (1980).
166. J. D. Watson, *Nature NB* **239,** 197 (1972).
167. B. W. Stillman, A. J. D. Bellett and A. J. Robinson, *Nature* **269,** 723 (1977).
168. M. D. Challberg and T. J. Kelly, Jr., *PNAS* **76,** 655 (1979).
169. J. M. Sogo, J. A. Garcia, M. A. Penalva and M. Salas, *Virology* **116,** 1 (1982).
170. J.-E. Ikeda, T. Enomoto and J. Hurwitz, *PNAS* **78,** 884 (1981).
171. J. H. Lichy, M. S. Horwitz and J. Hurwitz, *PNAS* **78,** 2678 (1981).
172. S. Pincus, W. Robertson and D. Rekosh, *NARes* **9,** 4919 (1981).
173. T. Enomoto, J. H. Lichy, J.-E. Ikeda and J. Hurwitz, *PNAS* **78,** 6779 (1981).
174. K. Watabe, M.-F. Shih, A. Sugino and J. Ito, *PNAS* **79,** 5245 (1982).

175. M. Shih, K. Watabe and J. Ito, *BBRC* **105**, 1031 (1982).
176. M. A. Penalva and M. Salas, *PNAS* **79**, 5522 (1982).
177. R. P. Mellado and M. Salas, *NARes* **11**, 7397 (1983).
178. A. Zaballos, M. Salas and R. P. Mellado, *Gene* **43**, 103 (1986).
179. L. Blanco and M. Salas, *PNAS* **82**, 6404 (1985).
180. M. D. Challberg, J. M. Ostrove and T. J. Kelly, Jr., *J. Virol.* **41**, 265 (1982).
181. R. A. Guggenheimer, K. Nagata, M. Kenny and J. Hurwitz, *JBC* **259**, 7815 (1984).
182. F. Tamanoi and B. W. Stillman, *PNAS* **79**, 2221 (1982).
183. J. Gutierrez, J. A. Garcia, L. Blanco and M. Salas, *Gene* **43**, 1 (1986).
184. B. G. M. van Bergen, P. A. van der Ley, W. van Driel, A. D. M. van Mansfeld and P. C. van der Vliet, *NARes* **11**, 1975 (1983).
185. G. D. Pearson, K.-C. Chow, R. E. Enns, K. G. Ahern, J. L. Corden and J. A. Harpst, *Gene* **23**, 293 (1983).
186. R. E. Enns, M. D. Challberg, K. G. Ahern, K.-C. Chow, C. Z. Mathews, C. R. Astell and G. D. Pearson, *Gene* **23**, 307 (1983).
187. R. T. Hay, N. D. Stow and I. M. McDougall, *JMB* **175**, 493 (1984).
188. M. Ruben, S. Bacchetti and F. Graham, *Nature* **301**, 172 (1983).
189. F. L. Graham, *EMBO J.* **3**, 2917 (1984).
190. R. T. Hay, *EMBO J.* **4**, 421 (1985).
191. K. Wang and G. D. Pearson, *NARes* **13**, 5173 (1985).
192. R. T. Hay, *JMB* **186**, 129 (1985).
193. A. Tolun, P. Alestrom and U. Pettersson, *Cell* **17**, 705 (1979).
194. B. W. Stillman, W. C. Topp and J. A. Engler, *J. Virol.* **44**, 530 (1982).
195. M. Shinagawa, T. Tshiyama, R. Padmanabhan, K. Fujinaga, M. Kamada and G. Sata, *Virology* **125**, 491 (1983).
196. C. Lally, T. Dorper, W. Groger, G. Antoine and E.-L. Winnacker, *EMBO J.* **3**, 333 (1984).
197. F. Tamanoi and B. W. Stillman, *PNAS* **80**, 6446 (1983).
198. M. D. Challberg and D. R. Rawlins, *PNAS* **81**, 100 (1984).
199. J. A. Garcia, M. A. Penalva, L. Blanco and M. Salas, *PNAS* **81**, 80 (1984).
200. D. R. Rawlins, P. J. Rosenfeld, R. J. Wides, M. D. Challberg and T. J. Kelly, Jr., *Cell* **37**, 309 (1984).
201. R. A. Guggenheimer, B. W. Stillman, K. Nagata, P. Tamanoi and J. Hurwitz, *PNAS* **81**, 3069 (1984).
202. A. W. M. Rijnders, B. G. M. van Bergen, P. C. van der Vliet and J. S. Sussenbach, *NARes* **11**, 8777 (1983).
203. K. Nagata, R. A. Guggenheimer and J. Hurwitz, *PNAS* **80**, 6177 (1983).
204. J.-E. Ikeda, T. Enomoto and J. Hurwitz, *PNAS* **79**, 2442 (1982).
205. J. H. Lichy, J. Field, M. S. Horwitz and J. Hurwitz, *PNAS* **79**, 5225 (1982).
206. J. A. Garcia, R. Pastrana, I. Prieto and M. Salas, *Gene* **21**, 65 (1983).
207. B. W. Stillman, F. Tamanoi and M. B. Mathews, *Cell* **31**, 613 (1982).
207a. K. Matsumoto, T. Saito, C. I. Kim, T. Ando and H. Hirokawa, *MGG* **196**, 381 (1984).
208. L. Blanco, J. A. Garcia, M. A. Penalva and M. Salas, *NARes* **11**, 1309 (1983).
209. K. Matsumoto, T. Saito and H. Hirokawa, *MGG* **191**, 26 (1983).
210. K. Watabe and J. Ito, *NARes* **11**, 8333 (1983).
211. B. R. Friefeld, J. H. Lichy, J. Hurwitz and M. S. Hurwitz, *PNAS* **80**, 1589 (1983).
212. J. Field, R. M. Gronostajski and J. Hurwitz, *JBC* **259**, 9487 (1984).
213. L. Blanco and M. Salas, *PNAS* **81**, 5325 (1984).

214. M. Dunsworth, R. E. Schell and A. J. Berk, *NARes* **8,** 543 (1980).
215. L. Blanco and M. Salas, *Virology* **153,** 179 (1986).
216. K. Watabe, M. S. Leusch and J. Ito, *BBRC* **123,** 1019 (1984).
217. L. Blanco, J. Gutierrez, J. M. Lazaro, A. Bernad and M. Salas, *NARes* **14,** 4923 (1986).
218. B. W. Stillman, *Cell* **35,** 7 (1983).
219. K. Nagata, R. A. Guggenheimer, T. Enomoto, J. H. Lichy and J. Hurwitz, *PNAS* **79,** 6438 (1982).
220. J. Nowock, U. Borgmeyer, A. W. Puschel, R. A. W. Rupp and A. E. Sippel, *NARes* **13,** 2045 (1985).
221. P. A. J. Leegwater, P. C. van der Vliet, R. A. W. Rupp, J. Nowock and A. E. Sippel, *EMBO J.* **5,** 381 (1986).
222. P. C. van der Vliet, D. van Dam and M. M. Kwant, *FEBS Lett.* **171,** 5 (1984).
223. G. J. M. Pruijn, W. van Driel and P. C. van der Vliet, *Nature* **322,** 656 (1986).
224. M.-A. Bjornsti, B. E. Reilly and D. L. Anderson, *PNAS* **78,** 5861 (1981).
225. T. I. Ponomareva, N. A. Grodnitskaya, E. E. Goldberg, N. M. Chaplygina, B. S. Naroditsky and T. I. Tikchonenko, *NARes* **6,** 3119 (1979).
226. H. Hirochika and K. Sakaguchi, *Plasmid* **7,** 59 (1982).
227. Y. Kikuchi, K. Hirai and F. Hishinuma, *NARes* **12,** 5685 (1984).
228. Y. Kikuchi, K. Hirai, N. Gunge and F. Hishinuma, *EMBO J.* **4,** 1881 (1985).
229. J. C. Stam, J. Kwakman, M. Meijer and A. R. Stuitje, *NARes* **14,** 6871 (1986).
230. R. J. Kemble and R. D. Thompson, *NARes* **10,** 8181 (1982).
231. C. D. Chase and D. R. Pring, *Plant Mol. Biol.* **6,** 53 (1986).
232. H. Hirochika, K. Nakamura and K. Sakaguchi, *EMBO J.* **3,** 761 (1984).
233. F. Sor, M. Wesolowski and H. Fukuhara, *NARes* **11,** 5037 (1983).
234. C. S. Levings III and R. R. Sederoff, *PNAS* **80,** 4055 (1983).
235. M. Paillard, R. R. Sederoff and C. S. Levings III, *EMBO J.* **4,** 1125 (1985).
236. F. Sanger, A. R. Coulson, T. Friedman, G. M. Air, B. G. Barrell, N. L. Brown, J. C. Fiddes, C. A. Hutchinson III, P. M. Slocombe and M. Smith, *JMB* **125,** 225 (1978).
237. G. N. Godson, B. G. Barrell, R. Staden and J. C. Fiddes, *Nature* **276,** 236 (1978).
238. P. D. Baas, *BBA* **825,** 111 (1985).
239. B. Francke and D. S. Ray, *JMB* **61,** 565 (1971).
240. J. E. Ikeda, A. Yudelevich and J. Hurwitz, *PNAS* **73,** 2669 (1976).
241. S. A. Langeveld, A. D. M. van Mansfeld, P. D. Baas, H. S. Jansz, G. A. van Arkel and P. J. Weisbeek, *Nature* **271,** 417 (1978).
242. A. D. M. van Mansfeld, S. A. Langeveld, P. J. Weisbeek, P. D. Baas, G. A. van Arkel and H. S. Jansz, *CSHSQB* **43,** 331 (1979).
243. F. Heidekamp, S. A. Langeveld, P. D. Baas and H. S. Jansz, *NARes* **8,** 2009 (1980).
244. F. Heidekamp, P. D. Baas and H. S. Jansz, *J. Virol.* **42,** 91 (1982).
245. C. Sumida-Yasumoto, J.-E. Ikeda, E. Benz, K. J. Marians, R. Vicuna, S. Sugrue, S. L. Zipursky and J. Hurwitz, *CSHSQB* **43,** 311 (1979).
246. J. F. Scott, S. Eisenberg, L. L. Bertsch and A. Kornberg, *PNAS* **74,** 193 (1977).
247. S. Eisenberg, J. Griffith and A. Kornberg, *PNAS* **74,** 3198 (1977).
248. K. Koths and D. Dressler, *JBC* **255,** 4328 (1980).
249. A. Aoyama, R. K. Hamatake and M. Hayashi, *PNAS* **80,** 4195 (1983).
250. M. J. Roth, D. R. Brown and J. Hurwitz, *JBC* **259,** 10556 (1984).
251. D. R. Brown, M. J. Roth, D. Reinberg and J. Hurwitz, *JBC* **259,** 10545 (1984).
252. S. Eisenberg and A. Kornberg, *JBC* **254,** 5328 (1979).
253. J.-E. Ikeda, A. Yudelevich, N. Shimamoto and J. Hurwitz, *JBC* **254,** 9416 (1979).

254. S. A. Langeveld, G. A. van Arkel and P. J. Weisbeek, *FEBS Lett.* **114,** 269 (1980).
255. A. D. M. van Mansfeld, H. A. A. M. van Teeffelen, P. D. Baas, G. H. Veeneman, J. H. van Boom and H. S. Jansz, *FEBS Lett.* **173,** 351 (1984).
256. S. Sanhueza and S. Eisenberg, *J. Virol.* **53,** 695 (1985).
257. A. D. M. van Mansfeld, H. A. A. M. van Teeffelen, P. D. Baas and H. S. Jansz, *NARes* **14,** 4229 (1986).
258. E. S. Tessman, *JMB* **17,** 218 (1966).
259. A. van der Ende, R. Teertstra and P. J. Weisbeek, *NARes* **10,** 6846 (1982).
260. P. Weisbeek, F. van Mansfeld, C. Kuhlemeier, G. van Arkel and S. Langeveld, *EJB* **114,** 501 (1981).
261. K. J. Marians, J.-E. Ikeda, S. Schlagman and J. Hurwitz, *PNAS* **74,** 1965 (1977).
262. R. L. Low, K. Arai and A. Kornberg, *PNAS* **78,** 1436 (1981).
263. S. A. Langeveld, A. D. M. van Mansfeld, J. M. de Winter and P. J. Weisbeek, *NARes* **7,** 2177 (1979).
264. S. Eisenberg, *J. Virol.* **35,** 409 (1980).
265. A. D. M. van Mansfeld, H. A. A. M. van Teeffelen, A. C. Fluit, P. D. Baas and H. S. Jansz, *NARes* **14,** 1845 (1986).
266. A. D. M. van Mansfeld, S. A. Langeveld, P. D. Baas, H. S. Jansz, G. A. van der Marel, G. H. Veeneman and J. H. van Boom, *Nature* **288,** 561 (1980).
267. F. Heidekamp, P. D. Baas, J. H. van Boom, G. H. Veeneman, S. L. Zipursky and H. S. Jansz, *NARes* **9,** 3335 (1981).
268. A. C. Fluit, P. D. Baas, J. H. van Boom, G. H. Veeneman and H. S. Jansz, *NARes* **12,** 6443 (1984).
269. D. R. Brown, T. Schmidt-Glenewinkel, D. Reinberg and J. Hurwitz, *JBC* **258,** 8402 (1983).
270. A. C. Fluit, P. D. Baas and H. S. Jansz, *Virology* **154,** 357 (1986).
271. A. C. Fluit, P. D. Baas and H. S. Jansz, *EJB* **149,** 579 (1985).
272. E. Linney and M. Hayashi, *Nature NB* **245,** 6 (1979).
273. S. A. Langeveld, A. D. M. van Mansfeld, A. van der Ende, J. H. van de Pol, G. A. van Arkel and P. J. Weisbeek, *NARes* **9,** 545 (1981).
274. S. Sanhueza and S. Eisenberg, *PNAS* **81,** 4285 (1984).
275. S. Eisenberg and M. Finer, *NARes* **8,** 5305 (1980).
276. A. D. M. van Mansfeld, H. A. A. M. van Teeffelen, J. Zandberg, P. D. Baas, H. S. Jansz, G. H. Veeneman and J. H. van Boom, *FEBS Lett.* **150,** 103 (1982).
277. A. S. Zolotukhin, Yu. F. Drygin and A. A. Bogdanov, *Bioorgan. Khim.* (Russ.) **10,** 1109 (1984).
278. A. S. Zolotukhin, Yu. F. Drygin and A. A. Bogdanov, *Dokl. Akad. Nauk SSSR* **283,** 213 (1985).
279. S. Eisenberg and R. Ascarelli, *NARes* **9,** 1991 (1981).
280. J. Colasanti and D. T. Denhardt, *J. Virol.* **53,** 807 (1985).
281. P. D. Baas, W. R. Teertstra, A. D. M. van Mansfeld, H. S. Jansz, G. A. van der Marel, G. H. Veeneman and J. H. van Boom, *JMB* **152,** 615 (1981).
282. Y.-C. Tse, K. Kirkegaard and J. C. Wang, *JBC* **255,** 5560 (1980).
283. M. A. Lovett and D. R. Helinksi, *JBC* **250,** 8790 (1975).
284. G. J. Warren, A. J. Twigg and D. J. Sherratt, *Nature* **274,** 259 (1978).
285. D. G. Guiney and D. R. Helinski, *JBC* **250,** 8796 (1975).
286. K. L. Zhuklys, Yu. F. Drygin and A. A. Bogdanov, *Bioorgan. Khim.* (Russ.) **10,** 567 (1984).
287. R. R. Koepsel, R. W. Murray, W. D. Rosenblum and S. A. Khan, *PNAS* **82,** 6845 (1985).

288. R. R. Koepsel, R. W. Murray and S. A. Khan, *PNAS* **83,** 5484 (1986).
289. C. R. Astell, M. Thomson, M. Merchlinsky and D. C. Ward, *NARes* **11,** 999 (1983).
290. C. R. Wobbe and S. Mitra, *PNAS* **82,** 8335 (1985).
291. C. R. Astell, M. Thomas, M. B. Chow and D. C. Ward, *CSHSQB* **47,** 751 (1982).
292. M. Chow, J. W. Bodnar, M. Polvino-Bodnar and D. C. Ward, *J. Virol.* **57,** 1094 (1986).
293. C. R. Astell, M. B. Chow and D. C. Ward, *J. Virol.* **54,** 171 (1985).
294. P. Tiollais, C. Pourcel and A. Dejean, *Nature* **317,** 489 (1985).
295. W. H. Gerlich and W. S. Robinson, *Cell* **21,** 801 (1980).
296. D. Ganem, L. Greenbaum and H. E. Varmus, *J. Virol.* **44,** 374 (1982).
297. K. L. Molnar-Kimber, J. Summers, J. M. Taylor and W. S. Mason, *J. Virol.* **45,** 165 (1983).
298. B. Weiser, D. Ganem, C. Seeger and H. E. Varmus, *J. Virol.* **48,** 1 (1983).
299. J.-M. Lien, C. E. Aldrich and W. S. Mason, *J. Virol.* **57,** 229 (1986).
300. J. Summers and W. S. Mason, *Cell* **29,** 403 (1982).
301. M. Buscher, W. Reiser, H. Will and H. Schaller, *Cell* **40,** 717 (1985).
302. K. L. Molnar-Kimber, J. W. Summers and W. S. Mason, *J. Virol.* **51,** 181 (1984).
303. C. Seeger, D. Ganem and H. E. Varmus, *Science* **232,** 477 (1986).
304. V. I. Agol, *Mol. Biol.* (Russ.) **20,** 309 (1986).

Addendum: Foreign Gene Expression in Plant Cells

A number of papers studying the expression of foreign genes in plants and plant cells have appeared since the submission of this review. There is now evidence that monocotyledonous plants can be transformed *in planta* by direct gene transfer as well as by agroinfection. Other reports extend the range of plant transformation systems and apply this technology to further study gene regulation in plant cells. Some of these papers are briefly discussed below.

1. *Transformation of monocots. Agrobacterium*-mediated transformation of maize was first suggested by the detection in inoculated plants of T-DNA-specific enzyme activities [A. C. F. Graves and S. L. Goldman, *Plant Mol. Biol.* **7**, 43 (1986)] and subsequently demonstrated by viral infection of plants inoculated with *Agrobacterium* strains carrying tandemly repeated copies of maize streak virus in their T-DNA [N. Grimsley, T. Hohn, J. W. Davies and B. Hohn, *Nature* **325**, 177 (1987)].

A *nos-neo* chimeric gene has been used to show stable transformation of rice protoplasts after direct gene transfer [H. Uchimiya, T. Fushimi, H. Hashimoto, H. Harada, K. Syono and Y. Sugawara, *MGG* **204**, 204 (1986)], while transient expression of the *cat* gene was demonstrated in rice, wheat, and sorghum protoplasts after DNA transfer by electroporation. [T.-M. Ou-Lee, R. Turgeon and R. Wu, *PNAS* **83,** 6815 (1986)]. Interestingly, the latter study also showed expression in these plant cells of another non-plant promoter, that present in the *copia* long terminal repeat of *Drosophila*.

Finally, transgenic plants have been obtained after injection of DNA into the young inflorescences of rye [A. de la Pena, H. Lorz and J. Schell, *Nature* **325**, 274 (1987)].

2. *Cocultivation of callus cells with Agrobacterium*. This system has been extended to the high frequency transformation of carrot embryogenic suspension cells and has allowed efficient regeneration of transgenic plants [R. J. Scott and J. Draper, *Plant Mol. Biol.* **8**, 265 (1987)].

3. *Luciferase genes as reporters for gene expression*. The expression of the firefly and bacterial (*Vibrio harveyi*) chimeric luciferase genes in tobacco and carrot cells has been demonstrated after *Agrobacterium*-mediated transfer and electroporation [D. W. Ow, K. V. Wood, M. DeLuca, J. R. De Wet, D. R. Helinski and S. H. Howell, *Science* **234**, 856 (1986); C. Koncz, O. Olsson, W. H. R. Langridge, J. Schell and A. A. Szalay, *PNAS* **84**, 131 (1987)]. This reporter gene, contrary to the other existing ones, allows nondestructive assay of gene activity, as its product (luciferase) mediates light emission when transformed cells are incubated with the proper substrates. Furthermore, since bacterial luciferase is composed of two subunits, these experiments also demonstrated correct assembly of two bacterial gene products in transgenic plant cells.

4. *Gene regulation in transformed plant cells*. Transposition of the maize transposable element *Ac* has been demonstrated in transgenic tobacco plants and appears to occur in a similar fashion, i.e., accompanied by short duplications at the site of insertion. This phenomenon enables the marking and cloning of genes in dicots as it has in maize [B. Baker, J. Schell, H. Lorz and N. Fedoroff, *PNAS* **83**, 4844 (1986)].

Previous studies reviewed in this paper describe the properties of light-regulated genes after chimeric gene transfer into plants. These chimeric genes consist of a *neo* coding sequence cloned downstream from a promoter region isolated from light-inducible genes. Further studies demonstrate that in addition to possessing the properties of an enhancer, a 247-bp sequence present in the 5′-noncoding region of the chlorophyll *a/b* binding protein gene acts as a silencer in the roots of transgenic plants, regardless of its orientation. This demonstrates negative regulation of gene expression in plants [J. Simpson, J. Schell, M. Van Montagu and L. Herrera-Estrella, *Nature* **323**, 551 (1986)].

It was shown earlier that antisense RNA (RNA with a polarity opposite to that of a naturally occurring mRNA molecule) is capable of inhibiting gene expression in a variety of prokaryotes and eukaryotes. Such information was also recently gained in the case of carrot protoplasts electroporated with a mixture of plasmids containing the *cat* gene, either in the correct orientation relative to a plant promoter, or cloned in the opposite orientation [stop codon downstream from the promoter and initiation codon upstream from the poly(A) signal]. These experiments show that the transient expression of the *cat* gene is strongly inhibited when a 100-fold excess antisense *cat* plasmid is electroporated together with the sense plasmid. Presumably, this inhibition occurs through the formation of sense–antisense double-stranded mRNA molecules unavailable for translation [J. R. Ecker and R. W. Davis, *PNAS* **83**, 5372 (1986)].

Index

A

B

C

O

P

S

T